Lecture Notes in Artificial Intelligence 12915

Subseries of Lecture Notes in Computer Science

Series Editors

Randy Goebel
University of Alberta, Edmonton, Canada
Yuzuru Tanaka
Hokkaido University, Sapporo, Japan
Wolfgang Wahlster
DFKI and Saarland University, Saarbrücken, Germany

Founding Editor

Jörg Siekmann
DFKI and Saarland University, Saarbrücken, Germany

More information about this subseries at http://www.springer.com/series/1244

Thierry Denœux · Eric Lefèvre ·
Zhunga Liu · Frédéric Pichon (Eds.)

Belief Functions:
Theory and Applications

6th International Conference, BELIEF 2021
Shanghai, China, October 15–19, 2021
Proceedings

Springer

Editors
Thierry Denœux (iD)
Université de Technologie de Compiègne
Compiègne, France

Eric Lefèvre (iD)
Université d'Artois
Béthune, France

Zhunga Liu (iD)
Northwestern Polytechnical University
Xi'an, China

Frédéric Pichon (iD)
Université d'Artois
Béthune, France

ISSN 0302-9743 ISSN 1611-3349 (electronic)
Lecture Notes in Artificial Intelligence
ISBN 978-3-030-88600-4 ISBN 978-3-030-88601-1 (eBook)
https://doi.org/10.1007/978-3-030-88601-1

LNCS Sublibrary: SL7 – Artificial Intelligence

This Springer imprint is published by the registered company Springer Nature Switzerland AG
The registered company address is: Gewerbestrasse 11, 6330 Cham, Switzerland

Preface

The theory of belief functions, also referred to as evidence theory or Dempster-Shafer theory, was first introduced by Arthur P. Dempster in the context of statistical inference, and was later developed by Glenn Shafer as a general framework for modeling epistemic uncertainty. These early contributions have been the starting points of many important developments not only in statistics but also in computer science and engineering. The theory of belief functions is now well established as a general framework for reasoning with uncertainty, and has well-understood connections to other frameworks such as probability, possibility, and imprecise probability theories. It has been applied in diverse areas such as machine learning, information fusion, and pattern recognition.

The series of biennial International Conferences on Belief Functions (BELIEF), sponsored by the Belief Functions and Applications Society, are dedicated to the confrontation of ideas, the reporting of recent achievements, and the presentation of the wide range of applications of this theory. The first edition of this conference series was held in Brest, France, in 2010. Later editions were held in Compiègne, France, in 2012, Oxford, UK, in 2014, Prague, Czech Republic, in 2016, and again in Compiègne, France, in 2018. The 6th International Conference on Belief Functions (BELIEF 2021) was held in Shanghai, China, during October 15–19, 2021, together with the 1st International Conference on Cognitive Analytics, Granular Computing, and Three-way Decisions (CCGT 2021). Such a joint meeting promotes interactions and discussions between different communities working on different aspects of uncertainty theories. It was held both onsite and online due to the COVID-19 situation.

This volume contains the proceedings of BELIEF 2021. It contains 30 accepted submissions, each reviewed by at least three reviewers. Original contributions were solicited on theoretical aspects (including, for example, mathematical foundations, statistical inference) as well as on applications in various areas including classification, clustering, data fusion, image processing, and so on.

We would like to thank all the people who made this volume and this conference possible: all contributing authors, the organizers, and the Program Committee members who helped to build such an attractive program. We are especially grateful to our four invited speakers, Chunlai Zhou (Renmin University) for his talk "Basic Utility Theory for Belief Functions", Deqiang Han (Xi'an Jiaotong University) for his talk "Learning-based Modelized Methods for Evidence Combination", Zengjing Chen (Shandong University) for his talk "A Central Limit Theorem for Sets of Probability Measures", and Van-Nam Huynh (Japan Advanced Institute of Science and Technology) for his talk "Machine Learning Coupled with Evidential Reasoning for User Preference". We would also like to thank all our generous sponsors: Elsevier and the International Journal of Approximate Reasoning, as well as the International Society for Information Fusion (ISIF). Furthermore, we would like to thank the editors of the

Springer-Verlag series Lecture Notes in Artificial Intelligence (LNCS/LNAI) and
Springer-Verlag for their dedication to the production of this volume.

August 2021   Thierry Denœux
                                                                    Eric Lefèvre
                                                                   Zhunga Liu
                                                                Frédéric Pichon

Organization

BELIEF-CCGT Conference Co-chairs

Thierry Denœux   Université de technologie de Compiègne, France
Duoqian Miao   Tongji University, China
Yiyu Yao   University of Regina, Canada

Program Committee Chairs

Zhunga Liu   Northwestern Polytechnical University, China
Frédéric Pichon   Université d'Artois, France

Steering Committee

Éric Lefèvre   Université d'Artois, France
Zhunga Liu   Northwestern Polytechnical University, China
David Mercier   Université d'Artois, France
Frédéric Pichon   Université d'Artois, France
Zhihua Wei   Tongji University, China
Xiaodong Yue   Shanghai University, China

Program Committee

Alessandro Antonucci   Dalle Molle Institute for Artificial Intelligence, Switzerland
Olivier Colot   Université de Lille, France
Ines Couso   University of Oviedo, Spain
Fabio Cuzzolin   Oxford Brookes University, UK
Yong Deng   University of Electronic Science and Technology of China, China
Thierry Denœux   Université de technologie de Compiègne, France
Sébastien Destercke   Université de technologie de Compiègne, France
Jean Dezert   ONERA, France
Didier Dubois   Toulouse Institute of Computer Science Research, France
Zied Elouedi   Institut Supérieur de Gestion de Tunis, Tunisia
Chao Fu   Hefei University of Technology, China
Ruobin Gong   Rutgers University, USA
Deqiang Han   Xi'an Jiaotong University, China
Van Nam Huynh   Japan Advanced Institute of Science and Technology, Japan
Radim Jiroušek   University of Economics, Prague, Czech Republic

Anne-Laure Jousselme	Centre for Maritime Research and Experimentation, Italy
John Klein	Université de Lille, France
Václav Kratochvíl	Institute of Information Theory and Automation, CAS, Czech Republic
Éric Lefèvre	Université d'Artois, France
Xinde Li	Southeast University, China
Liping Liu	University of Akron, USA
Zhunga Liu	Northwestern Polytechnical University, China
Liyao Ma	University of Jinan, China
Arnaud Martin	Université de Rennes 1, France
Ryan Martin	North Carolina State University, USA
David Mercier	Université d'Artois, France
Enrique Miranda	University of Oviedo, Spain
Serafín Moral	University of Granada, Spain
Frédéric Pichon	Université d'Artois, France
Benjamin Quost	Université de technologie de Compiègne, France
Emmanuel Ramasso	École nationale supérieure de mécanique et des microtechniques, France
Johan Schubert	Swedish Defence Research Agency, Sweden
Prakash Shenoy	University of Kansas, USA
Zhigang Su	Southeast University, China
Barbara Vantaggi	University of Rome La Sapienza, Italy
Jiang Wen	Northwestern Polytechnical University, China
Jian-Bo Yang	University of Manchester, UK
Zhi-jie Zhou	Rocket Force University of Engineering, China

Additional Reviewers

Anh Hoang
Vo Vinh
Yiru Zhang
Zuowei Zhang

Contents

Clustering

Fast Unfolding of Credal Partitions
in Evidential Clustering

Zuowei Zhang[1,2], Arnaud Martin[2(✉)], Zhunga Liu[1], Kuang Zhou[3],
and Yiru Zhang[4]

[1] School of Automation, Northwestern Polytechnical University, Xi'an, China
zuowei_zhang@mail.nwpu.edu.cn, liuzhunga@nwpu.edu.cn
[2] Univ Rennes, CNRS, IRISA, Rue E. Branly, 22300 Lannion, France
Arnaud.Martin@univ-rennes1.fr
[3] School of Mathematics and Statistics, Northwestern Polytechnical University,
Xi'an, China
kzhoumath@nwpu.edu.cn
[4] Department of Computer Science, St. Francis Xavier University,
Antigonish, Canada

Abstract. Evidential clustering, based on the notion of *credal partition*,
has been successfully applied in many fields, reflecting its broad appeal
and usefulness as one of the steps in exploratory data analysis. However,
it is time-consuming due to the introduction of meta-cluster, which is
regarded as a new cluster and defined by the disjunction (union) of several special (singleton) clusters. In this paper, a simple and fast method
is proposed to extract the credal partition structure in evidential clustering based on modifying the iteration rule. By doing so, the invalid
computation associated with meta-clusters is effectively eliminated. It is
superior to known methods in terms of execution time. The results show
the potential of the proposed method, especially in large data.

Keywords: Evidential clustering · Fast credal partition · Belief
functions · Uncertainty

1 Introduction

Clustering is an important branch in data mining and unsupervised machine
learning, which has extensive applications in various domains. It aims to find
groups or clusters of objects that are similar to one another but dissimilar to
objects in any other clusters [1]. Of course, a wide variety of methods for clustering (object and relational) data have been developed with different philoso-
phies [1–7]. Among them, *credal partitions* [2,3] based on the theory of belief
functions (TBF) [8,9] have attracted a lot of attention since it can provide an
efficient tool in characterizing uncertain and imprecise information, which can
help the users reduce the high risk of errors in some fields [10]. The TBF is a
theoretical framework for reasoning with uncertain information and has been
widely used, for example, in clustering [2–6] and decision-making [11,12].

© Springer Nature Switzerland AG 2021
T. Denœux et al. (Eds.): BELIEF 2021, LNAI 12915, pp. 3–12, 2021.
https://doi.org/10.1007/978-3-030-88601-1_1

A *credal partition* [2,3] can generate three types of clusters: singleton clusters, meta-clusters, and noise cluster. The meta-cluster is a new cluster with the same properties as a singleton cluster, and it is defined by the disjunction (union) of several singleton clusters [3–5]. As a result, the meta-cluster can be considered a transition cluster among these different close singleton clusters, aiming to represent the imprecision (partial ignorance) of clustering for the overlapping objects. In this case, the query object may belong to any singleton cluster and meta-cluster with different masses of belief. For instance, evidential *c*-means (ECM) [3], one of the most famous methods based on the notion of *credal partition*, was presented by Denœux and Masson. It is considered an evidential version of the fuzzy *c*-means (FCM) [13] and noise clustering (NC) [14] based on the TBF. Of course, many evidential clustering methods for relational data have also emerged [2,6].

We would agree that the meta-cluster introduced by *credal partitions* has dramatically enriched our extraction of different data structures. This flexibility allows us to gain a deeper insight into various data in the real world. However, this advantage takes a high cost since putting the frame under the power-set 2^Ω brings an exponentially increasing computational burden. Of course, these methods also reduce the computational cost by limiting the number of elements in meta-clusters [2,3,15]. But this also limits some possibilities. For example, we only allow the number of focal sets to be less than or equal to 3, but objects may be difficult to be assigned to 4 or 5 singleton clusters in some scenarios. The method we propose in this paper does not impose such a restriction but can reduce the computational complexity.

Motivating by the above considerations, this paper proposes a simple and fast method to extract the credal partition structure in evidential clustering, called Fast evidential clustering (FAST-EC). It aims to split the iterative rule of traditional evidential clustering into the following two steps:

1) The query set is only allowed to be iterated under the frame of discernment Ω until the robust singleton cluster centers are obtained. Then, the centers of the associated singleton clusters calculate that of meta-clusters.
2) The masses of beliefs of the objects belonging to different clusters are iterated again under the power-set 2^Ω. In this case, one only needs to iterate one time to capture the final credal partition.

The rest is organized as follows. The notion of credal partitions is briefly recalled in Sect. 2. The FAST-EC method is then introduced in Sect. 3. In Sect. 4, we conduct some experiments to study the performances of FAST-EC using some synthetic and real-world datasets. Section 5 concludes this paper.

2   Notion of Credal Partitions

One starts with a frame of discernment $\Omega = \{\omega_1, ..., \omega_c\}$ consisting of a finite discrete set of mutually exclusive and exhaustive hypotheses (clusters) The power-set of Ω denoted 2^Ω is the set of all the subsets of Ω. For example, if

$\Omega = \{\omega_1, \omega_2, \omega_3\}$, then $2^\Omega = \{\emptyset, \omega_1, \omega_2, \omega_3, \{\omega_1 \cup \omega_2\}, \{\omega_1 \cup \omega_3\}, \{\omega_2 \cup \omega_3\}, \Omega\}$. The singleton element (e.g., $\{\omega_1\}$) represents a specific cluster. Meta-clusters, unions of several singleton elements, represent partial ignorance (e.g., $\{\omega_1 \cup \omega_2\}$).

Suppose that one has a set $O = \{o_1, o_2, ..., o_n\}$ of n objects. A mass of belief is a function $m(\cdot)$ from 2^Ω to [0,1] satisfying $\sum_{A \in 2^\Omega} m(A) = 1$ and $m(\emptyset) = 0$. The subsets A of Ω such that $m(A) > 0$ are called the focal elements of $m(\cdot)$. The n-tuple $M = (m_1, m_2, ..., m_n)$ is called a *credal partition* [3].

Table 1. Illustration of credal partitions.

	$\emptyset$	$\{\omega_1\}$	$\{\omega_2\}$	Ω
m_1	1	0	0	0
m_2	0	0.8	0.2	0
m_3	0	0.1	0.9	0
m_4	0	0	0.1	0.9

Example 1: The 4-tuple $M = (m_1, m_2, m_3, m_4)$ with $\Omega = \{\omega_1, \omega_2\}$ is a simple illustration of credal partitions in Table 1. One can see that objects o_2 and o_3 are likely to belong to the cluster $\{\omega_1\}$ and $\{\omega_2\}$, respectively. In contrast, object o_1 has the whole mass of beliefs assigned to the outlier cluster (*i.e.*, $\emptyset$). The object o_4 has the most significant mass to Ω, indicating that it might belong to $\{\omega_1\}$ or $\{\omega_2\}$, and we cannot obtain the exact cluster information in this case.

3 Fast Evidential Clustering Method

The fast evidential clustering (FAST-EC) method aims to reduce the computational burden by changing the iteration rule of existing methods. The implementation process will be introduced in this Section and take evidential c-means [3], called FAST-ECM, as an example. The FAST-ECM mainly consists of two parts: 1) Disjunction of singleton cluster centers; 2) Extraction of the credal partition structure. These two parts are discussed in Subsects. 3.1 and 3.2, respectively.

3.1 Disjunction of Singleton Cluster Centers

Let us consider a query set $\mathcal{O} = \{o_1, o_2, ..., o_n\}$ of n data objects clustered with the frame of discernment $\Omega = \{\omega_1, ..., \omega_c\}$. Since we only need to obtain the centers of the c singleton clusters in this step, the meta-clusters and noise cluster do not need to be considered in this iteration process. Inspired by ECM, the m_{ij} can be redefined as follows:

$$m_{ij} = \frac{|A_j|^{-a/(\beta-1)} d_{ij}^{-2/(\beta-1)}}{\sum_{A_k \neq \emptyset} |A_k|^{-a/(\beta-1)} d_{ik}^{-2/(\beta-1)}}, \quad |A_j| = |A_k| = 1, \sum_{j=1}^{c} m_{ij} = 1. \quad (1)$$

It is easy to understand from Eq. (1) that the simplified mass of belief m_{ij} does not consider the meta-clusters and noise cluster during the iteration process. It can significantly reduce the computational burden since the meta-clusters will bring many invalid calculations of the distance between the object and the centers of meta-clusters. Thus, Eq. (1) can be further simplified as follows:

$$m_{ij} = \frac{d_{ij}^{-2/(\beta-1)}}{\sum d_{ik}^{-2/(\beta-1)}}, \ \sum_{j=1}^{c} m_{ij} = 1. \tag{2}$$

One may find that the simplified mass function m_{ij} is very similar to the membership function u_{ij} in fuzzy c-means (FCM) [13]. In fact, the initial iteration is a simple derivation of FCM under belief functions, and an FCM-like objective function with $\sum_{j=1}^{c} m_{ij} = 1$ can be defined as follows:

$$J_{FAST-EC} = \sum_{i=1}^{n} \sum_{A_j \subseteq \Omega, A_j \neq \emptyset} |A_j|^\alpha m_{ij}^\beta d_{ij}^2, \ |A_j| = 1. \tag{3}$$

where the exponent α allowing to control the degree of penalization. Parameter β has the same meaning as in FCM. d_{ij} is the Euclidean distance between the object o_i and the cluster A_j. Thus, the center vector v_j of the singleton cluster obtained by minimizing $J_{FAST-EC}$ is defined by:

$$v_j = \frac{\sum_{i=1}^{n} m_{ij}^\beta \mathbf{x}_i}{\sum_{i=1}^{n} m_{ij}^\beta}, \ \forall j = 1, ..., c \tag{4}$$

Here the singleton cluster centers may not be the same as that of ECM, but they are very similar. Since the meta-clusters and the noise cluster have a constraining behavior on the iteration, it may lead to a slight difference in the results compared with ECM. However, this difference is within reasonable limits.

3.2   Extraction of Credal Partition Structure

In this subsection, we will quickly extract the credal partition structure for evidential clustering with the frame of power-set 2^Ω. Since we have obtained the real centers of singleton clusters from Subsect. 3.1, the center $\overline{v}_j$ of the meta-cluster A_j ($|A_j| > 1$), as suggested in ECM, can be defined by:

$$\overline{v}_j = \frac{1}{|A_j|} \sum_{k=1}^{c} s_{kj} v_k \tag{5}$$

where $s_{kj} = 1$ if ω_k and $s_{kj} = 0$ otherwise. Then, one can obtain the real center matrix V of the clusters, including singleton and meta-clusters. Thus, the masses of belief of the query object $\mathbf{o}_i$ belonging to different clusters, as suggested in ECM, can be obtained by:

$$\begin{cases} m_{ij} = \dfrac{|A_j|^{-a/(\beta-1)} d_{ij}^{-2/(\beta-1)}}{\sum_{A_k \neq \emptyset} |A_k|^{-a/(\beta-1)} d_{ik}^{-2/(\beta-1)}}, \\ m_{i\emptyset} \triangleq 1 - \sum_{A_j \neq \emptyset} m_{ij}. \end{cases} \tag{6}$$

One needs to note that since the real centers are already known, it only needs to iterate once to get the final mass value of the query object belonging to different clusters. Thus, one can quickly extract the credal partition structure after obtaining the mass matrix $M = (m_1, m_2, ..., m_n)$.

Discussion: In previous work, we proposed a dynamic evidential clustering (DEC) method [5]. Both DEC and this work have adopted a similar way to obtain singleton cluster centers in the first step, but their goals are not the same. DEC strives to avoid unreasonable results that may arise from evidential clustering when the singleton cluster centers are very close to meta-clusters and improve execution efficiency. In this paper, we hope that different evidential clustering can quickly unfold under different adaptive scenarios. In the case of DEC, this work can be regarded as its first step. The flowchart of FAST-EC is presented in Fig. 1 to show how it works and illustrate the basic principle clearly.

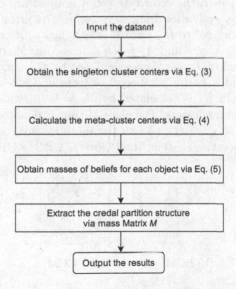

Fig. 1. Flowchart of the proposed FAST-EC.

3.3 Complexity Analysis in FAST-EC

To reduce the computational complexity of classical evidential clustering, FAST-EC splits the iteration rule into two steps: 1) Disjunction of singleton cluster

Table 2. Complexity comparison of some classical credal partitions.

ECM	FAST-ECM	CCM	FAST-CCM	MECM	FAST-MECM	DEC
$O(n2^c)$	$O(nc)$	$O(n2^c)$	$O(nc)$	$O(n2^c)$	$O(nc)$	$O(nc)$

centers; 2) Extraction of the credal partition structure. Since one only needs to iterate once to extract the credal partition structure after obtaining the real cluster centers, the computational complexity of FAST-EC mainly depends on the iteration of the singleton cluster centers. Specifically, the computational complexity is $O(nc)$, where n is the number of the objects in the query set, and c is the real number of singleton clusters. Thus, it reduces to a level similar to FCM [13], which is much lower than classical evidential clustering. Here we show the corresponding complexity of some credal partitions in evidential clustering methods, as shown in Table 2.

4   Experiment Applications

Two experiments have been done to test and evaluate the performances of the FAST-EC method and compared it to DEC [5], ECM [3], CCM [4] and MECM [6], corresponding to FAST-ECM, FAST-CCM, and FAST-MECM, respectively. In this paper, the error rate (R_e), imprecision rate (R_i) and accuracy (R_a) are used as evaluation indicators [4,5]. The error is counted for one object is explicitly belonged to $\{\omega_k\}$ but it is assigned to A_j with $\{\omega_k\} \cap A_j = \emptyset$. Given A_j with $\{\omega_k\} \cap A_j \neq \emptyset$ and $A_j \neq \{\omega_k\}$, it is considered as a meta-cluster. The error rate denoted by R_e (in %) is calculated by $R_e = n_e/n$, where n_e is the number of errors, and n is the number of objects under test. The imprecision rate denoted by R_i (in %) is calculated by $R_i = n_i/n$, where n_i is number of objects assigned to meta-clusters. The elapsed time is denoted by $T(s)$. For convenience, we denote $\omega_{k,t} = \{\omega_k, \omega_t\}$. All the methods are run in Matlab 2019A with Windows 10 operating system, Intel Core i7 CPU, 8 GB RAM.

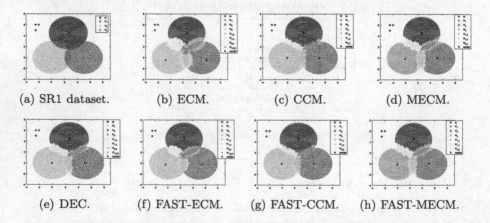

(a) SR1 dataset.   (b) ECM.   (c) CCM.   (d) MECM.

(e) DEC.   (f) FAST-ECM.   (g) FAST-CCM.   (h) FAST-MECM.

Fig. 2. Clustering results of SR dataset by different methods.

First, we consider a particular 3-class synthetic dataset with two dimensions under $\Omega = \{\omega_1, \omega_2, \omega_3\}$, which we called SR dataset. It consists of 933 data points, including 3 noisy data in the five-pointed star shape. The 310 objects in each class are evenly distributed in a circle with the radius of $r = 2$, and the centers of three circles are c_1: (0,0), c_2: (2.5,5), c_3: (5,0). Figure 2 and Table 3 correspond to the clustering results of the SR dataset by different methods. One can easily deduce such a conclusion from the results that FAST-EC can reduce computational complexity and improve clustering efficiency when the accuracy and the error rates are very similar. In addition, it can be found that FAST-EC as the first step of DEC, its execution time is faster than traditional evidential clustering methods, including ECM, CCM, and MECM.

Table 3. Statistics results of SR dataset by different methods.

	ECM	FAST-ECM	CCM	FAST-CCM	MECM	FAST-MECM	DEC
R_a	72.78	79.21	82.96	82.96	76.75	77.50	80.70
R_e	0	0	0	0	0.21	0.21	0.32
R_i	27.22	20.79	17.04	17.04	23.04	22.29	18.98
T	0.5650	0.0990	0.3806	0.0637	0.3296	0.0725	0.0649

Then, seven real datasets from UCI Machine Learning Repository [16] are employed to evaluate the performance of the proposed FAST-EC method. The results with real datasets for the different methods are reported in Table 4. One can find that the accuracy of FAST-EC is often higher than that of traditional methods while effectively reducing the imprecision rate since the FAST-EC can obtain more reasonable cluster centers in the iteration process. This is caused by the difference in the distribution of singleton cluster centers. In the iteration process of traditional methods, the singleton centers are often constrained by meta-clusters and then biased toward the centers of meta-clusters, determined by the objective function minimization rule. For example, given a dataset, we find that the distribution of singleton cluster centers obtained by FAST-EC is more reasonable than the distribution of singleton cluster centers obtained by traditional evidential clustering. Of course, this insignificant difference may not be easily observed. Furthermore, the execution time shows that FAST-EC is more efficient than traditional methods because it avoids the enormous computational burden caused by meta-clusters and obtains more robust cluster centers. In addition, we can find that FAST-EC and DEC have similar convergence speeds because they are both dedicated to reducing the computational complexity of traditional evidential clustering. The difference is that FAST-EC is only an improvement on the computational complexity of the original method and does not consider the flaws in the design of different evidential clustering. On the other hand, in DEC, we solved the design flaws of methods such as ECM and designed a dynamic frame framework for some imprecise objects.

Table 4. Statistics of clustering for UCI datasets by different methods (in %).

Data	Method	R_a	R_e	R_i	T
Contraceptive #Class.(3) #Attr.(9) #Inst.(1573)	ECM	34.60	57.91	7.49	2.4824
	FAST-ECM	36.05	57.36	6.59	0.0875
	CCM	35.17	56.48	8.35	1.5513
	FAST-CCM	35.64	56.89	7.47	0.0529
	MECM	35.17	56.48	8.35	0.4001
	FAST-MECM	35.64	56.89	7.47	0.0513
	DEC	30.28	56.82	12.90	0.1666
Satimage #Class.(7) #Attr.(36) #Inst.(6435)	ECM	66.48	28.08	5.44	21.8436
	FAST-ECM	66.55	29.90	3.45	0.5821
	CCM	60.80	30.54	8.66	15.4795
	FAST-CCM	55.60	34.10	10.30	0.3495
	MECM	53.85	37.95	8.20	13.4121
	FAST-MECM	54.95	34.85	10.20	0.3524
	DEC	63.37	24.07	12.56	0.5630
Spambase #Class.(2) #Attr.(57) #Inst.(4597)	ECM	61.78	33.85	4.37	5.5081
	FAST-ECM	62.01	33.07	4.92	0.3347
	CCM	63.15	34.07	2.78	3.2178
	FAST-CCM	62.87	34.35	2.78	0.3001
	MECM	62.32	33.26	4.42	3.1023
	FAST-MECM	62.32	33.61	4.07	0.3061
	DEC	64.15	31.79	4.06	0.1196
Abalone #Class.(3) #Attr.(8) #Inst.(4174)	ECM	36.23	36.22	27.55	6.9495
	FAST-ECM	39.88	40.35	19.77	0.2432
	CCM	48.11	45.18	6.71	4.0985
	FAST-CCM	44.87	42.43	12.70	0.1420
	MECM	46.05	42.57	11.38	3.2804
	FAST-MECM	41.30	41.02	17.68	0.1398
	DEC	47.37	42.13	10.50	0.1882
Segment #Class.(7) #Attr.(36) #Inst.(4174)	ECM	44.50	51.39	4.11	48.79
	FAST-ECM	46.28	51.30	2.42	0.5828
	CCM	54.23	41.53	4.24	18.9143
	FAST-CCM	57.89	37.86	4.24	0.3386
	MECM	42.38	49.74	7.88	8.3136
	FAST-MECM	46.13	45.58	8.29	0.3529
	DEC	45.60	49.17	5.23	0.3724
Magic #Class.(5) #Attr.(10) #Inst.(19020)	ECM	56.31	43.32	0.37	20.1879
	FAST-ECM	57.70	42.17	0.13	1.2250
	CCM	52.42	39.33	8.25	12.1652
	FAST-CCM	52.94	37.54	9.52	0.8094
	MECM	49.78	36.77	13.44	7.0703
	FAST-MECM	50.50	35.19	14.31	0.5823
	DEC	51.97	37.15	10.88	0.6273
Vehicle #Class.(4) #Attr.(18) #Inst.(846)	ECM	39.00	52.84	8.16	0.3703
	FAST-ECM	38.31	52.71	8.98	0.0863
	CCM	42.24	51.73	6.03	0.6578
	FAST-CCM	44.04	49.81	6.15	0.0818
	MECM	57.09	41.96	0.95	0.6491
	FAST-MECM	57.33	41.84	0.83	0.0824
	DEC	41.50	48.23	10.28	0.1803

5   Conclusion

This paper presented a simple and fast method to extract the credal partition structure in evidential clustering based on modifying the iteration rule. We found that more robust singleton cluster centers can be obtained while avoiding the computational burden caused by meta-clusters. The experiments show that FAST-EC can significantly reduce the computational complexity compared with traditional methods while getting more robust results. This will further expand the applications of credal partitions in the future.

Acknowledgement. This work was supported in part by the National Natural Science Foundation of China under Grants U20B2067, 61790552, 61790554, and in part by the Innovation Foundation for Doctor Dissertation of Northwestern Polytechnical University under Grant CX201953.

References

1. Liang, J., Bai, L., Dang, C., Cao, F.: The K-means-type algorithms versus imbalanced data distributions. IEEE Trans. Fuzzy Syst. **20**(4), 728–745 (2012)
2. Denæux, T., Masson, M.H.: EVCLUS: evidential clustering of proximity data. IEEE Trans. Syst. Man Cybern. Part B (Cybern.) **34**(1), 95–109 (2004)
3. Masson, M.H., Denoeux, T.: ECM: an evidential version of the fuzzy c-means algorithm. Pattern Recogn. **41**(4), 1384–1397 (2008)
4. Liu, Z.G., Pan, Q., Dezert, J., Mercier, G.: Credal c-means clustering method based on belief functions. Knowl.-Based Syst. **74**, 119–132 (2015)
5. Zhang, Z.W., Liu, Z., Martin, A., Liu, Z.G., Zhou, K.: Dynamic evidential clustering algorithm. Knowl.-Based Syst. **213**, 106643 (2021)
6. Zhou, K., Martin, A., Pan, Q., Liu, Z.G.: Median evidential c-means algorithm and its application to community detection. Knowl.-Based Syst. **74**, 69–88 (2015)
7. Zhang, Z.W., Liu, Z., Ma, Z., Zhang, Y., Wang, H.: A new belief-based incomplete pattern unsupervised classification method. IEEE Trans. Knowl. Data Eng. (2021). https://doi.org/10.1109/TKDE.2021.3049511
8. Shafer, G.: A Mathematical Theory of Evidence. Princeton University Press, Princeton (1976)
9. Smets, P.: The combination of evidence in the transferable belief model. IEEE Trans. Pattern Anal. Mach. Intell. **12**(5), 447–458 (1990)
10. Liu, Z.G., Pan, Q., Mercier, G., Dezert, J.: A new incomplete pattern classification method based on evidential reasoning. IEEE Trans. Cybern. **45**(4), 635–646 (2015)
11. Liu, Z.G., Pan, Q., Dezert, J., Martin, A.: Combination of classifiers with optimal weight based on evidential reasoning. IEEE Trans. Fuzzy Syst. **26**(3), 1217–1230 (2018)
12. Denoeux, T.: Decision-making with belief functions: a review. Int. J. Approx. Reason. **109**, 87–110 (2019)
13. Bezdek, J.C.: Pattern Recognition with Fuzzy Objective Function Algorithms. Springer, Heidelberg (2013)

14. Sen, S., Davé, R.N.: Clustering of relational data containing noise and outliers. In: 1998 IEEE International Conference on Fuzzy Systems Proceedings. IEEE World Congress on Computational Intelligence (Cat. No. 98CH36228), vol. 2, pp. 1411–1416. IEEE (1998)
15. Denoeux, T., Sriboonchitta, S., Kanjanatarakul, O.: Evidential clustering of large dissimilarity data. Knowl.-Based Syst. **106**, 179–195 (2016)
16. Asuncion, A., Newman, D.: UCI machine learning repository (2007)

Credal Clustering for Imbalanced Data

Zuowei Zhang[1,2](✉), Zhunga Liu[1], Kuang Zhou[3], Arnaud Martin[2], and Yiru Zhang[4]

[1] School of Automation, Northwestern Polytechnical University, Xi'an, China
zuowei_zhang@mail.nwpu.edu.cn, liuzhunga@nwpu.edu.cn
[2] Univ Rennes, CNRS, IRISA, Rue E. Branly, 22300 Lannion, France
Arnaud.Martin@univ-rennes1.fr
[3] School of Mathematics and Statistics, Northwestern Polytechnical University, Xi'an, China
kzhoumath@nwpu.edu.cn
[4] Department of Computer Science, St. Francis Xavier University, Antigonish, Canada

Abstract. Traditional evidential clustering tends to build clusters where the number of data for each cluster fairly close to each other. However, it may not be suitable for imbalanced data. This paper proposes a new method, called credal clustering (CClu), to deal with imbalanced data based on the theory of belief functions. Consider a dataset with C wanted classes, the credal c-means (CCM) clustering method is employed at first to divide the dataset into some (i.e., S ($S > C$)) clusters. Then these clusters are gradually merged following a given principle based on the density of meta-clusters and the associated singleton clusters. The merging is finished when C singleton wanted classes are obtained. During this merging procedure, the objects in each singleton cluster will be assigned to one new singleton class. Moreover, a weighted mean vector rule is developed to classify the objects in the unmerged meta-cluster to the associated new classes using the K-Nearest neighbor technique. Two experiments show that CClu can handle imbalanced datasets with high accuracy, and the errors are reduced by properly modeling imprecision.

Keywords: Evidential clustering · Belief functions · Imbalanced data · Credal c-means · K-NN

1 Introduction

Cluster analysis remains an important topic in data mining and machine learning. Clustering aims to group similar data and separate dissimilar data from a set into, what we call, clusters [1–4]. A recent credal partition [5,6] based on the theory of belief functions (TBF) is developed by Denœux and Masson. The TBF [7,8] provides an efficient tool to deal with uncertain and imprecise information, and it has been well applied in many fields, such as classification [9,10], clustering [11,12], information fusion [13]. The credal partition allows the

© Springer Nature Switzerland AG 2021
T. Denœux et al. (Eds.): BELIEF 2021, LNAI 12915, pp. 13–21, 2021.
https://doi.org/10.1007/978-3-030-88601-1_2

object to belong to any singleton cluster and the set of several clusters (called meta-cluster) with different belief degrees. By doing this, it can well represent the imprecision of some overlapping datasets. A wide variety of methods based on the credal partition for clustering object data have been developed, such as evidential c-means (ECM) [6], credal c-means (CCM) [11], and dynamic evidential clustering (DEC) [12]. However, the performance of these algorithms is often degraded when there is a very different number of data in each cluster, which we call imbalanced data [14]. Therefore, this paper proposes a new credal clustering (CClu) method for imbalanced datasets based on the TBF.

CClu mainly consists of three steps: 1) producing some sub-clusters, 2) merging these sub-clusters, and 3) classifying the objects in some sub-clusters (meta-clusters) that are not merged to obtain the final results. In the first step, CCM is employed to produce multiple sub-clusters, and the number of sub-clusters is bigger than the required number of classes[1]. The meta-cluster represents the overlapping zone of different sub-clusters. Thus, if several sub-clusters are really from a common class, these sub-clusters can be overlapped, and there usually exist many objects in the associated meta-clusters. In the second step, these sub-clusters are gradually merged until the final number of singleton classes is satisfied. This inherits the idea of hierarchical clustering to some extent [4]. Then, we will cautiously classify the objects in the sub-clusters (meta-clusters) that are not merged to the associated new singleton classes by a mean vector rule using the K-nearest neighbors (KNN) technique.

This paper is organized as follows. After a brief recall of credal c-means in Sect. 2, the new CClu method is presented in the Sect. 3. The proposed CClu is then tested in Sect. 4 and compared with several other clustering methods. The conclusion of this paper is finally given in Sect. 5.

2   Brief Recall of Credal c-means

Credal c-means clustering (CCM) [11] is developed based on the TBF, which extends the given framework $\Omega = \{\omega_1, ..., \omega_C\}$ to the power-set 2^{Ω}. For example, if $\Omega = \{\omega_1, \omega_2, \omega_3\}$, then $2^{\Omega} = \{\emptyset, \omega_1, \omega_2, \omega_3, \{\omega_1, \omega_2\}, \{\omega_1, \omega_3\}, \{\omega_2, \omega_3\}, \Omega\}$. CCM can produce three clusters: singleton clusters, meta-clusters, and noise cluster $\emptyset$. The meta-cluster is defined as the set of several singleton clusters and is considered as a kind of imprecise transition cluster among these different singleton clusters[2], which is used to represent partial ignorance. The mass value of the object $\mathbf{x}_i$ belonging to any singleton or meta-cluster $A_j \in 2^{\Omega}$, $j = 1, ..., 2^C$, is denoted as $m_{ij} \triangleq m_{\mathbf{x}_i}(A_j)$. It is determined by the Euclidean distance between the prototype vector of $\overline{\boldsymbol{v}}_j$ and $\mathbf{x}_i$. The center $\overline{\boldsymbol{v}}_j$ of the meta-cluster is the mean of that of the associated singleton clusters and defined by:

$$\overline{\boldsymbol{v}}_j = \frac{1}{|A_j|} \sum_{c=1}^{\mathcal{C}} s_{cj} \boldsymbol{v}_c, \quad \text{with} \quad s_{cj} = \begin{cases} 1 \text{ if } \{\omega_c\} \in A_j; \\ 0 \quad \text{otherwise.} \end{cases} \tag{1}$$

---

[1] For ease of description, we define the final clustering results as $\mathcal{C}$ wanted classes.
[2] We refer to the singleton clusters included in the meta-cluster as the associated ones.

where v_c is the center of the singleton cluster $\{\omega_c\}$, and $|A_j|$ is the cardinality of A_j, *i.e.*, the number of associated singleton clusters. In CCM, the mass of belief of the object belonging to a meta-cluster depends not only on the distance between the object and the center of the meta-cluster but also on the distance between the object and the centers of associated singleton clusters. The objective function of CCM denoted by J_{CCM} is designed according to these basic principles and given by:

$$J_{CCM}(M,V) = \sum_{i=1}^{N} \sum_{A_j \in S^{\Omega}} m_{ij}^{\beta} D_{ij}^2, \quad \sum_{A_j \in S^{\Omega}} m_{ij} = 1 \tag{2}$$

with

$$D_{ij}^2 = \begin{cases} \delta^2, & A_j = \emptyset; \\ d_{ij}^2, & |A_j| = 1; \\ \dfrac{\sum\limits_{\omega_c \in A_j} d_{ic}^2 + \gamma \, d_{ij}^2}{|A_j| + \gamma}, & |A_j| > 1. \end{cases} \tag{3}$$

where $M = (m_1, ..., m_N) \in R^{N \times 2^C}$ is the mass of belief matrix for all objects, and $V_{C \times p}$ is the matrix of clustering centers. N is the number of objects and p the number of attributes in the dataset. The parameters β, δ, γ, S^{Ω} have been discussed in detail [11].

3   Credal Clustering for Imbalanced Data

In this section, the CClu method is proposed to deal with imbalanced data based on the TBF. It mainly consists of three steps: 1) produce some sub-clusters and estimate the corresponding densities, 2) merge the sub-clusters with the given rule, and 3) classify the imprecise objects in some sub-clusters (meta-clusters) that are not merged to obtain the final clustering results.

3.1   Estimation of Cluster Density

For a C-class problem, CCM is employed first to generate sub-clusters (the number of singleton clusters $S > C$ and some meta-clusters) that are more than the actual number (C) of classes, where only two singleton clusters can be included in one meta-cluster. By doing this, the densities of these sub-clusters are then calculated based on the KNN. Here the density ρ_j of the cluster A_j (singleton cluster or meta-cluster) can be obtained with a common method as follows:

$$\rho_j = \left[\frac{1}{\mathcal{O}_j} \sum_{o=1}^{\mathcal{O}_j} \bar{d}_o \right]^{-1} \tag{4}$$

with

$$\bar{d}_o = \frac{1}{\mathcal{K}} \sum_{k=1}^{\mathcal{K}} d_{ok} \tag{5}$$

where $\mathcal{O}_j$ denotes the number of objects in the cluster A_j, and $\overline{d}_o$ is the distance of the nearest neighbor; d_{ok} is the Euclidean distance between $\mathbf{x}_o \in A_j$ ($o = 1, ..., \mathcal{O}$) and the k-th neighbor. Here the average is taken as the distance of the nearest neighbor to make the density more robust, and $\mathcal{K}$ is the number of considered neighbors. One can easily find that the density ρ_j is the reciprocal of the average of all objects in A_j and that the more dispersed (discrete) the objects in Λ_j, the smaller ρ_j is. In other words, ρ_j can characterize the degree of dispersion and also be used as a basis for decision-making on A_j.

3.2   Cluster Merging Rule

As we mentioned earlier, a meta-cluster is considered a locally imprecise cluster among the different associated singleton clusters. It means that the object may belong to any associated singleton clusters when the object is assigned to the meta-cluster. Hence, a meta-cluster is a good bridge between the associated singleton clusters. We can find that the objects in one (majority or minority) class may be clustered into several singleton clusters and meta-clusters in CClu. In this case, the densities of these singleton clusters would be very similar, and the density of the meta-cluster should be greater than or between the densities of the associated singleton clusters because the objects in the meta-cluster are usually distributed in the relative center of the (majority or minority) class. At the same time, if the density of the meta-cluster is less than that of the two associated singleton clusters, it means that the two singleton clusters belong to different classes. Therefore, there are three cases of density relations between the meta-cluster $\Lambda_i = \{\omega_k, \omega_t\}$ and the two associated singleton clusters (*i.e.*, $\{\omega_k\}$ and $\{\omega_t\}$) in CClu:

$\mathcal{C}1: \rho_{\omega_k} (\rho_{\omega_t}) \leq \rho_{\Lambda_i}$
$\mathcal{C}2: \rho_{\omega_k} (\rho_{\omega_t}) < \rho_{\Lambda_i} < \rho_{\omega_t} (\rho_{\omega_k})$
$\mathcal{C}3: \rho_{\Lambda_i} \leq \rho_{\omega_k} (\rho_{\omega_t})$

The meta-cluster and the associated singleton clusters satisfying $\mathcal{C}1$ and $\mathcal{C}2$ can be merged, and those satisfying $\mathcal{C}1$ should be merged first. The merging process also satisfies transitivity. That is, the newly merged sub-classes can be merged again if there are one or more sub-clusters in both of them.

3.3   Classification of Imprecise Objects

Objects in unmerged meta-clusters, called imprecise objects, will be classified into new classes or new meta-classes. Thus, a weighted mean vector method is developed to classify these imprecise objects to the associated classes using KNN. Thus, we first find the $\mathcal{K}$ neighbors of the object $\mathbf{x}_i$ from the known new singleton classes[3], denoted as $\mathbf{x}_k^{\{\widetilde{\omega}_i\}}$, $k = 1, ..., \mathcal{K}$. $\mathbf{x}_k^{\{\widetilde{\omega}_i\}}$ indicates that the neighbor $\mathbf{x}_k$ belongs to the class $\{\widetilde{\omega}_i\}$. For example, the meta-cluster $\{\omega_{1,5}\}$

---

[3] It should be noted that the $\mathcal{K}$ neighbors here have clear class information now.

includes two singleton clusters $\{\omega_1\}$ and $\{\omega_5\}$, where $\{\omega_1\}$ is merged into the real singleton class $\{\widetilde{\omega}_{i'}\}$, and $\{\omega_5\}$ is merged into $\{\widetilde{\omega}_{i''}\}$. In this case, the object $\mathbf{x}_i$ in $\{\omega_{1,5}\}$ will be classified into the singleton class $\{\widetilde{\omega}_{i'}\}$ or $\{\widetilde{\omega}_{i''}\}$, or the new meta-class $\{\widetilde{\omega}_{i',i''}\}$.

Since the distances between the object $\mathbf{x}_i$ in the unmerged meta-clusters and the neighbors $\mathbf{x}_k^{\{\widetilde{\omega}_i\}}$ $(k = 1, ..., \mathcal{K})$ are usually different, the bigger distance d_{ik} generally leads to the smaller discounting factor ϕ_{ik}. For $k = 1, ..., \mathcal{K}$, this factor ϕ_{nk} is simple defined by:

$$\phi_{ik} = \frac{d_{ik}^{-1}}{\sum\limits_{k=1}^{\mathcal{K}} d_{ik}^{-1}} \tag{6}$$

where d_{ik} is the Euclidean distance between $\mathbf{x}_i$ and its k-neighbor.

By doing this, one can obtain $\mathcal{K}$ vectors, named $\mathcal{V}_k^{\{\widetilde{\omega}_i\}}$, where $\mathbf{x}_i$ is the starting point, and $\mathbf{x}_k \in \{\widetilde{\omega}_i\}$ is the ending point. Assume that $\mathbf{x}_i$ may belong to the new classes $\widetilde{\omega}_c$ and $\widetilde{\omega}_{c'}$, we then can obtain two partial vectors, called Φ_c and $\Phi_{c'}$, for example, Φ_c is defined as follows:

$$\Phi_c = s_k \cdot \sum_{k=1}^{\mathcal{K}} \phi_{ik} \mathcal{V}_k^{\{\widetilde{\omega}_i\}}, \quad \text{with} \quad s_k = \begin{cases} 1 \text{ if } \mathbf{x}_k \in \{\widetilde{\omega}_c\}; \\ 0 \quad \text{otherwise.} \end{cases} \tag{7}$$

Based on the above analysis, one can easily find that the sum vector $\Phi = \sum\limits_{k=1}^{\mathcal{K}} \phi_{ik} \mathcal{V}_k^{\{\widetilde{\omega}_i\}}$ is mainly determined by the partial vector (i.e., Φ_c or $\Phi_{c'}$) of the real class to which the object $\mathbf{x}_i$ belongs. Thus, we define the weighted mean vector rule as follows:

$$\mathbf{x}_i \in \begin{cases} \{\widetilde{\omega}_c\}, & \text{if } \mathcal{L}_c < \mathcal{L}_{c'}, \; |\mathcal{L}_c - \mathcal{L}_{c'}| > \varphi; \\ \{\widetilde{\omega}_{c,c'}\}, & \text{if } \mathcal{L}_c \approx \mathcal{L}_{c'}, \; |\mathcal{L}_c - \mathcal{L}_{c'}| \leq \varphi; \\ \{\widetilde{\omega}_{c'}\}, & \text{if } \mathcal{L}_{c'} < \mathcal{L}_c, \; |\mathcal{L}_{c'} - \mathcal{L}_c| > \varphi. \end{cases} \tag{8}$$

where $\mathcal{L}_c$, for example, is the cosine distance (value) between Φ and Φ_c, i.e., $\mathcal{L}_c = \cos(\Phi, \Phi_c)$. φ is the meta-class parameter, and $\varphi \in [0, 2]$. In general, the larger φ is, the more the objects are assigned to the meta-classes.

Guideline for Choosing Parameters: In applications, the value of $\mathcal{S}$ $(\mathcal{C} < \mathcal{S} < N)$ cannot be too small, but too large requires a huge computational cost. $\mathcal{S} = 2\mathcal{C}$ is recommended as the default based on some experiments. The parameter φ should be tuned depending on the imprecision degree that one accepts. $\mathcal{K}$ is the number of neighbors, and we recommend $\mathcal{K} = 5$ as the default.

Computational Complexity: The computational burden mainly depends on two parts: producing some sub-clusters and calculating Euclidean distance for obtaining the cluster density. To reduce the burden, CCM is simplified here. We find the centers of singleton clusters first, and that of meta-clusters are then calculated. Thus, the complexity of CClu method is $O(N \cdot \mathcal{S})$.

The pseudo-code of CClu is presented in Algorithm 1 to show how CClu works and illustrate its basic principle clearly.

Algorithm 1. Credal clustering for imbalanced data.

Require: The query set: $\mathcal{X} = \{\mathbf{x}_1, ..., \mathbf{x}_N\}$; Given the parameters:
      $\mathcal{S}$: the number of sub-clusters, $\mathcal{S} = 2\mathcal{C}$;
      $\mathcal{K}$: the number of neighbors, $\mathcal{K} = 5$;
      φ: the meta-class parameter, $0 \leq \varphi \leq 0.4$.
Ensure: Clustering decision results.
  for $n = 1$ to N
  Generate $\mathcal{S}$ singleton clusters and some meta-clusters;
    for $j = 1$ to $2\mathcal{C} + C_{2\mathcal{C}}^2 + 1$
    Calculate the densities of different clusters using Eqs. (4) and (5);
    end
  Merge meta-clusters and related singleton clusters based on the cases $\mathcal{C}1$ and $\mathcal{C}2$;
  Generate new singleton classes and find imprecise objects;
  Calculate the vectors of imprecise objects using Eqs. (6), (7);
  Classify objects in unmerged meta-clusters using Eq. (8).
  end
  return Class label.

---

4   Experimental Applications

Two experiments have been done to test and evaluate the performances of the CClu method. Since the introduction of meta-cluster, the error rate (R_e), imprecision rate (R_i) and accuracy (R_a) are employed as indicators of performance [11,12]. The error is counted for one object is explicitly belonged to $\{\omega_c\}$ but it is clustered into A with $\{\omega_c\} \cap A = \emptyset$. Given A with $\{\omega_c\} \cap A \neq \emptyset$ and $A \neq \{\omega_c\}$, it is considered as a meta-class. The error rate denoted by R_e (in %) is calculated by $R_e = N_e/N$, where N_e is the number of errors, and N is the number of objects under test. The imprecision rate denoted by R_i (in %) is calculated by $R_i = N_i/N$, where N_i is number of objects assigned to meta-classes.

The first experiment is mainly used to clearly explain the use of CClu ($\varphi = 1$) in clustering imbalanced data on 2-class datasets. It has 2400 artificial data points which arise from a mixture of two bivariate Gaussian densities given by:

$$\frac{5}{6} \cdot \text{Gaussian} \begin{pmatrix} 6 \\ 0 \end{pmatrix} \begin{pmatrix} 5 & 0 \\ 0 & 5 \end{pmatrix} + \frac{1}{6} \cdot \text{Gaussian} \begin{pmatrix} 14 \\ 0 \end{pmatrix} \begin{pmatrix} 2 & 0 \\ 0 & 2 \end{pmatrix}$$

The generated dataset is shown in Fig. 1(a), and the clustering results of C-means [1], Fuzzy c-means (FCM) [2], and Credal c-means (CCM) [11] are shown in Fig. 1 (b)-(d), respectively.

The average error rates, imprecision rates, and accuracy are given in Table 1. One can see that error rates are lower while the accuracy of the CClu method is higher than the other methods. When $\mathcal{S}$ increases, one can get similar results, but at a higher computational cost. Thus, we should find a good compromise between the error rate and computational cost, and a large number of experiments we have done show that a small computational cost can be paid while achieving a lower error rate when $\mathcal{S} = 2\mathcal{C}$. So it is recommended as default in applications.

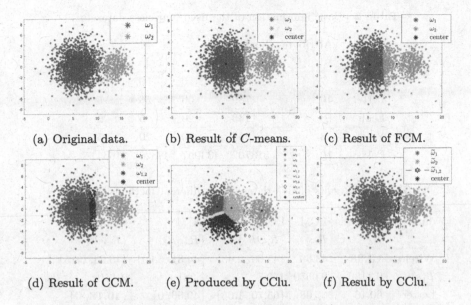

(a) Original data.   (b) Result of C-means.   (c) Result of FCM.

(d) Result of CCM.   (e) Produced by CClu.   (f) Result by CClu.

Fig. 1. The results of the 2-class dataset by different methods.

Table 1. Results of the 2-class dataset (in %).

	C-means	FCM	CCM	CClu
R_e	5.5	14.08	7.12	**1.21**
R_i	\	\	8.25	**1.04**
R_a	94.5	85.92	84.63	**97.75**

In the second experiment, we selected five UCI datasets (wireless, letter, ecoli, yeast, and Poker)[4]. There are the classes (1-class is complete, and 2, 3, 4-class randomly select 50 data objects) in Wireless Indoor Localization dataset and the classes (A-class is complete, and B, C, D-class randomly select 66, 36, 5 data objects) in Letter Image Recognition dataset, and the classes in Ecoli dataset (*i.e.*, cp, om, omL, and imL), Yeast dataset (*i.e.*, CYT, VAC, POX, and ERL) and Poker dataset (*i.e.*, 0, 1, 2, 3). The basic information is given in Table 2.

The clustering results with execution time of the real datasets by different methods have been shown in Table 3. We can see that the error rates of CClu are lower than other applied methods since the CClu reasonably employs multiple centers to represent one class and then merges the sub-clusters that originally belong to the same class. One can also find that CClu can deal well with the imbalanced dataset at the cost of an enormous computational burden. Hence, the CClu method is suitable for handling the case where high accuracy is required, but the computational speed is not very crucial.

---

[4] It is available: http://archive.ics.uci.edu/ml.

Table 2. Basic information of the used four-class datasets.

Name	Attributes	Instances			
		$\{\omega_1\}$	$\{\omega_2\}$	$\{\omega_3\}$	$\{\omega_4\}$
Wireless	7	500	50	50	50
Letter	16	789	66	36	5
Ecoli	8	143	20	5	2
Yeast	8	463	30	20	5
Poker	10	513702	433097	48828	21634

Table 3. Results of UCI datasets by different methods (in %).

data	C-means	FCM	ECM	CCM	CClu
	R_e	R_e	$\{R_e, R_i\}$	$\{R_e, R_i\}$	$\{R_e, R_i\}$
Wireless	**13.38**	43.69	$\{45.69, 2.43\}$	$\{44.31, 5.69\}$	$\{15.23, 0.15\}$
$T(s)$	0.1150	0.0550	1.787	2.381	0.4610
Letter	60.16	67.08	$\{65.70, 1.65\}$	$\{49.56, 0\}$	$\{\mathbf{40.18, 0}\}$
$T(s)$	0.1150	0.0540	2.149	5.382	0.5280
Ecoli	21.76	58.24	$\{55.29, 0.59\}$	$\{32.35, 0\}$	$\{\mathbf{17.65, 0}\}$
$T(s)$	0.114	0.115	0.892	1.239	0.456
Yeast	46.72	57.92	$\{59.02, 0\}$	$\{47.10, 0\}$	$\{\mathbf{16.99, 0}\}$
$T(s)$	0.115	0.053	0.202	0.314	2.012
Poker	73.92	68.18	$\{68.22, 0\}$	$\{51.43, 0\}$	$\{\mathbf{24.33, 0.002}\}$
$T(s)$	21.49	24.67	1014	1159	986.8

5 Conclusion

In this paper, we proposed a new CClu method for clustering imbalanced data based on the theory of belief functions. It mainly consists of three steps: 1) produce some sub-clusters and estimate the corresponding densities, 2) merge the sub-clusters with the given rule, and 3) classify the imprecise objects in some sub-clusters (meta-clusters). In CClu, the objects that are difficult to classify could be assigned to the meta-classes to reduce the risk of errors. Two experiments with artificial and real datasets are used to evaluate the performances of CClu by comparing it with other classical methods. The results show that CClu can efficiently improve the accuracy when clustering imbalanced data and capture the imprecision thanks to the meta-class.

Acknowledgements. This work was supported in part by the National Natural Science Foundation of China under Grants U20B2067, 61790552, 61790554, and in part by the Innovation Foundation for Doctor Dissertation of Northwestern Polytechnical University under Grant CX201953.

References

1. Jain, A.K.: Data clustering: 50 years beyond K-means. Pattern Recognit. Lett. **31**(8), 651–666 (2010)
2. Bezdek, J.: Pattern Recognition with Fuzzy Objective Function Algorithm. Plenum Press, New York (1981)
3. Davé, R.-N.: Clustering relational data containing noise and outliers. Pattern Recognit. Lett. **12**, 657–664 (1991)
4. Johnson, S.C.: Hierarchical clustering schemes. Psychometrika **32**(3), 241–254 (1967)
5. Denœux, T., Masson, M.H.: EVCLUS: evidential clustering of proximity data. IEEE Trans. Syst. Man Cybern. B Cybern. **34**(1), 95–109 (2004)
6. Masson, M.H., Denœux, T.: ECM: An evidential version of the fuzzy c-means algorithm. Pattern Recognit. **41**(4), 1384–1397 (2008)
7. Shafer, G.: A Mathematical Theory of Evidence. Princeton University Press, Princeton (1976)
8. Smets, P.: The combination of evidence in the transferable belief model. IEEE Trans. Pattern Anal. Mach. Intell. **12**(5), 447–458 (1990)
9. Huang, L., Liu, Z., Pan, Q., Dezert, J.: Evidential combination of augmented multi-source of information based on domain adaptation. Sci. China Inf. Sci. **63**(11), 1–14 (2020). https://doi.org/10.1007/s11432-020-3080-3
10. Liu, Z.G., Huang, L.Q., Zhou, K., Denœux, T.: Combination of transferable classification with multisource domain adaptation based on evidential reasoning. IEEE Trans. Neural. Netw. Learn. Syst. **32**, 2015–2029 (2020)
11. Liu, Z.G., Pan, Q., Dezert, J., Mercier, G.: Credal c-means clustering method based on belief functions. Knowl.-Based Syst. **74**, 119–132 (2015)
12. Zhang, Z.W., Liu, Z., Martin, A., Liu, Z.G., Zhou, K.: Dynamic evidential clustering algorithm. Knowl.-Based Syst. **213**, 106643 (2021)
13. Denœux, T.: Decision-making with belief functions: a review. Int. J. Approx. Reason. **109**, 87–110 (2019)
14. Liang, J.Y., Bai, L., Dang, C.: The K-means-type algorithms versus imbalanced data distributions. IEEE Trans. Fuzzy Syst. **20**(4), 728–745 (2012)

Evidential Weighted Multi-view Clustering

Kuang Zhou$^{(\boxtimes)}$, Mei Guo, and Ming Jiang

School of Mathematics and Statistics, Northwestern Polytechnical University,
Xi'an 710072, Shaanxi, People's Republic of China
kzhoumath@nwpu.edu.cn, xming@mail.nwpu.edu.cn

Abstract. Generally, the data to be clustered are from one single view. In real clustering applications, sometimes the data are insufficient so that it is difficult to learn an ideal cluster model. In such cases, multi-view data can be taken into consideration in the clustering task. However, the inconsistency cross views may increase the cluster uncertainty. In this research, a new clustering method for multi-view object data, called MvWECM (Multi-view Weighted Evidential C-Means) is introduced in the framework of belief functions. The proposed method can take consistency and diversity cross each view into account by incorporating the concept of view weights to measure the importance of each view. An objective function is defined to look for the best credal partitions over the different views. Experimental results on generated and UCI data sets show the advantage of the proposed method.

Keywords: Multi-view clustering · Belief functions · View weights

1 Introduction

Clustering is an unsupervised classification technique which has been applied to many fields, such as recommendation system [1], community detection [3] and so on [9]. Sometimes, data from one single view are too insufficient to study an ideal clustering model. Fortunately, nowadays multi-view data are very common in the real world. Generally, multi-view data can be obtained from multi-source, describing the same objection from distinctive aspects. For example, in the user clustering problem which is often required in personalized recommendation systems, the online user review data contains traveling histories, personal attributes, ratings in the form of texts and imagines, *etc.*

Some clustering approaches combine the multiple data into single-view data when dealing with multiple data, but this may lead to the loss of diversity information [4]. The multi-view clustering approach can be used to deal with this problem and it has been a popular topic in data mining [11]. Besides, some clustering algorithms based on multi-view data are proposed. Bickel et al. proposed partitioning multi-view clustering algorithms for text data [2]. Wang et al. proposed a novel multi-view clustering algorithm termed multi-view affinity

T. Denœux et al. (Eds.): BELIEF 2021, LNAI 12915, pp. 22–32, 2021.
https://doi.org/10.1007/978-3-030-88601-1_3

propagation based on max-product belief propagation [10]. Jiang et al. proposed a multi-view fuzzy clustering algorithm by weighting the importance of each view [5]. Zhang et al. proposed a Two-level Weighted Collaborative k-means algorithms [11], taking collaborative manner to keep the diversity and consistency among each view.

The diversity across each view may increase the cluster uncertainty. The theory of belief functions is an efficient mathematical tool for uncertain information representation and fusion. There are already some clustering approaches developed in this framework, such as the Evidential c-means (ECM) clustering [6], which can derive effective clustering results for uncertain data. In this paper, inspired by ECM and a multi-view evidenial clustering method called CEC [7], we propose a multi-view clustering approach with a collaborative strategy named Multi-view Weighted Evidential C-Means algorithm (MvWECM). Different from CEC which can not distinguish the importance of different views, in MvWECM we introduce the concept of view weights to qualify the contribution of each view to cluster structure. Experimental results on Gaussian and UCI data sets show the effectiveness of the proposed method.

This paper is organized as follows. The proposed MvWECM is introduced in Sect. 2. Experimental results are shown in Sect. 3. Conclusions are given in the final section.

2 Multi-view Weighted Evidential c-means

The MvWECM algorithm is introduced in this section. We assume that the number of clusters in each view is the same but the feature dimensions are different. The objective function of MvWECM will be defined first in Sect. 2.1, and then the optimization method will be described in Sect. 2.2.

2.1 The Objective Function

Notation $X = \{X[1], X[2], \cdots, X[T]\}$ denotes the dataset with T views, where $X[t] = \{x_1[t], x_2[t], \cdots, x_N[t]\}$ denots N samples of the t_{th} view. The dimensions of X in each view are denoted by $Q = \{q_1, q_2, \cdots, q_T\}$.

Let the discernment frame be $\Omega = \{\omega_1, \omega_2, \cdots, \omega_c\}$. Denote the weights of the T views by $W = (w[1], w[2], \cdots, w[T])$. Let $M[t] = \{m_1[t], m_2[t], \cdots, m_N[t]\}$ denote the BBAs in t_{th} view, where $m_i[t] = \{m_{ij}[t] \triangleq m_i(A_j)[t] | A_j \subseteq \Omega\}$ represents mass assignments for the cluster membership of sample $x_i[t]$. Let $V = \{V[1], V[2], \cdots, V[T]\}$ represent the cluster center sets for T views, where $V[t]$ is a matrix of size $(c \times q_t)$.

We aim to find out the multi-view optimal credal partial, cluster centers and view wights by minimizing the following objective function:

$$
J_{\text{MvWECM}} = \sum_{t=1}^{T} w[t] \left(\sum_{i=1}^{N} \sum_{\{j|A_j \subseteq \Omega, A_j \neq \emptyset\}} c_j^\alpha m_{ij}^2[t] d_{ij}^2[t] + \sum_{i=1}^{n} \delta^2[t] m_{i\emptyset}^2[t] \right)
$$

$$
+ \eta \sum_{t=1}^{T} \sum_{s \neq t}^{T} k[t,s] \sum_{i=1}^{N} \sum_{\{j|A_j \subseteq \Omega, A_j \neq \emptyset\}} (m_{ij}[t] - m_{ij}[s])^2 d_{ij}^2[t]
$$

$$
+ \beta \sum_{t=1}^{T} w[t] \log w[t], \tag{1}
$$

subject to:

$$
\sum_{t=1}^{T} w[t] = 1, \ w[t] \in [0,1], \ t = 1, 2, \cdots, T, \tag{2}
$$

and

$$
\sum_{\{j|A_j \subseteq \Omega, A_j \neq \emptyset\}} m_{ij}[t] + m_{i\emptyset}[t] = 1, \forall A_j \subseteq \Omega, i = 1, \cdots, N; t = 1, \cdots, T. \tag{3}
$$

In Eq. (1), notations s_{kj} and c_j^α have the same meaning as those in ECM algorithm. Parameters η and β are introduced to control the effect of disagreement among views and view weights respectively. $k[t,s] \in [0,1]$ is used to measure the collaborative strength of the s_{th} and t_{th} views. The greater value of $k[t,s]$ is, the stronger collaborative strength between s_{th} view and t_{th} one is. It's easy to get $k[t,t] = 0, t = 1 \cdots T$. Notation $d_{ij}[t]$ denotes the Euclidean distance between sample $x_i[t]$ and barycenter of focal set A_j. The objective function J_{MvWECM} consists of three parts. The first part is similar to ECM, which computes the sum of with-in weighted distance in each view. The second part qualifies the disagreement across multiple views. The last part is the entropy-based terms, reflecting the degree of view weights influence in Eq. (1).

2.2   Optimization

The Lagrange multiplier approach is adopted here to minimize the objective function J_{MvWECM}. The optimization process can be iteratively proceeded by the following three steps.

(1) Update the basic belief assignments

In this part, the BBA $M[t]$ was updated with view weight $w[t]$ and clustering centers $V[t]$ fixed. Aiming to minimize the constrained function with respect to the BBA matrix $M[t]$ in t_{th} view, $N \times T$ Lagrange multiplier $\lambda[t], (i = 1 \cdots N, t = 1 \cdots, T)$ are introduced into Lagrangian Function $L(M[t], \lambda[t])$, and we show it as follows:

$$L(\boldsymbol{M}[t], \lambda[t]) = J_{\text{MvWECM}}$$

$$-\sum_{t=1}^{T}\sum_{i=1}^{N}\lambda_i[t]\left(\sum_{\{j|A_j\subseteq\Omega, A_j\neq\emptyset\}} m_{ij}[t] + m_{i\emptyset}[t] - 1\right). \qquad (4)$$

Differentiating the above equation with respect to $m_{ij}[t]$, $m_{i\emptyset}[t]$ and $\lambda_i[t]$ and setting the partial differential equation into zero, we can obtain:

$$\frac{\partial L(\boldsymbol{M}[t], \lambda[t])}{\partial m_{ij}[t]} = 2w[t]c_j^\alpha m_{ij}[t]d_{ij}^2[t]$$

$$+ 2\eta\sum_{\substack{s\neq t}}^{T} k[t,s](m_{ij}[t] - m_{ij}[s])d_{ij}^2[t] - \lambda_i[t] = 0, \qquad (5)$$

$$\frac{\partial L(\boldsymbol{M}[t], \lambda[t])}{\partial m_{i\emptyset}[t]} = 2w[t]\delta[t]^2 m_{i\emptyset}[t] - \lambda_i[t] = 0 \qquad (6)$$

and

$$\frac{\partial L(\boldsymbol{M}[t], \lambda[t])}{\partial \lambda_i[t]} = -\sum_{\{j|A_j\subseteq\Omega, A_j\neq\emptyset\}} m_{ij}[t] - m_{i\emptyset}[t] + 1 = 0. \qquad (7)$$

From Eqs. (5) and (6), we can deduce

$$m_{ij}[t] = \frac{\eta\varphi_{ij}[t]}{w[t]c_j^\alpha + \eta\psi[t]} + \frac{\lambda_i[t]}{2(w[t]c_j^\alpha + \eta\psi[t])d_{ij}^2[t]} \qquad (8)$$

and

$$m_{i\emptyset}[t] = \frac{\lambda_i[t]}{2w[t]\delta^2[t]}, \qquad (9)$$

where

$$\psi[t] = \sum_{\substack{s\neq t}}^{T} k[t,s] \qquad (10)$$

and

$$\varphi_{ij}[t] = \sum_{\substack{s\neq t}}^{T} k[t,s]m_{ij}[s]. \qquad (11)$$

Equations (8) and (9) are substituted into Eq. (7), then Lagrange multiplier $\lambda_i[t]$ can be expressed by fixed parameters as follows:

$$\lambda_i[t] = \frac{1 - \sum\limits_{\{j|A_j\subseteq\Omega, A_j\neq\emptyset\}} \frac{\eta\varphi_{ij}[t]}{w[t]c_j^\alpha + \eta\psi[t]}}{\sum\limits_{\{j|A_j\subseteq\Omega, A_j\neq\emptyset\}} \frac{1}{2(w[t]c_j^\alpha + \eta\psi[t])d_{ij}^2[t]} + \frac{1}{2w[t]\delta^2[t]}}. \qquad (12)$$

Then Eq. (12) is substituted into Eqs. (5) and (6), the update rule of BBA $M[t]$ is the minimum point of J_{MvWECM} with other variables fixed, can be derived as follows:

$$m_{ij}[t] = \frac{\eta\varphi_{ij}[t]}{w[t]c_j^{\alpha} + \eta\psi[t]} + \frac{\left(1 - \sum\limits_{\{j|A_j \subseteq \Omega, A_j \neq \emptyset\}} \frac{\eta\varphi_{ij}[t]}{w[t]c_j^{\alpha} + \eta\psi[t]}\right) \frac{1}{(w[t]c_j^{\alpha} + \eta\psi[t])d_{ij}^2}}{\sum\limits_{\{j|A_j \subseteq \Omega, A_j \neq \emptyset\}} \frac{1}{(w[t]c_j^{\alpha} + \eta\psi[t])d_{ij}^2[t]} + \frac{1}{w[t]\delta^2[t]}}$$

$$\forall i = 1, 2 \cdots, N; \forall t = 1, \cdots, T,$$

(13)

and

$$m_{i\emptyset}[t] = 1 - \sum_{\{j|A_j \subseteq \Omega, A_j \neq \emptyset\}\}} m_{ij}[t].$$

(14)

(2) Update the view weights

In this part, view weight $w[t]$ is updated with BBA $M[t]$ and clustering center $V[t]$ fixed. Similar to the first step, the Lagrangian Function is given as follows:

$$L(w[t], \mu) = J_{\text{MvWECM}} - \mu\left(\sum_{t=1}^{T} w[t] - 1\right),$$

(15)

where μ is the Lagrangian multiplier. Similarly, differentiating the Lagrangian with respect to $w[t]$ and μ, we can obtain:

$$\frac{\partial L(w[t], \mu)}{\partial w[t]} = \sum_{i=1}^{N} \sum_{\{j|A_j \subseteq \Omega, A_j \neq \emptyset\}} c_j^{\alpha} m_{ij}^2[t] d_{ij}^2[t]$$

$$+ \sum_{i=1}^{N} \delta^2[t] m_{i\emptyset}^2[t] + \beta(1 + \log w[t]) - \mu = 0$$

(16)

and

$$\frac{\partial L(w[t], \mu)}{\partial \mu} = -\sum_{i=1}^{T} w[t] + 1 = 0.$$

(17)

From Eq. (16), we can deduce:

$$\Delta[t] + \beta(\log w[t] + 1) - \mu = 0,$$

(18)

where

$$\Delta[t] = \sum_{i=1}^{N} \sum_{\{j|A_j \subseteq \Omega, A_j \neq \emptyset\}} c_j^{\alpha} m_{ij}^2[t] d_{ij}^2[t] + \sum_{i=1}^{N} \delta^2[t] m_{i\emptyset}^2[t].$$

(19)

Equation (18) is equivalent to

$$w[t] = \exp\left\{\frac{-\Delta[t] - \beta}{\beta}\right\} \exp\left\{\frac{\mu}{\beta}\right\}. \tag{20}$$

Substituting the above formula into Eq. (2), we have

$$\sum_{t=1}^{T} \exp\left\{\frac{-\Delta[t] - \beta}{\beta}\right\} \exp\left\{\frac{\mu}{\beta}\right\} = 1. \tag{21}$$

From the above formula, we can get

$$\exp\left\{\frac{\mu}{\beta}\right\} = \frac{1}{\sum\limits_{t=1}^{T} \exp\left\{\frac{-\Delta[t]-\beta}{\beta}\right\}}. \tag{22}$$

Finally, $\exp\left\{\frac{\mu}{\beta}\right\}$ in Eq. (22) is substituted into Eq. (21), and the update rule of $w[t]$ can be derived as follows:

$$w[t] = \frac{\exp\left\{\frac{-\Delta[t]-\beta}{\beta}\right\}}{\sum\limits_{t=1}^{T} \exp\left\{\frac{-\Delta[t]-\beta}{\beta}\right\}}. \tag{23}$$

(3) Update the clustering centers

In this part, clustering center $V[t]$ is updated with variable BBA $M[t]$ and view weight $w[t]$ fixed. We can find the minimization of J_{MvWECM} with respect to $V[t]$ is an unconstrained optimization problem. The partial derivation of J_{MvWECM} with respect to the centers are given by

$$\frac{\partial J}{\partial \boldsymbol{v}_l[t]} = w[t] \sum_{i=1}^{N} \sum_{\{j|A_j \subseteq \Omega, A_j \neq \emptyset\}} c_j^{\alpha} m_{ij}^2[t] \frac{\partial d_{ij}^2[t]}{\partial \boldsymbol{v}_l[t]}$$

$$+ \eta \sum_{s \neq t}^{T} k[t, s] \sum_{i=1}^{N} \sum_{\{j|A_j \subseteq \Omega, A_j \neq \emptyset\}} (m_{ij}[t] - m_{ij}[s])^2 \frac{\partial d_{ij}^2[t]}{\partial \boldsymbol{v}_l[t]}, \tag{24}$$

where

$$\frac{\partial d_{ij}^2}{\partial \boldsymbol{v}_l[t]} = -2\frac{1}{c_j} s_{lj} \boldsymbol{x}_i[t] + 2\frac{1}{c_j^2} \sum_{k=1}^{c} s_{lj} s_{kj} \boldsymbol{v}_k[t], \forall l = 1\ldots c; \forall t = 1\ldots T. \tag{25}$$

Setting these derivatives to zeros, from Eqs. (24) and (25) we can get l linear equations of $\boldsymbol{v}_k[t]$:

$$
\sum_{i=1}^{N} \boldsymbol{x}_i[t] \sum_{\{j|w_l \in A_j\}} w[t]c_j^{\alpha-1}m_{ij}^2[t]
$$

$$
+ \sum_{i=1}^{N} \boldsymbol{x}_i[t] \sum_{s \neq t}^{T} \eta k[t,s] \sum_{\{j|w_l \in A_j\}} (m_{ij}[t] - m_{ij}[s])^2 \frac{1}{c_j}
$$

$$
= \sum_{k=1}^{c} \boldsymbol{v}_k[t] \sum_{i=1}^{N} \sum_{\{j|\{w_l,w_k\} \subseteq A_j\}} w[t]c_j^{\alpha-2}m_{ij}^2[t]
$$

$$
+ \sum_{k=1}^{c} \boldsymbol{v}_k[t] \sum_{s \neq t}^{T} \eta k[t,s] \sum_{i=1}^{N} \sum_{\{j|\{w_l,w_k\} \subseteq A_j\}} (m_{ij}[t] - m_{ij}[s])^2 \frac{1}{c_j^2}. \tag{26}
$$

Let $\boldsymbol{B}[t]$ be a matrix of size $(c \times q_t)$ that is defined by

$$
B_{lq}[t] = \sum_{i=1}^{N} x_{iq}[t] \sum_{\{j|\omega_l \subseteq A_j\}} c_j^{\alpha-1}m_{ij}^2[t]w[t], l = 1, \cdots, c; q = 1, \cdots, q_t, \tag{27}
$$

and define $\boldsymbol{B}[t,s]$ as a matrix of size $(c \times q_t)$:

$$
B_{lq}[t,s] = \sum_{i=1}^{N} x_{iq}[t] \sum_{\{j|\omega_l \in A_j\}} (m_{ij}[t] - m_{ij}[s])^2 \frac{1}{c_j}, l = 1, \cdots, c; q = 1, \cdots, q_t. \tag{28}
$$

Let $\boldsymbol{H}[t]$ be a matrix of size $(c \times c)$ that is defined by

$$
H_{lk}[t] = \sum_{i=1}^{N} \sum_{\{j|\{\omega_l,\omega_k\} \subseteq A_j\}} w[t]c_j^{\alpha-2}m_{ij}^2[t], \; l,k = 1 \cdots c, \tag{29}
$$

and $\boldsymbol{H}[t,s]$ be a matrix of size $(c \times c)$ given by

$$
H_{lk}[t,s] = \sum_{i=1}^{N} \sum_{\{j|\{\omega_l,\omega_k\} \subseteq A_j\}} (m_{ij}[t] - m_{ij}[s])^2 \frac{1}{c_j^2}, \; l,k = 1 \cdots c. \tag{30}
$$

With those notations, the update of cluster centers of the t_{th} view can be obtained by

$$
V[t] = \left(\boldsymbol{H}[t] + \sum_{s \neq t}^{T} \eta k[t,s]\boldsymbol{H}[t,s] \right)^{-1} \left(\boldsymbol{B}[t] + \sum_{s \neq t}^{T} \eta k[t,s]\boldsymbol{B}[t,s] \right). \tag{31}
$$

For clarify, the proposed MvWECM approach is summarized in Algorithm 1.

Algorithm 1. Multi-view Weighted Evidential C-Means clustering.

Input: : The data set $X = \{X[1], X[2], \cdots, X[T]\}$, parameters α,β and η, the number of clusters c, the maximum number of iteration n_{max}

1: **Initialization:**$n = 0$, $M[t] = M_0[t]$, $V[t] = V_0[t]$, $W = (1/T, \cdots, 1/T)$, $\forall t = 1 \cdots T$

2: **repeat**

3:      $n \leftarrow n + 1$

4:      Claulate the value of objective function J based on Eq. (1)

5:      $J_{Old} \leftarrow J$

6:      Update the basic belief assignment M based on Eqs. (13) and (14)

7:      Update the view wights W based on Eq. (23)

8:      Update the clustering center V based on Eq. (31)

9:      Calculate the value of objective function J based on Eq. (1)

10: **until** $|J - J_{Old}| < 10^{-3}$ or $n \geq n_{max}$

Output: The basic belief assignments M, the view wights W and the cluster centers V

---

3 Experiments

In this section, some experiments are performed on Gaussian and UCI data set Iris to evaluate the effectiveness of the proposed algorithm. Fused credal partitions derived from MvWECM and ECM are calculated by:

$$\overline{M}_{MvWECM} = \sum_{t=1}^{T} w_t \cdot M_{MvWECM}[t], \quad \overline{M}_{ECM} = \frac{1}{T} \sum_{t=1}^{T} M_{ECM}[t]. \quad (32)$$

Credal partitions are transformed into hard ones by maximizing the corresponding Pignistic probability [8]. Then, Adjusted Rand Index (ARI), Normalized Mutual Information (NMI) and Cluster Accuracy (CA) are adopted to measure the clustering performance.

A. Gaussian data set

The mean values and covariance matrices of Gaussian data are listed in Table 1. We set $c = 2$, $k[t,s] = 1(t \neq s)$, $\beta = 25$ and $\eta = 1$ in MvWECM , while other parameters are set as default.

Table 1. Distribution of Gaussian data set

View	Mean	Covariance	Size	Mean	Covariance	Size
1_{th} view	(4,4)	$\begin{bmatrix} 1 & 0 \\ 0 & 1 \end{bmatrix}$	200	(7,7)	$\begin{bmatrix} 1.2 & 0 \\ 0 & 1.2 \end{bmatrix}$	200
2_{th} view	(3,6)	$\begin{bmatrix} 2 & 0 \\ 0 & 2 \end{bmatrix}$	200	(5,7)	$\begin{bmatrix} 2 & 0 \\ 0 & 2 \end{bmatrix}$	200

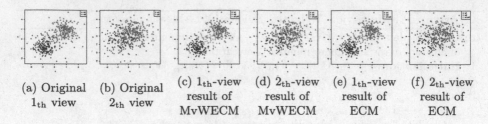

(a) Original
1_{th} view

(b) Original
2_{th} view

(c) 1_{th}-view
result of
MvWECM

(d) 2_{th}-view
result of
MvWECM

(e) 1_{th}-view
result of
ECM

(f) 2_{th}-view
result of
ECM

Fig. 1. Clustering results of 2-class Gaussian data set by different methods.

Table 2 shows that the proposed MvWECM has gained improvement compared with ECM in terms of all the three indexes. We can find that the proportion of sample in overlapping zones of cluster $\{\omega_1\}$ and $\{\omega_2\}$ in the second view is larger than that in the first view as Fig. 1.a - Fig. 1.b shows. It indicates that the first view is more important than the second one. As expected, the first view wight provided by MvWECM is larger than the second view one. There are lots of samples belonging to $\{\omega_1\}$ or $\{\omega_2\}$ are assigned to the meta-class $\{\omega_1, \omega_2\}$ by MvWECM as Fig. 1.c - Fig. 1.d shows, while the result of ECM in Fig. 1.e - Fig. 1.f not only has the uncertain partition but also shows that many samples are assigned incorrectly.

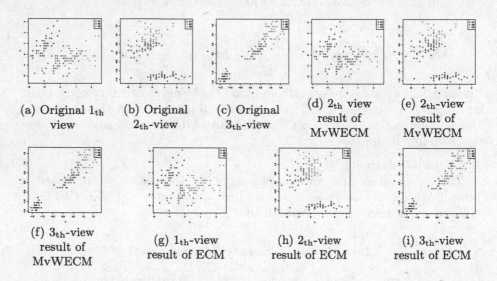

(a) Original 1_{th}
view

(b) Original
2_{th}-view

(c) Original
3_{th}-view

(d) 2_{th} view
result of
MvWECM

(e) 2_{th}-view
result of
MvWECM

(f) 3_{th}-view
result of
MvWECM

(g) 1_{th}-view
result of ECM

(h) 2_{th}-view
result of ECM

(i) 3_{th}-view
result of ECM

Fig. 2. Clustering results of Iris data set transformed by different methods.

Table 2. The ARI, NMI and CA values of clustering results on Gaussian data set .

Index	MvWECM			ECM		
	ARI	NMI	CA	ARI	NMI	CA
Value	0.87	0.80	0.94	0.85	0.77	0.93

B. Iris data set

The Iris data are divided into 3 views by features, such that the distributions of each view are are (a) Sepal.Length, Sepal.Width; (b) Sepal.Width, Petal.Length; (c) Petal.Length, Petal.Width respectively. We set $c = 3$, $k[t, s] = 1(t \neq s)$, $\beta = 25$, $\eta = 1$ in MvWECM, while other parameters are set as default.

Table 3 shows that, especially, AC obtained by MvWECM has outperformed 6% over ECM. Figure 2.a - Fig. 2.c show that $\{\omega_1\}$ is easier to identify, however, $\{\omega_2\}$ and $\{\omega_3\}$ are more difficult to divide depending on the first and the second views but easier on 3_{th} view. The view weights given by MvWECM are $[0.011, 0.037, 0.952]$, those correspond the importance of the view we found on the benchmark data set in advance. The results in Fig. 2.d - Fig. 2.i show that the samples in $\{\omega_2\}$ and $\{\omega_3\}$ are divided as much as possible by MvWECM than ones by ECM. MvWECM has gained improvement compared with ECM. This can be attributed to the fact that it can mine and measure the contribution of each view to cluster structure by weighting views, meanwhile it preserves the diversity of each view when finding the mutual information from different views.

Table 3. The ARI, NMI and CA values of clustering results on Iris data set.

Index	MvWECM			ECM		
	ARI	NMI	CA	ARI	NMI	CA
Value	0.58	0.61	0.71	0.53	0.60	0.67

4 Conclusion

In this study, we proposed a novel multi-view clustering method called MvWECM in the framework of belief functions. The innovation of this approach is that can take advantage of multi-view data and consider the weights of different views. The uncertain cluster structure is modeled by credal partitions. Experimental results show that MvWECM is effective on both Gaussian and UCI data sets.

Acknowledgements. This work was supported by the National Natural Science Foundation of China (No. 61701409), the Natural Science Basic Research Plan in Shaanxi Province of China (No.2018JQ6005), the Aero Science Foundation of China (No.20182053023), the Science Research Plan of China (Xi'an) Institute for Silk Road Research (2019ZD02), and the Fundamental Research Funds for the Central Universities of China (No. 310201911cx041).

References

1. Ahmed, M., Imtiaz, M.T., Khan, R.: Movie recommendation system using clustering and pattern recognition network. In: 2018 IEEE 8th Annual Computing and Communication Workshop and Conference (CCWC), pp. 143–147. IEEE (2018)
2. Bickel, S., Scheffer, T.: Multi-view clustering. In: ICDM, vol. 4, pp. 19–26. Citeseer (2004)
3. Ferreira, L.N., Zhao, L.: Time series clustering via community detection in networks. Inf. Sci. **326**, 227–242 (2016)
4. Huang, D., Lai, J., Wang, C.D.: Ensemble clustering using factor graph. Pattern Recogn. **50**, 131–142 (2016)
5. Jiang, Y., Chung, F.L., Wang, S., Deng, Z., Wang, J., Qian, P.: Collaborative fuzzy clustering from multiple weighted views. IEEE Trans. Cybern. **45**(4), 688–701 (2014)
6. Masson, M.H., Denoeux, T.: ECM: an evidential version of the fuzzy c-means algorithm. Pattern Recogn. **41**(4), 1384–1397 (2008)
7. Qiao, Y., Li, S., Denœux, T.: Collaborative evidential clustering. In: Kearfott, R.B., Batyrshin, I., Reformat, M., Ceberio, M., Kreinovich, V. (eds.) IFSA/NAFIPS 2019 2019. AISC, vol. 1000, pp. 518–530. Springer, Cham (2019). https://doi.org/10.1007/978-3-030-21920-8_46
8. Smets, P.: Decision making in the TBM: the necessity of the pignistic transformation. Int. J. Approx. Reason. **38**(2), 133–147 (2005)
9. Tsai, C.F., Hung, C.: Cluster ensembles in collaborative filtering recommendation. Appl. Soft Comput. **12**(4), 1417–1425 (2012)
10. Wang, C.D., Lai, J.H., Philip, S.Y.: Multi-view clustering based on belief propagation. IEEE Trans. Knowl. Data Eng. **28**(4), 1007–1021 (2015)
11. Zhang, G.Y., Wang, C.D., Huang, D., Zheng, W.S., Zhou, Y.R.: Tw-co-k-means: two-level weighted collaborative k-means for multi-view clustering. Knowl.-Based Syst. **150**, 127–138 (2018)

Unequal Singleton Pair Distance for Evidential Preference Clustering

Yiru Zhang[1,2(✉)] and Arnaud Martin[3]

[1] St. Francis Xavier University, Antigonish, Canada
[2] Hainan University, Haikou, China
[3] University Rennes, CNRS, IRISA, DRUID, Lannion, France

Abstract. Evidential preference based on belief function theory has been proposed recently, simultaneously characterizing preference information with uncertainty and imprecision. However, traditional distances on belief functions do not adapt to some intrinsic properties of preference relations, especially when indifference relation is taken into comparison, therefore may cause inconsistent results in preference-based applications. In order to solve this issue, Unequal Singleton Pair (USP) distance has been proposed previously, with applications limited in preference aggregation. This paper explores forward the effectiveness of USP distance in preference clustering, especially confronting multiple conflicting sources. Moreover, a combination strategy for multiple conflicting sources of preference is proposed. The experiments on synthetic data show that USP distance can effectively improve the clustering results in Adjusted Rand Index (ARI).

Keywords: Belief function theory · Preference clustering · Distance

1 Introduction

With the blossomy development of the digital world, there are various ways to describe one's preference information, such as binary choice (like, dislike), rated with ranks, scores, even colors. Indeed, it is challenging to accurately and effectively cluster preferences, and data quality is one problem. Low quality may be caused by uncertainty, conflicts, incompleteness, or other flaws. We refer to such preference data as "imperfect" in this paper.

Many works have been devoted to modeling imperfect preferences. For example, fuzzy preference [11], possibilistic model [1], probabilistic model [8] and Plackett-Luce model [9] have been proposed to deal with preference with uncertainty and have gained success in various scenarios of applications. However, these methods are usually limited to uncertainty information with uncertainty by proportional or probabilistic values by imposing distribution assumptions.

Dissimilarity measures play an important role in preference analysis, notably in preference aggregation [5,13] and preference learning [7,14] applications. The former application concerns combining multiple preferences into a consensus one,

© Springer Nature Switzerland AG 2021
T. Denœux et al. (Eds.): BELIEF 2021, LNAI 12915, pp. 33–43, 2021.
https://doi.org/10.1007/978-3-030-88601-1_4

while the latter one concentrates on machine learning over preference information, usually applied in ranking problems [4]. Preference clustering is a mission in preference learning, aiming at categorize the preference information based on their similarities, often applied in recommendation systems and community detection tasks [6,16,17]. Indeed, some preference aggregation strategies are intrinsically identical to the minimization of distance sums, as demonstrated in a work of Viappini [15].

Naturally, dissimilarity measure methods in BFT come into the focus for evidential preferences. Even though many dissimilarity measures have been proposed in BFT, they are proved not suitable for evidential preference in [19] because of the conflicts between inherent properties of preference relations. An important one is the equal dissimilarity value between singletons. Formally, in a framework of discernment (FoD) $\Omega = \{\omega_1, \omega_2, \ldots, \omega_H\}$, $\forall \omega_p, \omega_q \in \Omega$, to the limit of our knowledge, the dissimilarity function $d(\cdot)$ over two singletons $d(\omega_p, \omega_q)$ is a constant, usually normalized as 1. However, dissimilarity between singletons should be naturally discriminated in preference relation. For example, the dissimilarity d_Δ between three binary preference relations "strict prefer to" (denoted as $\succ$), "indifferent to" (denoted as $\approx$), and "inverse strict prefer to" (or "preferred by", denoted as $\prec$) is naturally $d_\Delta(\succ, \prec) > d_\Delta(\succ, \approx)$ while all dissimilarity measures in BFT output $d_\Delta(\succ, \prec) = d_\Delta(\succ, \approx)$. This valuation set ignores the intermediate role of "indifference" between the two directions of "strict preference", which is detrimental in distance based applications with weak preferences. Zhang *et al.* [19] discussed negative consequences of such valuation in preference aggregation application and proposed Unequal Singleton Pair (USP) distance, solving the issue by discriminating the dissimilarity between different singleton pairs with other important properties in BFT still guaranteed.

The effectiveness of USP distance in evidential preference aggregation has already been demonstrated [19], while not applied in evidential preference clustering.

In this paper, we study USP distance in evidential preference clustering applications. The evidential preferences are obtained from conflicting preference sources over identical alternative pairs. In our method, the conflicts between multiple sources are interpreted as the ignorance of an agent. The experiments show that the clustering results are improved by applying USP distance in terms of Adjust Rand Index (ARI).

The paper is organized as follows: in Sect. 2, basic notions on belief functions as well as evidential preference model are introduced, followed by the calculation tutorial of USP distance and clustering model in Sect. 3. Afterward, the comparison experiments of clustering with other distances are depicted in Sect. 4. Conclusion and discussions are given finally in Sect. 5.

2 Preliminary

2.1 Belief Functions

Let $\Omega = \{\omega_1, \ldots, \omega_H\}$ be a finite set representing all possible status of a categorical attribute, the uncertainty and imprecision of this attribute is expressed by Basic Belief Assignment (BBA).

Definition 1. *(Basic Belief Assignment (BBA))* A *Basic Belief Assignment* (BBA) on Ω is a function $m : 2^{\Omega} \to [0,1]$ such that:

$$m(\emptyset) = 0 \text{ and } \sum_{X \subseteq \Omega} m(X) = 1. \tag{1}$$

The subsets X of Ω such that $m(X) > 0$ are called *focal elements*, while the finite set Ω is called *the framework of discernment (FoD)*. Ω is also considered as *total ignorance* since it represents all the possibilities. A BBA representing *total ignorance* $(m(\Omega) = 1)$ is also called a *vacuous* BBA. A BBA is *simple supported* if a non-zero value is assigned only to one singleton and Ω. Besides, a BBA m is called *categorical* on element $X, X \in 2^{\Omega}$ if $m(X) = 1$, denoted as X^0. We refer to a categorical BBA on one singleton as *categorically simple supported*.

2.2 Evidential Preference Model

Preference modeling is usually based on order theory. In this paper, we use the widely accepted notions in studies of preferences from [12].

Definition 2. *(Preference relation)* Between any two alternatives a_i, a_j, only three exclusive relations possibly exist $\{\succ, \approx, \sim\}$, defined from binary relation R, with $\neg$ denoting logic negation, as:

Strict preference: $a_i \succ a_j$ iff $a_i R a_j$ and $a_j \neg R a_i$;
Indifference: $a_i \approx a_j$ iff $a_i R a_j$ and $a_j R a_i$;
Incomparability: $a_i \sim a_j$ iff $a_i \neg \succ a_j$ and $a_i \neg \prec a_j$ and $a_i \neg \approx a_j$.

Definition 3. *(Preference Structure)* A preference structure is a collection of binary relations defined on the set $\mathcal{A}$ and such that:

- for each couple (a_i, a_j), $a_i, a_j \in \mathcal{A}$, at least one relation is satisfied;
- for each couple (a_i, a_j), $a_i, a_j \in \mathcal{A}$, if one relation is satisfied, any other relation cannot be satisfied.

The evidential preference model is originally proposed by [10] on weak orders and extended to quasi orders with the consideration of *incomparability* by [18].

Definition 4. *(Evidential preference)* For any alternative pair $a_i, a_j \in \mathcal{A}$, four relations are possible. Therefore, the preference FoD Ω_{ij}^{pref} is defined as:

$$\Omega_{ij}^{pref} = \{\omega_{ij}^R | R \in \{\succ, \prec, \approx, \sim\}\}. \tag{2}$$

The degree of uncertainty on preference relation is represented by values on singletons. The imprecision is characterized by values on union sets.

With the combination rules in the framework of BFT, the evidential preference model is effective in group decision-making with imperfect preference information sources, as systematically discussed in [19].

3   Clustering Model for Evidential Preferences with Unequal Singleton Pair (USP) Distance

In this section, we introduce the clustering model over evidential preference with USP distance, followed by a brief tutorial for calculating USP distance.

3.1   Strategy of Reasoning and Clustering

The reasoning strategy is designed with the procedure depicted in Fig. 1, where σ denotes a preference structure, u an agent, and D the matrix of pairwise distances.

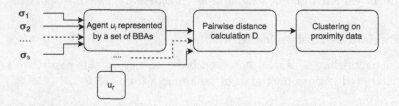

Fig. 1. Strategy of clustering

In a case that an identical agent u's $(u \in \mathcal{U})$ preference is expressed by multiple sources, agent u is therefore represented by a list of pairwise evidential preferences obtained by the combination of multiple sources. Afterward, pairwise distances between different agents are calculated for the clustering process. In this strategy, three main steps are included:

1. Combination of multiple conflicting preference sources for one agent;
2. Calculation of distances between different agents;
3. Clustering over agents based on the proximity distances.

In the following parts, we introduce the combination of multiple preferences and the calculation of distances, while the clustering method is out of the scope because any clustering method for proximity data is available.

3.2   Evidential Preference Reasoning and Combination

Evidential preferences are reasoned from conventional crisp preference information, wildly conflicting preferences from multiple sources. We develop an evidential preference reasoning strategy for multiple (conflicting) sources.

Given multiple preference structures (from multiple sources) $S = \{\sigma_1, \sigma_2, \ldots\}$
for one agent u on identical alternative set $\mathcal{A}$, the average mutual conflicting
(AMC) κ_{AMC} among S is defined as:

$$\kappa_{AMC} = \frac{1}{\binom{|S|}{2}} \sum_{\substack{\sigma_p, \sigma_q \in S \\ p < q}} d(\sigma_p, \sigma_q), \tag{3}$$

where d denotes a distance function for preference orders (rather than pair-wise
preferences). In this paper, we apply Fagin's distance [3], which is an extended
version of Kendall's distance.

The BBAs' values are obtained by mean rule combination with normalization
of AMC. For agent u's opinion between a_i and a_j, denote the crisp preference
from source s as $\sigma_s(a_i, a_j)$. The BBA m_{ij}^u representing agent u's evidential preference opinion between a_i, a_j is calculated by:

$$m_{ij}^u(X) = \begin{cases} \frac{1 - \kappa_{AMC}}{|S|} \sum_{s \in S} m_{ij}^s(X), & \text{if } X \neq \Omega; \\ \kappa_{AMC}, & X = \Omega, \end{cases}$$

where m_{ij}^s is categorical as it comes from a crisp preference source without
uncertainty nor imprecision.

The distance between two agents u_r and u_l are calculated by the mean value
of their pairwise preference distance, defined as:

$$d(u_r, u_l) = \frac{1}{\binom{|\mathcal{A}|}{2}} \sum_{\substack{a_i, a_j \in \mathcal{A} \\ i < j}} d_{BFT}(m_{ij}^{u_r}, m_{ij}^{u_l}), \tag{4}$$

where d_{BFT} denotes a distance function for BBAs in the theory of belief func-
tions, $\binom{|\mathcal{A}|}{2}$ the combination number of 2 elements out of $\mathcal{A}$. In our work, we
apply USP distance as introduced below to avoid the flaw mentioned in the
Sect. 1.

3.3   USP Distance

Unequal Singleton Pair distance is originally proposed to solve a flaw existing
in all dissimilarity measures in BFT. Before USP distance, all measures value
the dissimilarity between singletons equally. In a FoD $\Omega = \{\omega_1, \omega_2, \ldots, \omega_H\}$, the
dissimilarity between any two different singletons is a constant (normalized as
1), formally, $\forall \omega_m, \omega_n \in \Omega, \omega_m \neq \omega_n$:

$$d(\{\omega_m\}^0, \{\omega_n\}^0) \equiv 1. \tag{5}$$

USP distance, which is an extensive version of Jousselme distance, can solve this flaw. given for two BBAs m_1 and m_2 in Ω, USP distance is defined by:

$$d_{USP}(m_1, m_2) = \sqrt{(\boldsymbol{m_1} - \boldsymbol{m_2})^T \boldsymbol{\Sigma}(\boldsymbol{m_1} - \boldsymbol{m_2})}, \tag{6}$$

where Σ denotes the similarity matrix between elements in 2^Ω. In Jousselme distance, Σ is a Jaccard matrix defined on the structure of elements, while in USP distance, Σ is defined by resemblance $resemb$ and entirety $entire$ of the two elements. The value of resemblance and entirety are calculated by the difference in the similarity between singleton pairs.

Here we give a tutorial for USP distance calculation. Define a set of elements in 2^Ω, $W = \{X_1, X_2, \ldots, X_M\}$, therefore $W \subseteq 2^\Omega$. Denote $resemb(W)$ for $resemb(X_1, X_2, \ldots, X_M)$ and $entire(W)$ for $entire(X_1, X_2, \ldots, X_M)$ to simplify the expression. The size of W is defined by the number of elements $X \in 2^\Omega$, denoted by $|W|$. Singletons in W is defined by the union of all elements in W, formally:

$$\cup W = \bigcup_{X_i \in W} X_i. \tag{7}$$

To guarantee the uniqueness of the solution, the entirety value of a singleton is set as 1. Denote the subset of W by W_{sub}, $entire(W)$ is defined as a generalized version of cardinal function on the union sets:

$$entire(W) = \sum_{\omega \in \cup W} entire(\omega) + \sum_{t=1}^{|2^\Omega|} \sum_{\substack{W_{sub} \subseteq W \\ |W_{sub}|=t}} resemb(W_{sub}) \times (-1)^t. \tag{8}$$

To simplify the calculation, we assume that the resemblance values are nonzero only between two singletons and the entirety of a singleton is 1, formally:

$$resemb(W) = 0, \quad \forall W \subseteq 2^\Omega, |W| \geq 3, \tag{9}$$

$$entire(\omega) = 1, \quad \forall \omega \in \Omega. \tag{10}$$

Inserting above equations into Eq. (8), we have:

$$entire(X, Y) = \sum_{\omega \in X \cup Y} entire(\omega) - \sum_{\substack{\omega_m \in X \\ \omega_n \in Y \\ m \neq n}} resemb(\omega_m, \omega_n). \tag{11}$$

Hence, the similarity between two elements X and Y is calculated by:

$$sim(X_1, X_2) = \frac{\displaystyle\sum_{\substack{\omega_m \in X_1 \\ \omega_n \in X_2 \\ m \neq n}} resemb(\omega_m, \omega_n)}{\displaystyle\sum_{\omega \in X_1 \cup X_2} entire(\omega) - \sum_{\substack{\omega_m \in X_1 \\ \omega_n \in X_2 \\ m \neq n}} resemb(\omega_m, \omega_n)}. \tag{12}$$

To guarantee Eq. (9), the following constraint can be deduced:

$$\sum_{\substack{\omega_m,\omega_n\in\Omega \\ \omega_m\neq\omega_n}} sim(\omega_m,\omega_n) \leq 1. \tag{13}$$

3.4   Value Setting of USP Distance for Evidence Preference

Assume that similarities between categorical BBA representing preferences are:

$$\begin{aligned} d_\Delta(\omega^\succ,\omega^\approx) = d_\Delta(\omega^\prec,\omega^\approx) \quad &= x; \\ d_\Delta(\omega^\succ,\omega^\prec) = 1. \end{aligned} \tag{14}$$

Assume $resemb(\omega^\succ,\omega^\approx) = p$, from Eq. (12), we get:

$$p = \frac{2x}{1+x} \tag{15}$$

In this work, we take the extreme value as in [19], shown in Table 1, with which the similarity matrix Sim over 2^Ω can be obtained by Eq. (12).

Table 1. Similarity between singletons

sim	$\omega^\succ$	$\omega^\prec$	$\omega^\approx$
$\omega^\succ$	1	0	1/3
$\omega^\prec$	0	1	1/3
$\omega^\approx$	1/3	1/3	1

For preference structures, by applying Eq. (4), the USP distance degrades to Fagin's distance. Due to the space limitations, the proof will be provided in an extended version.

4   Experiments

In this paper, we show our first experiments on synthetic data generated by Algorithm 1. The implementation is realised by Python 3.7, based on iBelief package[1]. After calculation of pairwise distance over agents, a proximity measure applicable clustering method is used. In this paper, EkNNclus [2] is chosen as the clustering learner. Parameter selection in EkNNclus is not in the scope of this paper. In this paper, we directly set the number of clusters as in the data generation process.

---

[1] https://github.com/jusdesoja/iBelief_python.

Algorithm 1. Generate conflicting preference sources in $|C|$ clusters

Require: Cluster number $|C|$	4: **for** t in $1:T$ **do**
Switch time T	5: randomly generate index i, j;
neighbour size N	6: exchange ranking order of
Alternative size in each order $|A|$	a_i, a_j in σ_c to making a new
Ensure: $|C|$ clusters of preferences	order;
1: Initialise $|C|$ preference structures as	7: **end for**
centroids	8: **end for**
2: **for** each centroid σ_c **do**	9: **end for**
3: **for** n in $1:N$ **do**	

Confronting multiple preference sources, several methods are respectively compared with the average of Euclidean distance and Fagin (Kendall) distance. Clustering results are evaluated by Adjusted Rand Index and Silhouette score, depicted in Fig. 2. To avoid random errors, the average value of 20 times experiments is calculated.

Two sets of experiments are conducted to demonstrate the effectiveness of USP distance in preference clustering. The first one is done with two conflicting sources, while the neighborhood size of preferences over 10 items increment, depicted in Fig. 2. The second one is done with 8 clusters of preferences, with number of sources varying from 1 to 10 with step 2, depicted in Fig. 3.

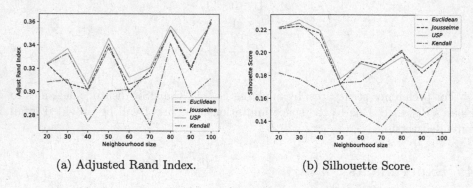

(a) Adjusted Rand Index.   (b) Silhouette Score.

Fig. 2. Clustering results with different neighbourhood size

It can be easily observed from Fig. 2 that USP distance outperforms other distances, especially in terms of ARI. The advantages of USP distance are obtained by moderating the dissimilarity between $\prec, \succ$ and $\prec, \approx$, which respects better the natural definition of the preference relations. From Fig. 3, the result is consistent with Experiment 1, that USP distance out performs other in ARI while worse in sihouette score. Moreover, with one source, experiments with Jousslem distance, USP distance and Kendall distance return identical clustering results.

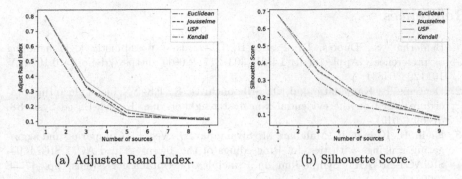

(a) Adjusted Rand Index.   (b) Silhouette Score.

Fig. 3. Clustering results with different number of sources

This proves the assertation that Jousselme distance and USP distance degrade to Kendall distance confronting conventional preferences in total orders. We also observe that both ARI and silhouette score dramatically decrease with the number of conflicting sources augmenting. This is due to the fact that the alternative space is small (with only 10 alternatives), therefore one pair of conflicting preference already takes a big portion in all preference structure. In deed, with. 10 sources of conflicting sources, two agents often become identical after the combination step. The results with 10 sources are similar in all distances, because the data is barely separable at this stage.

5   Discussion and Conclusion

This paper explores the usage of a previously proposed distance, Unequal Singleton Pair (USP) distance, into clustering applications over evidential preferences. A combination rule for multiple preference sources is also proposed by interpreting the conflicts as imprecision. By applying USP distance over evidential preferences, clustering results are improved in terms of ARI.

Compared with the simple average strategy, evidential reasoning with USP distance can moderate the conflict between different information sources. Unfortunately, this also causes some side effects on the clustering mission: The clustering results are improved while the clustering quality is jeopardized in terms of silhouette scores.

Despite that USP distance is empirically proven useful, its effectiveness over incomplete preference structure remains suspicious. In the evidential preference model, missing information is usually modeled by total ignorance, which is equivalent to complete imprecision. However, pieces of missing information are measured as identical by USP distance, making them easily clustered into one identical group. Such a phenomenon is ridiculously against logical facts. To correctly clustering incomplete data within BFT is in the scope of our future work.

References

1. Benferhat, S., Dubois, D., Prade, H.: Towards a possibilistic logic handling of preferences. Appl. Intell. **14**(3), 303–317 (2001). https://doi.org/10.1023/A:1011298804831
2. Denœux, T., Kanjanatarakul, O., Sriboonchitta, S.: EK-NNclus: a clustering procedure based on the evidential k-nearest neighbor rule. Knowl.-Based Syst. **88**, 57–69 (2015)
3. Fagin, R., Kumar, R., Mahdian, M., Sivakumar, D., Vee, E.: Comparing and aggregating rankings with ties. In: Proceedings of the Twenty-Third ACM SIGMOD-SIGACT-SIGART Symposium on Principles of Database Systems, pp. 47–58 (2004)
4. Fürnkranz, J., Hüllermeier, E.: Preference learning and ranking by pairwise comparison. In: Fürnkranz, J., Hüllermeier, E. (eds.) Preference Learning, pp. 65–82. Springer, Heidelberg (2010). https://doi.org/10.1007/978-3-642-14125-6_4
5. Jabeur, K., Martel, J.M., Khélifa, S.B.: A distance-based collective preorder integrating the relative importance of the group's members. Group Decis. Negot. **13**(4), 327–349 (2004). https://doi.org/10.1023/B:GRUP.0000042894.00775.75
6. Kamis, N.H., Chiclana, F., Levesley, J.: Preference similarity network structural equivalence clustering based consensus group decision making model. Appl. Soft Comput. **67**, 706–720 (2018)
7. Kamishima, T., Akaho, S.: Efficient clustering for orders. In: Zighed, D.A., Tsumoto, S., Ras, Z.W., Hacid, H. (eds.) Mining Complex Data, vol. 165, pp. 261–279. Springer, Heidelberg (2009). https://doi.org/10.1007/978-3-540-88067-7_15
8. Lu, T., Boutilier, C.: Vote elicitation with probabilistic preference models: empirical estimation and cost tradeoffs. In: Brafman, R.I., Roberts, F.S., Tsoukiàs, A. (eds.) ADT 2011. LNCS (LNAI), vol. 6992, pp. 135–149. Springer, Heidelberg (2011). https://doi.org/10.1007/978-3-642-24873-3_11
9. Luce, R.D.: Individual Choice Behavior: A Theoretical Analysis. Courier Corporation, North Chelmsford (2012)
10. Masson, M.-H., Destercke, S., Denoeux, T.: Modelling and predicting partial orders from pairwise belief functions. Soft. Comput. **20**(3), 939–950 (2014). https://doi.org/10.1007/s00500-014-1553-9
11. Orlovsky, S.: Decision-making with a fuzzy preference relation. Fuzzy Sets Syst. **1**(3), 155–167 (1978)
12. Öztürké, M., Tsoukiàs, A., Vincke, P.: Preference modelling. In: Multiple Criteria Decision Analysis: State of the Art Surveys. ISORMS, vol. 78, pp. 27–59. Springer, New York (2005). https://doi.org/10.1007/0-387-23081-5_2
13. Roy, B., Slowinski, R.: Criterion of distance between technical programming and socio-economic priority. RAIRO-Oper. Res. **27**(1), 45–60 (1993)
14. Tasgin, M., Bingol, H.O.: Community detection using preference networks. Physica A **495**, 126–136 (2018)
15. Viappiani, P.: Characterization of scoring rules with distances: application to the clustering of rankings. In: The Twenty-Fourth International Joint Conference on Artificial Intelligence (IJCAI 2015), pp. 104–110 (2015)
16. Wang, Y., Zhou, J.T., Li, X., Song, X.: Effective user preference clustering in web service applications. Comput. J. **63**(11), 1633–1643 (2020)

17. Yang, Y., Hooshyar, D., Jo, J., Lim, H.: A group preference-based item similarity model: comparison of clustering techniques in ambient and context-aware recommender systems. J. Ambient. Intell. Humaniz. Comput. 11(4), 1441–1449 (2018). https://doi.org/10.1007/s12652-018-1039-1
18. Zhang, Y., Bouadi, T., Martin, A.: Preference fusion and Condorcet's paradox under uncertainty. In: 20th International Conference on Information Fusion (2017)
19. Zhang, Y., Bouadi, T., Wang, Y., Martin, A.: A distance for evidential preferences with application to group decision making. Inf. Sci. 568, 113–132 (2021)

Transfer Learning

Transfer Evidential C-Means Clustering

Lianmeng Jiao[✉], Feng Wang, and Quan Pan

School of Automation, Northwestern Polytechnical University, Xi'an 710072, China
{jiaolianmeng,quanpan}@nwpu.edu.cn, fengwang@mail.nwpu.edu.cn

Abstract. Clustering is widely used in text analysis, natural language processing, image segmentation and other data mining fields. ECM (evidential c-means) is a powerful clustering algorithm developed in the theoretical framework of belief functions. Based on the concept of credal partition, it extends those of hard, fuzzy, and possibilistic clustering algorithms. However, as a clustering algorithm, it can only work well when the data is sufficient and the quality of the data is good. If the data is insufficient and the distribution is complex, or the data is sufficient but polluted, the clustering result will be poor. In order to solve this problem, using the strategy of transfer learning, this paper proposes a transfer evidential c-means (TECM) algorithm. TECM employs the historical clustering centers in source domain as the reference to guide the clustering in target domain. In addition, the proposed transfer clustering algorithm can adapt to situations where the number of clusters in source domain and target domain is different. The proposed algorithm has been validated on synthetic and real-world datasets. Experimental results demonstrate the effectiveness of transfer learning in comparison with ECM and the advantage of credal partition in comparison with TFCM.

Keywords: Evidential c-means · Clustering · Transfer learning

1 Introduction

Many current clustering algorithms such as c-means can produce good clustering results only under the premise of sufficient data. However, in practice, there are problems such as the data is insufficient and the distribution is complex, or the data is sufficient but polluted. One way to solve this problem is transfer learning. Transfer learning is applying knowledge learned in one domain or task to a different but related domain or task. At present, research on transfer learning mainly focuses on classification, while research on clustering is very limited despite the wide range of real-world clustering applications. Over the last decade or so, there are some studies on transfer learning for clustering. According to

This work was supported by the National Natural Science Foundation of China under Grant 61801386 and Grant 61790552, and the China Postdoctoral Science Foundation under Grant 2019M653743 and Grant 2020T130538.

© Springer Nature Switzerland AG 2021
T. Denœux et al. (Eds.): BELIEF 2021, LNAI 12915, pp. 47–55, 2021.
https://doi.org/10.1007/978-3-030-88601-1_5

the transfer method, they can be roughly divided into four categories [1]: (1) instance-based method, in which it assumes that certain parts of the data in source domain can be reused for learning in target domain by reweighting; (2) feature-representation-based method [2,3], in which the intuitive idea behind this case is to learn a "good" feature representation for target domain; (3) parameter-based method [4–10], in which it assumes that the source domain and the target domain share some parameters or prior distributions of the hyperparameters of the models; (4) relational-knowledge-based method [11], in which it deals with transfer learning for relational domains. Among them, parameter-based transfer clustering is a research hotspot, such as transfer fuzzy c-means [4] and transfer possibilistic c-means [7] have been proposed. The core idea of these algorithms are using the clustering centers in source domain to guide the clustering in target domain.

ECM (evidential c-means) [12] is a powerful clustering algorithm developed in the theoretical framework of belief functions, it extends those of hard, fuzzy, and possibilistic clustering algorithms. ECM is based on a new concept of partition, referred to as a credal partition. This is done by allocating, for each object, a mass of belief, not only to single clusters, but also to any subset of the set of clusters $\Omega = [w_1, ..., w_c]$. In order to enhance its performance in insufficient data situations, in this paper, transfer learning for ECM is exploited. Firstly, on the basis of retaining the structure of the classical ECM objective function, considering the similarity between the clustering centers of the source domain and the target domain, the clustering centers of the source domain and the clustering center correlation matrix are introduced into the new objective function. Then, the iterative algorithm for solving the objective function is derived.

The rest of the paper is organized as follows. In Sect. 2, ECM clustering algorithm is briefly reviewed. In Sect. 3, TECM clustering algorithm is proposed. Experimental results are reported and discussed in Sect. 4. Finally, conclusions are given in Sect. 5.

2 Evidential C-Means

In [13], it is proposed to represent partial knowledge regarding the class membership of an object i by a basic belief assignment (bba) m_i on the set $\Omega = [w_1, ..., w_c]$. Based on this representation, it is possible to model all situations ranging from complete ignorance to full certainty concerning the class of i. For each object i, $m_{ij} = m_i(A_j)$ is low when the distance d_{ij} between i and the focal set A_j is high. Like in fuzzy clustering, each class w_k is represented by a $v_k \in R^p$. We propose to associate to each subset A_j of Ω the barycenter $\overline{v}_j$ of the centers associated to the classes composing A_j. Some notations that will be used are introduced.

$$s_{kj} = \begin{cases} 1 & \text{if } w_k \in A_j \\ 0 & else, \end{cases} \tag{1}$$

and we compute the barycenter $\overline{v}_j$ associated to A_j by $\overline{v}_j = \frac{1}{c_j} \sum_{k=1}^{c} s_{kj} v_k$, where c_j denotes the cardinal of A_j.

Then, we propose to look for the credal partition $M = \{m_1, ..., m_n\} \in R^{n \times 2^c}$ and the matrix V of size $(c * p)$ of clustering centers by minimizing the following objective function:

$$J_{ECM}(M,\, V) = \sum_{i=1}^{n} \sum_{\{j/A_j \neq \emptyset, A_j \subseteq \Omega\}} c_j^{\alpha} m_{ij}^{\beta} \|x_i - \overline{v}_j\|^2 + \sum_{i=1}^{n} \delta^2 m_{i\emptyset}^{\beta},$$

$$\text{s.\,t.} \sum_{\{j/A_j \subseteq \Omega, A_j \neq \emptyset\}} m_{ij} + m_{i\emptyset} = 1 \quad \forall i = 1,\, n, \tag{2}$$

where $m_{i\emptyset}$ denotes $m_i(\emptyset)$, δ controls the amount of data considered as outliers, α and β are weighting exponents.

3   Transfer Evidential C-Means Clustering

3.1   Objective Function

Adopting the transfer method similar to that in [4], in this paper, we construct a novel objective function by utilizing the historical matrix of barycenter $\widetilde{V}_k$ in source domain, and propose the transfer evidential c-means clustering algorithm correspondingly. The objective function of TECM is defined as follows:

$$\min J_{TECM} = \sum_{i=1}^{n} \sum_{\{j/A_{t,j} \neq \emptyset, A_{t,j} \subseteq \Omega\}} c_j^{\alpha} m_{ij}^{\beta} \|x_i - \overline{v}_j\|^2 + \sum_{i=1}^{n} \delta^2 m_{i\emptyset}^{\beta}$$

$$+ \lambda \sum_{\{k/A_{s,k} \neq \emptyset, A_{s,k} \subseteq \Omega\}} \sum_{\{j/A_{t,j} \neq \emptyset, A_{t,j} \subseteq \Omega\}} c_j^{\alpha} r_{kj}^{\gamma} \|\widetilde{v}_k - \overline{v}_j\|^2, \tag{3}$$

$$\text{s.\,t.} \sum_{\{j/A_{t,j} \subseteq \Omega, A_{t,j} \neq \emptyset\}} m_{ij} + m_{i\emptyset} = 1 \quad \forall i = 1,\, n, \qquad \sum_{(j/A_{t,j} \neq \emptyset, A_{t,j} \in \Omega)} r_{kj} = 1,$$

where r_{kj} denotes the similarity between the barycenter $\overline{v}_j$ in the target domain and the barycenter $\widetilde{v}_k$ in the source domain, λ is a balance coefficient of transfer learning and γ is a weighting exponent.

For Eq. (3), the following explanations are given.

1) The first term and the second term in Eq. (3) are directly inherited from the classical ECM, which is mainly used to learn from the data available in the target domain.
2) The third one is used to learn the knowledge from the source domain. In this term, r_{kj} denotes the similarity between the jth barycenter in the target domain and the kth barycenter in the source domain; this term implies that if the jth barycenter in the target domain and the kth barycenter in the source domain are more similar, the jth barycenter in the target domain will learn more knowledge from the kth barycenter in the source domain.

3.2   Derivation

By minimizing (3) and using the Lagrange optimization, one may derive the following update equations for the credal partition M and the clustering center correlation matrix R:

$$m_{ij} = \frac{c_j^{-\alpha/(\beta-1)} \, ||x_i - \overline{v}_j||^{-2/(\beta-1)}}{\sum_{A_{t,l}\neq\emptyset} c_l^{-\alpha/(\beta-1)} ||x_i - \overline{v}_l||^{-2/(\beta-1)} + \delta^{-2/(\beta-1)}}, \tag{4}$$

$$r_{kj} = \frac{c_j^{-\alpha/(\gamma-1)} ||\widetilde{v}_k - \overline{v}_j||^{-2/(\gamma-1)}}{\sum_{A_{t,l}\neq\emptyset} c_l^{-\alpha/(\gamma-1)} ||\widetilde{v}_k - \overline{v}_l||^{-2/(\gamma-1)}} \tag{5}$$

Now it is considered that M and R are fixed. The minimization of J_{TECM} with respect to V is an unconstrained optimization problem. The partial derivatives of J_{TECM} with respect to the centers are set to zero:

$$\frac{\partial J_{TECM}}{\partial v_l} = 0, \tag{6}$$

namely,

$$\sum_i x_i \sum_{A_{t,j}\neq\emptyset} c_j^{\alpha-1} m_{ij}^\beta s_{lj} + \lambda \sum_{A_{s,k}\neq\emptyset} \widetilde{v}_k \sum_{A_{t,j}\neq\emptyset} c_j^{\alpha-1} r_{kj}^\gamma s_{lj}$$
$$= \sum_z v_z \sum_i \sum_{A_{t,j}\neq\emptyset} c_j^{\alpha-2} m_{ij}^\beta s_{lj} s_{zj} + \lambda \sum_z v_z \sum_{A_{s,k}\neq\emptyset} \sum_{A_{t,j}\neq\emptyset} c_j^{\alpha-2} r_{kj}^\gamma s_{lj} s_{zj}. \tag{7}$$

Let B_1, B_2, H_1 and H_2 be matrixs defined by

$$B_{1_{lq}} = \sum_{i=1}^n x_{iq} \sum_{A_{t,j}\neq\emptyset} c_j^{\alpha-1} m_{ij}^\beta s_{lj} = \sum_{i=1}^n x_{iq} \sum_{w_l\in A_{t,j}} c_j^{\alpha-1} m_{ij}^\beta, \tag{8}$$

$$B_{2_{lq}} = \sum_{A_{s,k}\neq\emptyset} \widetilde{v}_{kq} \sum_{A_{t,j}\neq\emptyset} c_j^{\alpha-1} r_{kj}^\gamma s_{lj} = \sum_{A_{s,k}\neq\emptyset} \widetilde{v}_{kq} \sum_{w_l\in A_{t,j}} c_j^{\alpha-1} r_{kj}^\gamma, \tag{9}$$

$$H_{1_{lz}} = \sum_i \sum_{A_{t,j}\neq\emptyset} c_j^{\alpha-2} m_{ij}^\beta s_{lj} s_{zj} = \sum_i \sum_{A_{t,j}\supseteq\{w_z,w_l\}} c_j^{\alpha-2} m_{ij}^\beta, \tag{10}$$

$$H_{2_{lz}} = \sum_{A_{s,k}\neq\emptyset} \sum_{A_{t,j}\neq\emptyset} c_j^{\alpha-2} r_{kj}^\gamma s_{lj} s_{zj} = \sum_{A_{s,k}\neq\emptyset} \sum_{A_{t,j}\supseteq\{w_z,w_l\}} c_j^{\alpha-2} r_{kj}^\gamma, \tag{11}$$

where $l = 1, c$, $q = 1, p$, and $z = 1, c$. With these notations, V is solution of the following linear system:

$$B_1 + \lambda B_2 = (H_1 + \lambda H_2)V, \tag{12}$$

which can be solved using a standard linear system solver. Based on the above analysis, the proposed TECM is presented in Algorithm 1.

Algorithm 1. Transfer Evidential C-means Clustering

Input: samples in target domain:$\{x_1,...x_n\}$, barycenter in source domain: $\{\tilde{v}_1,...\tilde{v}_k\}$, clustering number: c, weighting exponent: $\alpha \geqslant 0$, $\beta > 1$, $\gamma > 1$, distance to the empty set: $\delta > 0$, termination threshold: ε, balance coefficient of transfer learning: λ.

Output: credal partition M.

1: initial clustering centers V_0.
2: $t \leftarrow 0$
3: **repeat**
4:      $t \leftarrow t+1$
5:      compute credal partition M_t using (4);
6:      compute clustering center correlation matrix R_t using (5);
7:      compute B_1, B_2, H_1, H_2 using(8)(9)(10)(11);
8:      compute clustering centers V_t using (12);
9: **until** $(|J_{TECM}(t) - J_{TECM}(t-1)| < \varepsilon)$

---

4   Experimental Results

In this section, the proposed algorithm will be extensively evaluated on synthetic and real-world datasets. The experiment is divided into two parts, the first part is comparing with ECM algorithm to illustrate the effectiveness of the transfer learning, the second part is comparing with TFCM proposed in [4], the advantage of credal partition is illustrated.

4.1   Comparison of TECM and ECM

In this section, we will verify the effectiveness of the transfer learning by comparing the clustering effects of ECM. The indices used for performance evaluation are accuracy (ac), Adjusted Rand Index (ARI), Normalized Mutual Information (NMI) and Davies-Bouldin Index (DB). The definations of these indices are shown in [2, 6, 10].

Synthetic Dataset. Due to space constraints, we only consider a more general case of different clustering number here. T-1 and S-1 represent the datasets in target and source domain. Parameters used to generate datasets are given in Table 1. From Table 1, we can see that the first three clusters in S1 have similar but different means and covariances as T1, which means that S1 is a source domain dataset containing useful information for clustering dataset T1 in the target domain.

TECM is repeated 100 times on T-1. The means of ac, ARI, NMI, and DB values on the target dataset T-1 obtained by the proposed TECM algorithm using different parameter setting (λ) based on the knowledge extracted from the S-1 are shown in Fig. 1.

From the experimental results, it can be seen that when choosing appropriate λ, the clustering effect using TECM compared with ECM $(\lambda = 0)$ has large

Table 1. Parameters used to generate the synthetic datasets T-1 and S-1.

T-1	Mean	Covariance	Size
Cluster 1	[−4 4]	[6 0; 0 6]	10
Cluster 2	[1 13]	[6 0; 0 6]	10
Cluster 3	[4 6]	[6 0; 0 6]	10
S-1	Mean	Covariance	Size
Cluster 1	[−5 6]	[5 0; 0 5]	1000
Cluster 2	[0 15]	[5 0; 0 5]	1000
Cluster 3	[5 5]	[5 0; 0 5]	1000
Cluster 4	[−5 −5]	[5 0; 0 5]	1000

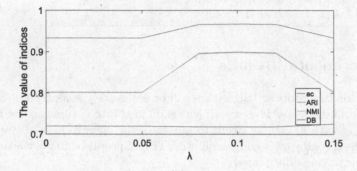

Fig. 1. The means of ac, ARI, NMI, and DB values on the target dataset T-1 obtained by the proposed TECM algorithm using different parameter setting (λ).

improvement. The values of three external indexs all increase and the value of internal index DB decreases when the appropriate λ is selected, this means that even if the clustering number of source domain and target domain is different, some structures in target domain are similar to that in source domain, the clustering effect will also be improved.

It also can be seen from the experimental results that as the value of λ increases, the clustering performance first becomes better and then becomes worst. When the value of λ is small, the effect of transfer learning is not obvious; when the value of λ is too large, negative transfer will occur. Therefore, it is very important to choose the appropriate λ. As for how to select the appropriate λ, firstly, the data of source domain and target domain are normalized. Then, the appropriate λ is selected in range of $[0, 1]$ through the grid search strategy.

Texture Image Segmentation. The experimental dataset used in this section is Brodatz texture image segmentation dataset. Specifically, six basic textures: D6, D11, D46, D93, D96, and D101, in this repository are used to synthesize the texture images acting as the source or target dataset in our experiment. The size of the composite texture image has been resized to 90 pixels by 90 pixels.

In order to simulate the real dataset environment, Gaussian noise is added to texture images in target domain. In the experiment, (a, c) in Fig. 2 represent the image in source domain and (b, d) in Fig. 2 represent the corresponding image in target domain. In the first group, the clustering number K of source domain and target domain is the same, while the latter group is different.

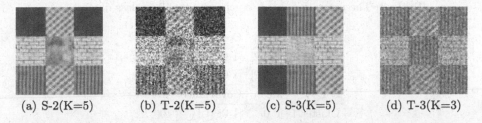

(a) S-2(K=5)   (b) T-2(K=5)   (c) S-3(K=5)   (d) T-3(K=3)

Fig. 2. Texture images in source domain and target domain.

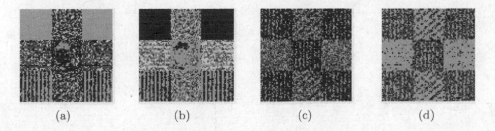

(a)   (b)   (c)   (d)

Fig. 3. The segmentation results of T-2 and T-3.

The results of image segmentation in target domain are shown in Fig. 3, where (a) is segmentation result of T-2 without using the transfer knowledge and (b) is the segmentation result of T-2 using the transfer knowledge from S-2, (c) is segmentation result of T-3 without using the transfer knowledge and (d) is the segmentation result of T-3 using the transfer knowledge from S-3.

It can be seen from the segmentation results that the segmentation effect of texture image is improved to a certain extent after using the transfer knowledge, and the texture information in the image is also clearer.

4.2   Comparison of TECM and TFCM

Two-Class Dataset. To illustrate the advantage of TECM over TFCM, let us consider the following Two-class dataset. The parameters used to generate the Two-class dataset are shown in the Table 2. The number of classes in TFCM and TECM are all set to 2 (w_1, w_2). But in TECM, the barycenters consist of w_1, w_2, $w_1 \cup w_2$ and empty set.

The clustering result of T-4 is shown as Fig. 4. In TECM, since no samples belong to the empty set, there are three barycenters in clustering result of TECM. The samples at the junction of two classes are separately clustered. Such clustering result is richer and more accurate compared with the clustering result of TFCM.

Table 2. The parameters used to generate the Two-class dataset.

T-4	Mean	Covariance	Size
Cluster 1	[0 0.2]	[1 0; 0 1]	10
Cluster 2	[1 0.2]	[1 0; 0 1]	10
S-4	Mean	Covariance	Size
Cluster 1	[0 0]	[1 0; 0 1]	100
Cluster 2	[1 0]	[1 0; 0 1]	100

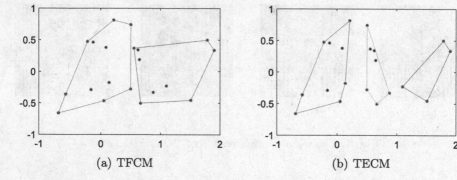

(a) TFCM   (b) TECM

Fig. 4. Clustering results of T-4 using TECM (a) and TFCM (b)

5   Conclusion

In this paper, a transfer clustering algorithm based on ECM is proposed. In this method, classical ECM is applyed to source data to gain the clustering centers of source domain. Then, the clustering centers of source domain and the target data are used to structure novel objective function. The experimental results show that transfer learning is useful, and the comparsion with TFCM illustrates the advantage of TECM. Although the proposed TECM clustering algorithm has demonstrated their promising performance, more works can be further addressed about this research topic. One important work is how to tune the tradeoff λ adaptively.

References

1. Pan, S.J., Yang, Q.: A survey on transfer learning. IEEE Trans. Knowl. Data Eng. **22**(10), 1345–1359 (2010)
2. Yang, L., Jing, L., Liu, B., Yu, J.: Common latent space identification for heterogeneous co-transfer clustering. Neurocomputing **269**, 29–39 (2017)
3. Dai, W., Yang, Q., Xue, G.R., Yu, Y.: Self-taught clustering. In: Proceedings of the 25th International Conference on Machine Learning, pp. 200–207. ACM, New York (2008)
4. Deng, Z., Jiang, Y., Chung, F.L., Choi, K.S., Wang, S.: Transfer prototype-based fuzzy clustering. IEEE Trans. Fuzzy Syst. **24**(5), 1210–1232 (2015)
5. Wang, R., Zhou, J., Liu, X., Han, S., Wang, L., Chen, Y.: Transfer clustering based on Gaussian mixture model. In: Proceedings of IEEE Symposium Series on Computational Intelligence, pp. 2522–2526. IEEE, Xiamen (2019)
6. Qin, J., Zhang, Y., Jiang, Y., Hang, W.: Transfer fuzzy clustering based on self-constraint of multiple medoids. J. Shandong Univ. (Eng. Sci.) **49**(2), 107–115 (2019)
7. Gargees, R., Keller, J.M., Popescu, M.: TLPCM: Transfer learning possibilistic c-means. IEEE Trans. Fuzzy Syst. (2020). In press
8. Cheng, Y., Jiang, Y., Qian, P., Wang, S.: A maximum entropy clustering algorithm based on knowledge transfer and its application to texture image segmentation. CAAI Trans. Intell. Syst. **12**(2), 179–187 (2017)
9. Sun, S., Jiang, Y., Qian, P.: Transfer learning based maximum entropy clustering. In: Proceedings of 4th IEEE International Conference on Information Science and Technology, pp. 829–832. IEEE, Shenzhen (2014)
10. Qian, P., et al.: Cluster prototypes and fuzzy memberships jointly leveraged cross-domain maximum entropy clustering. IEEE Trans. Cybern. **46**(1), 181–193 (2015)
11. Yu, L., Dang, Y., Yang, G.: Transfer clustering via constraints generated from topics. In: Proceedings of IEEE International Conference on Systems, Man, and Cybernetics, pp. 3203–3208. IEEE, Seoul (2012)
12. Masson, M.H., Denoeux, T.: ECM: an evidential version of the fuzzy c-means algorithm. Pattern Recogn. **41**(4), 1384–1397 (2008)
13. Denoeux, T., Masson, M.H.: EVCLUS: evidential clustering of proximity data. IEEE Trans. Syst. Man Cybern. Part B (Cybern.) **34**(1), 95–109 (2004)

Evidential Clustering Based on Transfer Learning

Kuang Zhou[1(✉)], Mei Guo[1], and Arnaud Martin[2]

[1] School of Mathematics and Statistics, Northwestern Polytechnical University,
Xi'an 710072, Shaanxi, People's Republic of China
kzhoumath@nwpu.edu.cn
[2] Univ Rennes, CNRS, IRISA, DRUID, Rue E. Branly, 22300 Lannion, France
Arnaud.Martin@univ-rennes1.fr

Abstract. Clustering is an essential part of data mining, which can be used to organize data into sensible groups. Among the various clustering algorithms, the prototype-based methods have been most popularly applied due to the easy implementation, simplicity and efficiency. However, most of them such as the c-means clustering are no longer effective when the data is insufficient and uncertain. While the data for the current clustering task may be sparse, there is usually some useful knowledge available in the related scenes. Transfer learning can be adopted to address such cross domain learning problems by using information from data in a related domain and transferring that data/knowledge to the target task. The inconsistency between different domains can increase the uncertainty in the data. To handle the insufficiency and uncertainty problems in the clustering task simultaneously, a prototype-based evidential transfer clustering algorithm, named transfer evidential c-means (TECM), is introduced in the framework of belief functions. The proposed algorithm employs the cluster prototypes of the source data as references to guide the clustering process of the target data. The experimental studies are presented to demonstrate the advantages of TECM in both synthetic and real-world data sets.

Keywords: Belief functions · Clustering · Transfer learning · Uncertainty · Source domain

1 Introduction

Clustering is an unsupervised technique aiming to classify patterns into groups [6,10]. It has been widely used in many fields such as image segmentation, market research and data analysis. Traditional clustering methods, such as c-means, usually work well when the data are sufficient. However, in real world, uncertain and noisy data are omnipresent. Moreover, sometimes we can not get enough data to train a fine clustering model. To address the problems of a lack of information and data impurity, several advanced cluster models have been developed,

© Springer Nature Switzerland AG 2021
T. Denœux et al. (Eds.): BELIEF 2021, LNAI 12915, pp. 56–65, 2021.
https://doi.org/10.1007/978-3-030-88601-1_6

such as semi-supervised learning [1], multi-view clustering [7], transfer learning [2,4] and so on.

Transfer learning can learn an effective model for the target domain by effectively leveraging useful information from the source domain [2]. Figure 1 illustrates a situation where transfer learning is useful. As we can see, it is difficult to obtain an ideal partition for the target data (the left figure) as they are too sparse. However, if information from the source domain (the right figure) is considered, more promising clustering results can be expected. In general, two kinds of information can be transferred from the source to the target domain: raw data or knowledge [9]. Due to the necessity of privacy protection in some applications, such as users' personal information, the original raw data in the source domain are not always accessible. Thus to employ some advanced knowledge from the source domain instead of raw data is more practical. For example, in the clustering task, the cluster prototypes of the source data (red triangles in the right figure) can be regarded as good references for the target domain.

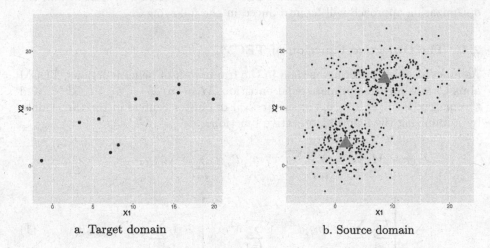

a. Target domain   b. Source domain

Fig. 1. An example where transfer learning is required for the clustering task. (Color figure online)

The available knowledge in the source domain can help us improve the cluster model, but the inconsistency between information from the two domains may increase the uncertainty in the data. The theory of belief functions is an efficient mathematical tool for uncertain information representation and fusion. The concept of credal partitions defined in the framework of belief functions is first proposed by Denœux et al. [3] to deal with the uncertain cluster structure, and following many evidential clustering methods have been designed and widely applied [5,8]. In this paper, we combine the idea of evidential clustering and transfer learning to develop a new clustering method, named transfer evidential c-means (TECM), for insufficient and uncertain data. It first identifies cluster prototypes based on the source domain, which are then transferred into the target domain to guide the clustering procedure. The experimental results on generated and UCI data show the effectiveness of the proposed method.

The remainder of this paper is organized as follows. The proposed TECM algorithm is presented in detail in Sect. 2. Numerical experiments are conducted in Sect. 3. Conclusions are drawn in the final section.

2   Transfer Evidential c-means

Inspired by the idea of evidential clustering and transfer learning, in this section we will introduce the transfer evidential c-means (TECM) clustering algorithm.

Denote the n data samples in the target domain by $\boldsymbol{X} = \{\boldsymbol{x}_1, \boldsymbol{x}_2, \cdots, \boldsymbol{x}_n\}$ and assume that there are c clusters. The frame of discernment is $\Omega = \{\omega_1, \cdots, \omega_c\}$. The available supervised knowledge in a related domain is represented by prototypes $\boldsymbol{V}^{(s)} = \{\boldsymbol{v}_1^{(s)}, \boldsymbol{v}_2^{(s)}, \cdots, \boldsymbol{v}_c^{(s)}\}$. The superscript (s) indicates that the prototypes are from the source domain. The objective function of TECM and the optimization approach will be introduced in the following.

2.1   The Objective Function of TECM

As an evidential clustering method in the framework of belief functions, TECM aims to look for the optimal credal partition $\boldsymbol{M} = (\boldsymbol{m}_1, \cdots, \boldsymbol{m}_n) \in \mathbb{R}^{n \times 2^c}$ and the matrix $\boldsymbol{V} = (\boldsymbol{v}_1, \cdots, \boldsymbol{v}_c)$ of size $(c \times p)$ of cluster centers in the target data by minimizing the following objective function:

$$J_{\text{TECM}}(\boldsymbol{M}, \boldsymbol{V}) = \sum_{i=1}^{n} \sum_{\substack{A_j \subseteq \Omega \\ A_j \neq \emptyset}} c_j^\alpha m_{ij}^\beta d_{ij}^2 + \sum_{i=1}^{n} \delta^2 m_{i\emptyset}^\beta$$

$$+ \beta_1 \left[\sum_{i=1}^{n} \sum_{\substack{A_j \subseteq \Omega \\ A_j \neq \emptyset}} c_j^\alpha m_{ij}^\beta d_{ij}^{2(s)} + \sum_{i=1}^{n} \delta^2 m_{i\emptyset}^\beta \right] + \beta_2 \sum_{k=1}^{c} ||\boldsymbol{v}_k^{(s)} - \boldsymbol{v}_k||^2, \qquad (1)$$

subject to:

$$\sum_{A_j \subseteq \Omega, A_j \neq \emptyset} m_{ij} + m_{i\emptyset} = 1, \qquad (2)$$

where m_{ij} denotes $m_i(A_j)$ and $m_{i\emptyset}$ denotes $m_i(\emptyset)$. $c_j = |A_j|$ denotes the cardinal of A_j. d_{ik} denotes the distance between $\boldsymbol{x}_i$ and the barycenter (prototype, denoted by $\overline{\boldsymbol{v}}_k$) associated with A_k:

$$d_{ik}^2 = ||\boldsymbol{x}_i - \overline{\boldsymbol{v}}_k||^2, \qquad (3)$$

where prototype $\overline{v}_k$ can defined mathematically by:

$$\overline{\boldsymbol{v}}_k = \frac{1}{c_k} \sum_{h=1}^{c} s_{hk} \boldsymbol{v}_h, \quad \text{with} \quad s_{hk} = \begin{cases} 1 & \text{if } \omega_h \in A_k \\ 0 & \text{else} \end{cases}. \qquad (4)$$

Notation v_h denotes the center of samples in cluster ω_h. Parameters α, β and δ control the degree of penalization for imprecise classes with high cardinality, the fuzziness of the partition, and the amount of data considered as outliers respectively. These parameters have the same meaning as those in ECM [8].

The objective functions in Eq. (1) has four terms. The first two terms are directly inherited from ECM, which are mainly used to learn from the target data. The third and fourth terms enable the model to learn with the knowledge from the source domain, where the knowledge in the form of cluster prototypes is available for the clustering task. β_1 and β_2 are nonnegative parameters which can balance the influence of data in the target domain and knowledge in the source domain. In the experiments, we suggest the default values for these parameter $\alpha = 1, \beta = 2, \beta_1 = \beta_2 = 1, \delta = 10$.

2.2 Optimization

To minimize the objective function J_{TECM}, the Lagrange multipliers method is adopted. First, consider that the prototype sets in the target domain, V, is fixed. To solve the constrained minimization problem with respect to the membership matrix M, n Lagrange multipliers $\lambda_i (i = 1, \cdots, n)$ are introduced and the Lagrangian can be written as:

$$L(M; \lambda_1, \cdots, \lambda_n) = J_{\text{TECM}} - \sum_{i=1}^{n} \lambda_i \left(\sum_{\substack{A_j \subseteq \Omega \\ A_j \neq \emptyset}} m_{ij} + m_{i\emptyset} - 1 \right). \tag{5}$$

Differentiating the Lagrangian with respect to m_{ij}, $m_{i\emptyset}$, and λ_i and setting the derivatives to zero, the necessary condition of optimality for M can be got as:

$$\frac{\partial L}{\partial m_{ij}} = c_j^\alpha \beta m_{ij}^{\beta-1} \left(d_{ij}^2 + \beta_1 d_{ij}^{2(s)} \right) - \lambda_i = 0. \tag{6}$$

$$\frac{\partial L}{\partial m_{i\emptyset}} = \beta m_{i\emptyset}^{\beta-1} \left(\delta^2 + \beta_1 \delta^2 \right) - \lambda_i = 0. \tag{7}$$

$$\frac{\partial L}{\partial \lambda_i} = \sum_{\substack{A_j \subseteq \Omega \\ A_j \neq \emptyset}} m_{ij} + m_{i\emptyset} - 1 = 0. \tag{8}$$

From Eqs. (6) and (7), it is easy to obtain

$$m_{ij} = \left(\frac{\lambda_i}{c_j^\alpha \beta \left(d_{ij}^2 + \beta_1 d_{ij}^{2(s)} \right)} \right)^{1/(\beta-1)}. \tag{9}$$

$$m_{i\emptyset} = \left(\frac{\lambda_i}{\beta(\delta^2 + \beta_1 \delta^{2(s)})} \right)^{1/(\beta-1)}. \tag{10}$$

Substituting Eqs. (9) and (10) into Eq.(8), we can get

$$\left(\frac{\lambda_i}{\beta} \right)^{1/(\beta-1)} = \frac{1}{\sum\limits_{\substack{A_j \subseteq \Omega \\ A_j \neq \emptyset}} \Delta_{ij} + \left(\frac{1}{\delta^2 + \beta_1 \delta^{2(s)}} \right)^{\frac{1}{\beta-1}}}, \tag{11}$$

where

$$\Delta_{ij} = \left(\frac{1}{c_j^\alpha \left(d_{ij}^2 + \beta_1 d_{ij}^{2(s)} \right)} \right)^{\frac{1}{\beta-1}}. \tag{12}$$

Return in Eqs. (9) and (10),

$$m_{ij} = \frac{\left(1/\left(c_j^\alpha \left(d_{ij}^2 + \beta_1 d_{ij}^{2(s)} \right) \right) \right)^{\frac{1}{\beta-1}}}{\sum\limits_{\substack{A_k \subseteq \Omega \\ A_k \neq \emptyset}} \left(1/\left(c_k^\alpha \left(d_{ik}^2 + \beta_1 d_{ik}^{2(s)} \right) \right) \right)^{\frac{1}{\beta-1}} + \left(\frac{1}{\delta^2 + \beta_1 \delta^2} \right)^{\frac{1}{\beta-1}}}, \tag{13}$$

and

$$m_{i\emptyset} = \frac{\left(\frac{1}{\delta^2 + \beta_1 \delta^2} \right)^{\frac{1}{\beta-1}}}{\sum\limits_{\substack{A_k \subseteq \Omega \\ A_k \neq \emptyset}} \left(1/\left(c_k^\alpha \left(d_{ik}^2 + \beta_1 d_{ik}^{2(s)} \right) \right) \right)^{\frac{1}{\beta-1}} + \left(\frac{1}{\delta^2 + \beta_1 \delta^2} \right)^{\frac{1}{\beta-1}}}. \tag{14}$$

Next we consider that the credal membership matrix M is fixed. It is easy to see the minimization of J_{TECM} with respect to V is an unconstrained optimization problem. The partial derivatives of J_{TECM} with respect to the prototypes of the specific classes in the target domain can be given by:

$$\frac{\partial J_{\text{TECM}}}{\partial v_l} = \sum_{i=1}^n \sum_{\substack{A_j \subseteq \Omega \\ A_j \neq \emptyset}} c_j^\alpha m_{ij}^\beta \frac{\partial d_{ij}^2}{\partial v_l} - 2\beta_2(v_l^{(s)} - v_l), \tag{15}$$

$$\frac{\partial d_{ij}^2}{\partial v_l} = 2 (x_i - \overline{v}_j) \left(-s_{lj} \frac{1}{c_j} \right), \tag{16}$$

where $\overline{v}_j$ is defined by Eq. (4). Thus we have:

$$\frac{\partial J_{\text{TECM}}}{\partial v_l} = -2 \sum_{i=1}^n \sum_{\substack{A_j \subseteq \Omega \\ A_j \neq \emptyset}} c_j^{\alpha-1} m_{ij}^\beta s_{lj} \left(x_i - \frac{\sum_{k=1}^c s_{kj} v_k}{c_j} \right)$$

$$- 2\beta_2 \left(v_l^{(s)} - v_l \right). \tag{17}$$

Setting these derivatives to zero, we can get l linear equations of v_k:

$$\sum_i x_i \sum_{\substack{A_j \subseteq \Omega \\ A_j \neq \emptyset}} c_j^{\alpha-1} m_{ij}^{\beta} s_{lj} = \sum_{k=1}^{c} v_k \sum_{i=1}^{n} \sum_{\substack{A_j \subseteq \Omega \\ A_j \neq \emptyset}} c_j^{\alpha-2} m_{ij}^{\beta} s_{kj} s_{lj}$$

$$- \beta_2 \left(v_l^{(s)} - v_l \right). \tag{18}$$

Let B be a matrix of size $(c \times p)$, and it can be defined by:

$$B_{lq} = \sum_{i=1}^{n} x_{iq} \sum_{\substack{A_j \subseteq \Omega \\ A_j \neq \emptyset}} c_j^{\alpha-1} m_{ij}^{\beta} s_{lj} = \sum_{i=1}^{n} x_{iq} \sum_{A_j \ni \omega_l} c_j^{\alpha-1} m_{ij}^{\beta}, \tag{19}$$

and H be a matrix of size $(c \times c)$ given by:

$$H_{lk} = \sum_{i=1}^{n} \sum_{\substack{A_j \subseteq \Omega \\ A_j \neq \emptyset}} c_j^{\alpha-2} m_{ij}^{\beta} s_{lj} s_{kj} = \sum_{i} \sum_{A_j \supseteq \{\omega_k, \omega_l\}} c_j^{\alpha-2} m_{ij}^{\beta}. \tag{20}$$

Let I be the $(c \times c)$ identity matrix. The prototype matrix v can be got by solving the following linear system:

$$(H + \beta_2 I) v = B + \beta_2 v^{(s)}. \tag{21}$$

3 Experiments

Some experiments are provided in this section. Generated Gaussian data and some UCI data sets are considered to show the performance of the proposed evidential transfer clustering method. In all experiments, the credal partitions provided by ECM and TECM are transformed into hard partitions by using maximum the corresponding Pignistic probability [11]. The parameters in ECM and TECM are all set as default ($\alpha = 1, \beta = 2, \beta_1 = \beta_2 = 1, \delta = 10$). Then, the Adjusted Rand Index (ARI) and Normalized Mutual Information (NMI) to measure closeness of a hard partition to the ground truth are adopted as performance index.

3.1 Gaussian Data Sets

As mentioned, TECM has advantages in the situation when the data in the target domain are insufficient and uncertain to train a good model. This experiment is to illustrate the application scope of TECM. Assume that source data and target data both follow two-dimensional Gaussian distribution. The mean values and covariance matrices of the source data and target data are listed in Table 1.

There are three clusters in both the target data and the source data. Denote the number of data samples in each cluster of the target and source domain by

Table 1. Distributions of source data and target data.

Mean	Covariance	Mean	Covariance
$\mu_1^{(s)} = [2, 4]$	$\begin{bmatrix} 10 & 0 \\ 0 & 10 \end{bmatrix}$	$\mu_1 = [3, 4]$	$\begin{bmatrix} 10 & 0 \\ 0 & 11 \end{bmatrix}$
$\mu_2^{(s)} = [9, 15]$	$\begin{bmatrix} 25 & 0 \\ 0 & 7 \end{bmatrix}$	$\mu_2 = [10.5, 12.5]$	$\begin{bmatrix} 25 & 0 \\ 0 & 7 \end{bmatrix}$
$\mu_3^{(s)} = [8, 30]$	$\begin{bmatrix} 30 & 0 \\ 0 & 20 \end{bmatrix}$	$\mu_3 = [9, 29]$	$\begin{bmatrix} 30 & 0 \\ 0 & 19.5 \end{bmatrix}$

n_t and n_s respectively. As mentioned, when n_t is small, it is difficult to cluster the samples in the target domain correctly.

The experiment is designed by increasing n_t gradually (from 10 to 500) and applying both ECM and TECM algorithms. For each n_t, Gaussian data are generated 100 times under the fixed parameters in Table 1. ECM and TECM algorithms are evoked each time. Noted that here in TECM the prototypes of clusters in the source domain are got by evoking c-means clustering method on the source data. The average values of ARI and NMI are reported and the results are shown in Fig. 2. As can be seen, the clustering results obtained by TECM is significantly better than those by ECM in terms of both ARI and NMI.

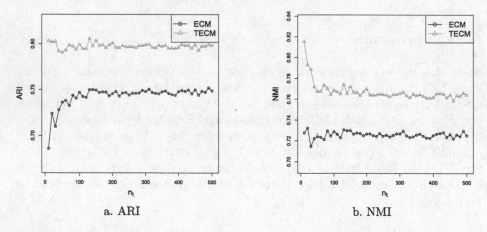

a. ARI   b. NMI

Fig. 2. The ARI and NMI value of the clustering results on Gaussian data.

3.2   Iris Data

This experiment is to show the effects of the prototypes available in the source domain on the clustering performance for the target data. We consider the Iris data set consisting of 50 samples from each of three species of Iris. Four features are measured from each sample: Sepal.Length (SL), Sepal.Width (SW),

Petal.Length (PL) and Petal.Width (PW). The four features are divided into two parts $FT1$ and $FT2$. The six cases are listed in Table 2. The samples with features in $FT1$ are regarded as the target data to be clustered.

Table 2. The feature division for Iris data.

Case	$FT1$	$FT2$	Case	$FT1$	$FT2$
Case 1	SL, SW	PL, PW	Case 4	SW, PL	SL, PW
Case 2	SL, PL	SW, PW	Case 5	SW, PW	PL, PL
Case 3	SL, PW	SW, PL	Case 6	PL, PW	SL, SW

In order to generate the prototypes in the source domain which are required before evoking TECM, for each case we first apply c-means clustering on the samples with features in $FT2$ and get the best hard partition for the 150 samples. Then the following two schemes are designed to get the prototypes $v_k^{(s)}$:

Scheme A: By the feature mean of samples in each group with feature set $FT2$;
Scheme B: By the feature mean of samples in each group with feature set $FT1$;

The methods with two schemes are termed by TECM-A and TECM-B respectively. We can see that in Scheme B the prototypes are from the target data (with $FT1$) based on a clustering rule learned with the source domain (with $FT2$), while in Scheme A the prototypes are from the source data (with $FT2$) based on a clustering rule learned with the same domain (with $FT2$). The ARI and NMI for the results by ECM, TECM-A and TECM-B are displayed in Fig. 3. From the figure we can see:

- For TECM-B, it performs better than ECM in all the cases except Cases 1 and 4, where the behavior of the two methods (TECM-B and ECM) is similar.
- For TECM-A, it performs worse than TECM-B in all the cases. It is not better than ECM in Cases 1, 3, 4, and 5.

In TECM-B, the transferred knowledge of prototypes have the same feature set as the target samples (this corresponds to the illustrative example in the introduction). The results show that the clustering performance is indeed improved by the use of information from the source. On the contrary, in TECM-A, the feature sets in the source and target domain are different. The knowledge from the source has a negative influence on the performance of transfer clustering in this situation. The imperfect matching between information provided by the two domains can degrade the clustering performance. We will study how to avoid such kind of negative transfer in the future.

3.3 UCI Data Sets

Three UCI data sets are used in this experiment: Seeds, Wine and Karate Club network. The first two data sets are object data while the last is a graph data.

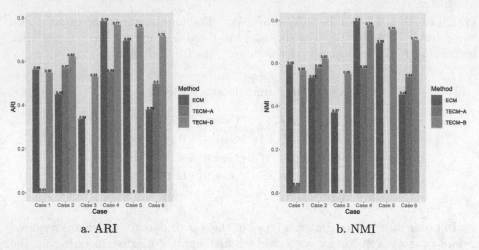

a. ARI   b. NMI

Fig. 3. The ARI and NMI for the clustering results on Iris set by TECM and ECM.

Table 3. The ARI and NMI of clustering results on UCI data sets.

Dataset	$nf = 2$				$nf = 3$			
	ARI		NMI		ARI		NMI	
	ECM	TECM	ECM	TECM	ECM	TECM	ECM	TECM
Seeds	0.4748	0.5074	0.4685	0.4999	0.5046	0.5276	0.4907	0.5177
Wine	0.3497	0.3938	0.357	0.3838	0.3233	0.3785	0.3419	0.3694
Karate	0.2636	1	0.3173	1	0.7717	1	0.7329	1

The number of samples of Seeds data is 210 while 178 for Wine data. The Karate Club network is a graph with 34 nodes and 78 edges.

There are 7 features in Seeds data and 13 for Wine data. In the experiment, nf features are randomly selected from the original data to form the target data set. For Karate graph, the vector embedding is first calculated by spectral decomposition of its adjacency matrices [12]. The embedding dimension is set to nf. Then ECM and TECM algorithms are used. We note here as the benchmarks for these data sets are known, in TECM the average values of the samples in the target data are directly used to simulate the prototypes in the source domain.

The ARI and NMI values of the clustering results provided by ECM and TECM are listed in Table 3. In all the experiments, the results by TECM are better than those by ECM as they generally have higher ARI and NMI values. This confirms the advantages of the evidential transfer clustering approach when there is some available positive transferred knowledge in the source domain.

4   Conclusion

In this study, the concept of knowledge transfer has been used to develop an evidential transfer clustering method named TECM for the application of clustering task when the target data are uncertain or insufficient. The proposed TECM algorithm can learn from not only the data of the target domain but also from the knowledge of the source domain in the form of prototypes as well. The experimental results on generated data and UCI data have demonstrated the effectiveness of TECM algorithm compared with ECM which is without the transfer learning ability.

In TECM, the number of clusters in the source domain and in the target domain is assumed to identical, which may be difficult to satisfy in real applications. How to deal with the case when the number of classes in the two domains is different will be studied in the future.

Acknowledgements. This work was supported by the National Natural Science Foundation of China (No. 61701409), the Natural Science Basic Research Plan in Shaanxi Province of China (No. 2018JQ6005), the Aero Science Foundation of China (No. 20182053023), the Science Research Plan of China (Xi'an) Institute for Silk Road Research (2019ZD02), and the Fundamental Research Funds for the Central Universities of China (No. 310201911cx041).

References

1. Bai, L., Liang, J., Cao, F.: Semi-supervised clustering with constraints of different types from multiple information sources. IEEE Trans. Pattern Anal. Mach. Intell. (2020)
2. Deng, Z., Jiang, Y., Chung, F.L., Ishibuchi, H., Choi, K.S., Wang, S.: Transfer prototype-based fuzzy clustering. IEEE Trans. Fuzzy Syst. **24**(5), 1210–1232 (2015)
3. Denœux, T., Masson, M.H.: EVCLUS: evidential clustering of proximity data. IEEE Trans. Syst. Man Cybern. Part B (Cybern.) **34**(1), 95–109 (2004)
4. Gargees, R., Keller, J.M., Popescu, M.: TLPCM: transfer learning possibilistic c-means. IEEE Trans. Fuzzy Syst. **29**, 940–952 (2020)
5. Gong, C., Su, Z.G., Wang, P.H., Wang, Q.: An evidential clustering algorithm by finding belief-peaks and disjoint neighborhoods. Pattern Recogn. **113**, 107751 (2021)
6. Jain, A.K.: Data clustering: 50 years beyond k-means. Pattern Recogn. Lett. **31**(8), 651–666 (2010)
7. Liu, X., et al.: Efficient and effective regularized incomplete multi-view clustering. IEEE Trans. Pattern Anal. Mach. Intell. (2020)
8. Masson, M.H., Denœux, T.: ECM: an evidential version of the fuzzy c-means algorithm. Pattern Recogn. **41**(4), 1384–1397 (2008)
9. Qian, P., et al.: Cluster prototypes and fuzzy memberships jointly leveraged cross-domain maximum entropy clustering. IEEE Trans. Cybern. **46**(1), 181–193 (2015)
10. Saxena, A., et al.: A review of clustering techniques and developments. Neurocomputing **267**, 664–681 (2017)
11. Smets, P.: Decision making·in the TBM: the necessity of the pignistic transformation. Int. J. Approx. Reason. **38**(2), 133–147 (2005)
12. Sussman, D.L., Tang, M., Fishkind, D.E., Priebe, C.E.: A consistent adjacency spectral embedding for stochastic blockmodel graphs. J. Am. Stat. Assoc. **107**(499), 1119–1128 (2012)

Ensemble of Adapters for Transfer Learning Based on Evidence Theory

Ying Lv[1] , Bofeng Zhang[1](✉) , Xiaodong Yue[1,2](✉) , Zhikang Xu[1] ,
and Wei Liu[1]

[1] School of Computer Engineering and Science, Shanghai University,
Shanghai 200444, China
{lvying2016,bfzhang,yswantfly,xuzhikangnba}@shu.edu.cn
[2] Shanghai Institute for Advanced Communication and Data Science,
Shanghai University, Shanghai, China

Abstract. Transfer learning hopes to borrow transferable knowledge from source domain (related domain) to build up an adapter for target domain. Since the adapter is built on the source domain, the robustness and generalization of a single adapter are more likely to be limited. To further improve the performance of the adapter, in this paper, we propose a parallel ensemble strategy based on evidence theory. Specifically, firstly, we quantify an adaptation degree for instances of source domain based on evidence theory. Secondly, we redefine Determinantal Point Processes (DPP) sampling with adaptation degree, and use the improved DPP sampling to generate k different subsets. Finally, we select and combine the base adapters that are trained by the subsets. In the proposed ensemble strategy, the adaptation degree can ensure the higher transferability of the base adapters, DPP sampling can increase the diversity among the base adapters. Thus, the ensemble strategy can reduce the conflict between accuracy and diversity, and improve the robustness and generalization of the adapters. Numerical experiments on real-world applications are given to comprehensively demonstrate the effectiveness and efficiency of our proposed ensemble strategy. The results show that the ensemble strategy can improve transfer performance.

Keywords: Transfer learning · Domain adaptation · Ensemble learning · Evidence theory · Determinantal point processes

1 Introduction

In the field of machine learning research, supervised learning methods have already witnessed outstanding performance in many applications. The key point of supervised learning is to collect sufficient labeled data sets for model training, which also limits the usage of supervised learning in the scenarios lack of training data. Furthermore, data annotating is usually a time-consuming, labor-expensive, or even unrealistic task.

T. Denœux et al. (Eds.): BELIEF 2021, LNAI 12915, pp. 66–75, 2021.
https://doi.org/10.1007/978-3-030-88601-1_7

To settle this situation, transfer learning (TL) [25,34,35] is a promising methodology, which aims to build an efficient model for the target domain by making use of labeled instances from other related source domains. Existing TL methods can be divided into four types, namely instance-based [3,5], feature-based [4,14], model-based [13,19], and deep learning-based [1,31]. Their fundament thought is to discover transferable knowledge through maximizing the consistency between the source domain and target domain, and transfer the knowledge to the model of target domain.

However, in transfer learning, most methods focus on improving the performance of the single adapter. Because the adapter is trained on source domain, a single adapter could be limited on robustness and generalization. For example, when the distance is far between source domain and target domain, the source domain exists a lot of outliers. These outliers lead easily to negative transfer for a single adapter.

To tackle this problem, in this paper, we propose a parallel ensemble strategy based on evidence theory. Specifically, firstly, we design a measure criterion, based on evidence theory, to quantify the adaptation degree of instances of source domain. Secondly, we redefine Determinantal Point Processes (DPP) sampling [22] with adaptation degree, and utilize improved DPP to sample from source domain for generating k different subsets. Finally, we train the base adapters by the subsets, and use the adaptation degree of subsets to select and combine the base adapters. In our ensemble strategy, the adaptation degree can ensure that the base adapters is more suited to target domain. In addition, according to the property of our improved DPP sampling, we can obtain the subsets with greater diversity and higher transferability. Specifically, the sampling process with DPP determines that the probability of two locally adjacent elements occurring at the same time is relatively small. The property can increase the diversity of subset. Thus, our ensemble strategy can reduce the conflict between diversity and accuracy, and improve the robustness and generalization of adapters. The contributions of this paper are summarized as follows.

- Proposing a parallel ensemble strategy that improves the robustness and generalization of the adapter in transfer learning.
- Quantifying an adaptation degree that source domain transfers to target domain.
- Improving the DPP sampling with adaptation degree for reducing the conflict between accuracy and diversity.

2 Evidence Theory

Evidence theory can be considered as a generalized probability [6,27]. It can use Dempster's rule to finish possibility reasoning [7,10,11]. Based on this view, Denoeux et al. combine the evidence theory with machine learning and designs some supervised and unsupervised algorithms that can solve the problem of imprecise information to improve the robustness of algorithms. Such as, Evidential K-NN classification [8], Evidential Linear Discriminant Analysis [26],

Evidential Neural Network Classifier [9] and multiple evidential clustering algorithms [12]. In this section, we recall mass function and Dempster's rule from evidence theory.

Let Ω be a finite set that includes all possible answers in decision problem. The Ω is called the frame of discernment. In the classification problems, the Ω can be regarded as the label space. We denote the power-set as 2^{Ω} and the cardinality of power-set is $2^{|\Omega|}$.

The mass function $m(\cdot)$ is the Basic Possibility Assignment (BPA) that represents support degree of evidence, and $m(\cdot)$ is a mapping from 2^{Ω} to the interval [0,1]. It satisfies the condition as follows:

$$\begin{cases} \sum_{A \in 2^{\Omega}} m(A) = 1 \\ m(\emptyset) = 0 \end{cases} \tag{1}$$

Dempster's rule reflects the combined effect of evidence. Let m_1 and m_2 be two mass functions induced by independent items of evidence. They can be combined using Dempster's rule to form a new mass function defined as:

$$(m_1 \oplus m_2)(A) = \frac{1}{1 - \kappa} \sum_{B \cap C = A} m_1(B) m_2(C) \tag{2}$$

where $A \subseteq \Omega$, $A \neq \emptyset$ and $(m_1 \oplus m_2)(\emptyset) = 0$. $\oplus$ is the combination operator of Dempster's rule. k is the degree of conflict between m_1 and m_2.

$$\kappa = \sum_{B \cap C = \emptyset} m_1(B) m_2(C) \tag{3}$$

3   Adapters Ensemble Based on Evidence Theory

3.1   Estimating Adaptation Degree Based on Evidence Theory

In this section, a measure criterion is designed, based on evidence theory, to quantify the adaptation degree of source domain for target domain.

In evidence theory, mass function $m(\Omega|x; \Phi)$ can be interpreted as a degree of knowing nothing for classification results based on evidence set, in which Ω can be considered as a label space in classification task and Φ denotes the evidence set. For transfer learning, when x is from the source domain and evidence set Φ is from the target domain, $m(\Omega|x; \Phi)$ can represent the unknown degree of x about target domain classification task. Thus, $m(\Omega|x\Phi)$ can reflect the adaptation degree of x for target domain task. Based on this view, in our work, we utilize evidence theory to design $m(\Omega|x; \Phi)$ for estimating the adaptation degree of source domain.

We first obtain evidence set Φ from a little of labeled target domain. To this end, the objective function is defined as

$$\Phi = \underset{\Phi}{\arg\min} f\left(x^s, \Phi \subset \mathcal{D}_l^t\right), \tag{4}$$

in which x^s is an instance of source domain, $\mathcal{D}_l^t$ is a little of labeled target domain, and the function $f(\cdot)$ measures the discrepancy between x^s and the evidence set Φ in a reproducing kernel Hilbert Space (RKHS) $\mathcal{H}$,

$$f\left(x^s, \Phi\right) = \left\| \varphi\left(x^s\right) - \frac{1}{|\Phi|} \sum_{e \in \Phi} \varphi(e) \right\|_{\mathcal{H}}^2, \tag{5}$$

where $\varphi : \mathcal{X} \mapsto \mathcal{H}$ is the feature mapping. $|\Phi|$ is the number of elements in evidence set. The optimal evidence set Φ in Equation (4) can be solved by the greedy search algorithm on a little of label target domain.

In evidence theory, the evidence set Φ can be viewed as a set of different granular evidence

$$\Phi = \{\Phi_1, \ldots, \Phi_n\}, \tag{6}$$

where $\Phi_c = \{(e_1, z_1 = c), \ldots, (e_m, z_m = c)\}$ is a set in which the labels of evidence is equal to c.

Then, based on the evidence set Φ, the mass functions are defined as

$$\begin{aligned} \mathrm{m}\left(\Omega|x^s; \Phi\right) &= \bigoplus_{\Phi_c \subseteq \Phi} m\left(\Omega|x^s; \Phi_c\right) \\ &= \frac{1}{\kappa} \prod_{c=1}^{n} m\left(\Omega|x^s; \Phi_c\right) = \frac{1}{\kappa} \prod_{c=1}^{n} \prod_{e \in \Phi_c} m\left(\Omega|x^s; e\right), \end{aligned} \tag{7}$$

where x^s is an instance of source domain, e denotes an element of evidence set. The orthogonal sum $\bigoplus$ represents the combination operator of Dempster's rule. κ is the degree of conflict between evidence. It can be interpreted as a normalizing factor.

$$\kappa = \sum_{c=1}^{n} (m\left(z = c|x^s; \Phi_c\right) \prod_{j \neq c} m\left(\Omega|x^s; \Phi_c\right)) + \prod_{c=1}^{n} m\left(\Omega|x^s; \Phi_c\right), \tag{8}$$

3.2 Improving DPP Sampling with Adaptation Degree

In this section, we modified the L-matrix in the Determinantal Point Processes (DPP) sampling using the adaptation degree.

Definition 1. *k-DPP sampling* [20]. *Suppose any sampled subset C^S consists of k instances from source domain, the k-DPP of sampling is defined by the following probability measure of subset selection with L-ensemble,*

$$P^k(C^S) = \frac{\det\left(L_{C^S}\right)}{\sum\limits_{C' \subseteq C \wedge |C'| = k} \det\left(L_{C'}\right)} \tag{9}$$

where $|C^S| = k$ and L_{C^S} is the $k \times k$ submatrix of L indexed by C^S.

Suppose $L = \sum_{i=1}^{M} \lambda_i v_i v_i^T$, λ_i refers to the eigenvalue corresponding to the eigenvector v_i, the probability of selecting a k-size subset C^S is

$$P^k(C^S) = \frac{\det(L_{C^S})}{\sum\limits_{C' \subseteq C \wedge |C'|=k} \det(L_{C'})} = \frac{\prod_{c_i \in C^S} \lambda_i}{\sum\limits_{C' \subseteq C \wedge |C'|=k} \left\{ \prod\limits_{c_j \in C'} \lambda_j \right\}}. \tag{10}$$

According to the definition above, the correlation matrix L determines the k-DPP sampling probability. The different correlation matrices have different sampling properties. For improving the diversity and adaptation of subsets, we redefine the matrix L.

(Matrix L Construction). According to [21], we can decompose the matrix L as a Gram matrix $L = B^T B$, and reformulate the matrix L with adaptation degree. Suppose each vector B_i in B has the form of $B_i = m(\Omega|x_i^s; \Phi) \cdot \phi_i$, in which $m(\Omega|x_i^s; \Phi)$ is the adaptation degree of ith element and ϕ_i is the normalized feature vector of the ith element. We rewrite the matrix L as

$$L = [L_{ij}]_{1 \le i,j \le M}, \ L_{ij} = (m(\Omega|x_i^s)) \cdot \phi_i^T \cdot \phi_j \cdot (m(\Omega|x_j^s)) \tag{11}$$

We can see that the inner product $\phi_i^T \phi_j \in [-1, +1]$ indicates the similarity between the elements i and j. Thus, L_{ij} consists of the adaptation degree and similarity of the pair of elements. Denoting the similarity $s_{ij} = \phi_i^T \cdot \phi_j$, we rewrite

$$L = \left\{ L_{ij} = m(\Omega|x_i^s) \cdot s_{ij} \cdot m(\Omega|x_j^s) \,|\, 1 \le i \le M, 1 \le j \le M \right\} \tag{12}$$

In our improved DPP sampling, the $m(\Omega|x_i^s)$ can ensure that the sampling instances are more transferability, and the s_{ij} can ensure that the instances are diverse. Thus, we can obtain the subset with greater diversity and higher transferability.

3.3   Adapters Selection and Ensemble

Let us consider a simple scenario of transfer learning with a large number of instances of labeled source-domain D^s, a small number of instances of labeled target-domain D_i^t that are available.

Firstly, we single out a subset T_i from source domain D^s, according to the improved DPP sampling with adaptation degree. Then, by repeating the process of DPP sampling, we can obtain a set of candidate subsets $\{T_1, T_2, \ldots, T_N\}$.

Secondly, we select the transfer data from the candidate subsets by

$$T = T \cup \{T_i | \, \text{Ent}(T_i) \le \beta\}, \tag{13}$$

in which $Ent\,(T_i)$ represents the confidence of subset in transfer process. The higher value of $\text{Ent}(T_i)$ represents that the subset T_i is more suited to target domain. The confidence is defined as

$$\text{Ent}\,(T_i) = -\frac{1}{|T_i|}\sum_{x \in T_i}(m(\Omega|x))\log(m(\Omega|x)). \tag{14}$$

Finally, according to each selected subset T_i, we train a set of base adapters $\{f_1, f_2, \ldots, f_N\}$ for target domain, and we use the confidence of subsets to calculate the weight of base classifier.

$$w_i = \frac{\text{Ent}\,(T_i)}{\sum_{i=1}^{|T|}\text{Ent}\,(T_i)}. \tag{15}$$

And we combine the base adapters by

$$H(x) = \sum_{i=1}^{|T|} w_i f_i(x). \tag{16}$$

The final decision is made by

$$H(\boldsymbol{x}) = \arg\max_c \sum_{i=1}^{|T|} w_i f_i^c(x). \tag{17}$$

where c is the class of instance x.

4   Experiments

In this section, we implement the experiment to validate the effectiveness of the proposed ensemble strategy on text and image datasets.

4.1   Data Preparation

Text Dataset: The Amazon product reviews dataset [2] is commonly used for the task of sentiment classification in transfer learning. The reviews are about four product domains: book (B), dvd (D), electronic (E) and kitchen appliance (K). We create 12 transfer learning subproblems by combining in a pairwise manner. For example, the $B \to D$ means that the book is as source domain and the dvd as the target domain.

Image Dataset: The Office+Caltech dataset [17] includes 10 classes. It is commonly used for the task of visual object recognition in transfer learning. It includes four domains: Amazon (A), Webcam (W), DSLR (D), and Caltech(C). We create 12 subproblems by combining in a pairwise manner.

4.2   Experimental Setting

Our Method, in our ensemble strategy, we choose the weighted logistic regression as the base adapter, in which the weights depend on adaptation degree.

Baseline Methods, we overall evaluate the performance of the method through comparing with **three deep learning-based transfer learning methods**: DANN [16], DDC [30], WDGAL [28] and **eight traditional transfer learning methods**: KMM [18], TCA [24],GFK [17],JDA [23], CORAL [29], MTLF [33], SCA [15], EasyTL [32] on text and image datasets.

4.3   Experimental Results

In the experiment of the overall evaluation, as shown in Table 1 and Table 2, the average accuracy improvement of our method is 2.41% and 2.49% respectively compared to the best baseline on the two types of datasets. These results indicate that our method achieves statistically superior performance against other comparing methods. Specifically, we observe from Table 1 that our method outperforms other well-established methods on 11 transfer learning subproblems of Amazon reviews. As shown in Table 2, our method outperforms other well-established transfer learning methods on eight subproblems of Office+Caltech. These results indicate that the proposed ensemble strategy is effective on text and image datasets.

In summary, our method achieves highly competitive performance against other comparing methods. These results clearly validate the effectiveness that the ensemble strategy based on evidence theory.

Table 1. Accuracy % on the Amazon product reviews

Data	Our method	TCA	CORAL	GFK	JDA	KMM	MTLF	SCA	EasyTL	WDGAL
$B \rightarrow D$	**85.62**	77.76	70.76	75.76	77.26	83.76	68.59	81.56	79.80	83.05
$B \rightarrow E$	**84.29**	75.54	66.21	72.00	75.93	79.02	69.63	78.08	79.70	83.28
$B \rightarrow K$	**85.55**	78.74	70.00	73.50	78.09	75.90	72.74	79.09	80.90	85.45
$D \rightarrow B$	**86.78**	76.05	73.05	71.85	77.65	80.5	70.70	82.35	79.90	80.72
$D \rightarrow E$	**81.62**	76.38	68.70	68.96	76.03	68.51	71.90	78.82	80.80	83.58
$D \rightarrow K$	**85.66**	79.34	71.96	75.70	78.29	76.45	74.18	80.39	82.00	86.24
$E \rightarrow B$	**82.85**	73.35	69.90	72.60	72.65	73.7	69.20	77.00	75.00	77.22
$E \rightarrow D$	**83.44**	73.66	65.71	71.11	72.16	77.86	70.73	77.26	75.30	78.28
$E \rightarrow K$	**89.32**	79.74	72.35	76.20	80.14	80.39	71.36	84.63	84.90	88.16
$K \rightarrow B$	**82.06**	73.05	67.45	73.75	75.05	74.25	66.04	78.90	76.50	77.16
$K \rightarrow D$	**83.30**	77.26	68.61	74.21	77.56	75.96	70.31	77.46	76.30	79.89
$K \rightarrow E$	**87.62**	78.74	75.68	76.58	80.32	85.00	68.58	85.65	82.50	86.29
Average	**84.84**	76.63	70.03	73.52	76.76	77.61	70.33	80.10	79.47	82.43

Table 2. Accuracy % on the Office+Caltech datasets with DeCAF6 Features

Data	Our method	TCA	CORAL	GFK	JDA	KMM	DANN	SCA	EasyTL	DCC
$A \to C$	**89.21**	78.98	83.88	76.85	75.07	83.08	87.80	78.81	81.66	85.00
$A \to D$	**82.20**	84.71	80.25	79.62	78.34	83.44	82.46	85.35	84.07	89.00
$A \to W$	**86.29**	74.92	74.58	68.47	70.85	74.24	77.81	75.93	72.88	86.10
$C \to A$	**95.29**	89.67	89.98	88.41	89.67	91.23	93.27	89.46	90.50	91.90
$C \to W$	**92.21**	77.29	78.64	80.68	80.00	80.34	89.47	85.42	75.59	85.40
$D \to A$	**94.10**	89.77	85.50	85.80	88.31	84.34	84.70	89.98	83.40	89.50
$D \to C$	**89.20**	79.96	79.25	74.09	73.91	71.86	82.21	78.09	74.09	81.10
$D \to W$	**95.58**	98.64	99.66	98.64	98.31	98.98	98.95	98.64	93.11	98.20
$W \to A$	**94.06**	84.45	77.14	75.26	80.27	72.81	82.98	86.12	74.53	84.90
$W \to C$	**88.79**	77.74	74.98	74.80	72.93	67.14	81.30	74.80	67.31	78.00
Average	**90.69**	84.96	84.15	81.04	83.02	83.85	87.67	85.88	81.24	88.20

5 Conclusion

To tackle the problem of limited performance of a single adapter in transfer learning, we proposed a novel ensemble strategy based on evidence theory. In the proposed ensemble strategy, the adaptation degree can ensure the high transferability of the base adapters, the DPP sampling can increase the diversity among the base adapters. Thus, the ensemble strategy can reduce the conflict between accuracy and diversity, and improve the robustness and generalization of the adapters. Experimental results on a large number of real-world datasets with text and image demonstrate that the proposed ensemble strategy achieves highly competitive performance against other state-of-the-art transfer learning methods. Moving forward, we believe that our ensemble strategy can potentially be used in other tasks, such as cross-modal medical image classification, etc.

References

1. Bengio, Y.: Deep learning of representations for unsupervised and transfer learning. In: Proceedings of ICML Workshop on Unsupervised and Transfer Learning, pp. 17–36 (2012)
2. Blitzer, J., Dredze, M., Pereira, F.: Biographies, bollywood, boom-boxes and blenders: Domain adaptation for sentiment classification. In: Proceedings of the 45th Annual Meeting of the Association of Computational Linguistics, pp. 440–447 (2007)
3. Chen, M., Weinberger, K.Q., Blitzer, J.: Co-training for domain adaptation. In: Advances in Neural Information Processing Systems, pp. 2456–2464 (2011)
4. Courty, N., Flamary, R., Tuia, D., Rakotomamonjy, A.: Optimal transport for domain adaptation. IEEE Trans. Pattern Anal. Mach. Intell. **39**(9), 1853–1865 (2016)
5. Dai, W., Yang, Q., Xue, G.R., Yu, Y.: Boosting for transfer learning. In: Proceedings of the 24th International Conference on Machine Learning, pp. 193–200 (2007)

6. Dempster, A.P., et al.: Upper and lower probabilities generated by a random closed interval. Ann. Math. Stat. **39**(3), 957–966 (1968)
7. Denœux, T.: Reasoning with imprecise belief structures. Int. J. Approx. Reason. **20**(1), 79–111 (1999)
8. Denoeux, T.: A k-nearest neighbor classification rule based on dempster-shafer theory. In: Yager, R.R., Liu, L. (eds.) Classic Works of the Dempster-Shafer Theory of Belief Functions, pp. 737–760. Studies in Fuzziness and Soft Computing, vol. 219. Springer, Berlin, Heidelberg (2008). https://doi.org/10.1007/978-3-540-44792-4_29
9. Denœux, T.: Logistic regression, neural networks and dempster-shafer theory: a new perspective. Knowl.-Based Syst. **176**, 54–67 (2019)
10. Denoeux, T.: Distributed combination of belief functions. Inf. Fusion **65**, 179–191 (2021)
11. Denoeux, T., Shenoy, P.P.: An interval-valued utility theory for decision making with dempster-shafer belief functions. Int. J. Approx. Reason. **124**, 194–216 (2020)
12. Denoeux, T., Sriboonchitta, S., Kanjanatarakul, O.: Evidential clustering of large dissimilarity data. Knowl.-Based Syst. **106**, 179–195 (2016)
13. Duan, L., Xu, D., Tsang, I.W.H.: Domain adaptation from multiple sources: a domain-dependent regularization approach. IEEE Trans. Neural Netw. Learn. Syst. **23**(3), 504–518 (2012)
14. Fernando, B., Habrard, A., Sebban, M., Tuytelaars, T.: Unsupervised visual domain adaptation using subspace alignment. In: Proceedings of the IEEE International Conference on Computer Vision, pp. 2960–2967 (2013)
15. Ghifary, M., Balduzzi, D., Kleijn, W.B., Zhang, M.: Scatter component analysis: a unified framework for domain adaptation and domain generalization. IEEE Trans. Pattern Anal. Mach. Intell. **39**(7), 1414–1430 (2017)
16. Ghifary, M., Kleijn, W.B., Zhang, M.: Domain adaptive neural networks for object recognition. In: Pham, D.-N., Park, S.-B. (eds.) PRICAI 2014. LNCS (LNAI), vol. 8862, pp. 898–904. Springer, Cham (2014). https://doi.org/10.1007/978-3-319-13560-1_76
17. Gong, B., Shi, Y., Sha, F., Grauman, K.: Geodesic flow kernel for unsupervised domain adaptation. In: 2012 IEEE Conference on Computer Vision and Pattern Recognition, pp. 2066–2073. IEEE (2012)
18. Huang, J., Gretton, A., Borgwardt, K., Scholkopf, B., Smola, A.J.: Correcting sample selection bias by unlabeled data. In: Advances in Neural Information Processing Systems, pp. 601–608 (2007)
19. Karbalayghareh, A., Qian, X., Dougherty, E.R.: Optimal bayesian transfer learning. IEEE Trans. Signal Process. **66**(14), 3724–3739 (2018)
20. Kulesza, A., Taskar, B.: k-DPPs: fixed-size determinantal point processes. In: Proceedings of the 28th International Conference on International Conference on Machine Learning, pp. 1193–1200 (2011)
21. Kulesza, A., Taskar, B.: Learning determinantal point processes (2011)
22. Kulesza, A., Taskar, B., et al.: Determinantal point processes for machine learning. Found. Trends® Mach. Learn. **5**(2–3), 123–286 (2012)
23. Long, M., Wang, J., Ding, G., Sun, J., Yu, P.S.: Transfer feature learning with joint distribution adaptation. In: Proceedings of the IEEE International Conference on Computer Vision, pp. 2200–2207 (2013)
24. Pan, S.J., Tsang, I.W., Kwok, J.T., Yang, Q.: Domain adaptation via transfer component analysis. IEEE Trans. Neural Netw. **22**(2), 199–210 (2010)
25. Pan, S.J., Yang, Q.: A survey on transfer learning. IEEE Trans. Knowl. Data Eng. **22**(10), 1345–1359 (2009)

26. Quost, B., Denœux, T., Li, S.: Parametric classification with soft labels using the evidential em algorithm: linear discriminant analysis versus logistic regression. Adv. Data Anal. Classif. **11**(4), 659–690 (2017)
27. Shafer, G.: A mathematical theory of evidence turns 40. Int. J. Approx. Reason. **79**, 7–25 (2016)
28. Shen, J., Qu, Y., Zhang, W., Yu, Y.: Wasserstein distance guided representation learning for domain adaptation. In: AAAI (2018)
29. Sun, B., Feng, J., Saenko, K.: Return of frustratingly easy domain adaptation. In: Thirtieth AAAI Conference on Artificial Intelligence (2016)
30. Tzeng, E., Hoffman, J., Zhang, N., Saenko, K., Darrell, T.: Deep domain confusion: Maximizing for domain invariance. arXiv preprint arXiv:1412.3474 (2014)
31. Venkateswara, H., Chakraborty, S., Panchanathan, S.: Deep-learning systems for domain adaptation in computer vision: learning transferable feature representations. IEEE Signal Process. Mag. **34**(6), 117–129 (2017)
32. Wang, J., Chen, Y., Yu, H., Huang, M., Yang, Q.: Easy transfer learning by exploiting intra-domain structures. In: 2019 IEEE International Conference on Multimedia and Expo (ICME), pp. 1210–1215 (2019)
33. Xu, Y., et al.: A unified framework for metric transfer learning. IEEE Trans. Knowl. Data Eng. **29**(6), 1158–1171 (2017)
34. Zhang, J., Li, W., Ogunbona, P., Xu, D.: Recent advances in transfer learning for cross-dataset visual recognition: a problem-oriented perspective. ACM Comput. Surv. (CSUR) **52**(1), 1–38 (2019)
35. Zhuang, F., et al.: A comprehensive survey on transfer learning. Proc. IEEE **109**(1), 43–76 (2020)

Classification

Improving Micro-Extended Belief Rule-Based System Using Activation Factor for Classification Problems

Long-Hao Yang[1,3,4（✉）], Jun Liu[3], Ying-Ming Wang[1,2], Hui Wang[3], and Luis Martínez[3,4]

[1] Decision Sciences Institute, Fuzhou University, Fuzhou, People's Republic of China
[2] Key Laboratory of Spatial Data Mining and Information Sharing of Ministry of Education, Fuzhou University, Fuzhou, People's Republic of China
[3] School of Computing, Ulster University, Northern Ireland, UK
{j.liu,h.wang}@ulster.ac.uk
[4] Department of Computer Science, University of Jaén, Jaén, Spain
martin@ujaen.es

Abstract. The *micro-extended belief rule-based system* (Micro-EBRBS) is an advanced rule-based system and has shown its superior ability in solving big data problems. To overcome the activation rule incompleteness and inconsistency of Micro-EBRBS, a new concept, named *activation factor* (AF), is introduced to revise the calculation of individual matching degree and, furthermore, an AF-based inference (AFI) method is proposed for improving Micro-EBRBS. A comparative analysis study is conducted using three classification datasets. Results demonstrate that the proposed AFI method can not only improve the accuracy of Micro-EBRBS, but also reduce the number of failed data in the process of rule inference.

Keywords: Extended belief rule-based system · Activation factor · Classification · Rule inference.

1 Introduction

The rule-based system is an artificial intelligent (AI) methodology that applies rules to manage quantitative data and qualitative knowledge. These rules usually take form of "IF statements THEN consequents", which constitutes a kernel component of the rule-based system, namely rule base. Compared to black box approaches, the rule-based system fosters the understanding of the underlying

Supported by the National Natural Science Foundation of China (Nos. 72001043 and 61773123), the Natural Science Foundation of Fujian Province of China (No. 2020J05122), the Humanities and Social Science Foundation of the Ministry of Education of China (No. 20YJC630188), the Chengdu International Science Cooperation Project (No. 2020-GH02-00064-HZ), and the Spanish Ministry of Economy and Competitiveness through the Spanish National Research Project PGC2018-099402-B-I00.

T. Denœux et al. (Eds.): BELIEF 2021, LNAI 12915, pp. 79–86, 2021.
https://doi.org/10.1007/978-3-030-88601-1_8

problem and the reasons behind intelligent decisions and predictions, which has been always one of the most promising directions to achieve transparent and explainable AI [1].

As one of advanced rule-based systems, the *micro-extended belief rule-based system* (Micro-EBRBS) [2] was recently proposed on the basis of the EBRBS [3], which is useful to handle uncertain information in both rule antecedent and consequent owing to belief structures. Comparing to other rule-based systems, Micro-EBRBS has shown its high efficiency and excellent accuracy in handling big data classification problems.

However, the current version of Micro-EBRBS usually suffers from the activation rule incompleteness and inconsistency problems because it is a data-driven decision model [4], where the former occurs due to the situation that Micro-EBRBS is unable to activate rules for producing an output class; the latter happens when Micro-EBRBS fails to produce a confident and accurate result because two or more rules with different consequents are activated.

Therefore, in the present work, a new concept, named *activation factor* (AF), is provided to revise the calculation of individual matching degrees, together with an AF-based inference (AFI) method for improving Micro-EBRBS. In order to validate the effectiveness of the proposed AFI method, three classification datasets are used to provide a case study and comparative anslysis for the improved Micro-EBRBS.

The remainder of the paper is organized as follows: Sect. 2 reviews the basic of Micro-EBRBS. Section 3 introduces the proposed AFI method. Section 4 provides the case study of classification problems for Micro-EBRBS. Finally, Sect. 5 concludes this work.

2   Basics of Micro-EBRBS

2.1   Micro-EBRB and Its Construction Scheme

Micro-extended belief rule base (Micro-EBRB) is the rule-base of Micro-EBRBS and it consists of a series of extended belief rules with exclusive division domains [2]. Suppose that there are M antecedent attributes $U_i(i = 1, ..., M)$ with J_i referential values $A_{i,j}(j = 1, ..., J_i)$ and one consequent attribute D with N consequents $D_n(n = 1, ..., N)$. The extended belief rule related to division domain $D(A_{i,j_i}; i = 1, ..., M)$ is written as follows:

$$R_{j_1 \cdots j_M} : IF\ U_1\ is\ \{A_{1,j}, \alpha_{1,j}^{j_1 \cdots j_M}; j = 1, ..., J_1\} \wedge \cdots \wedge U_M\ is\ \{A_{M,j},$$
$$\alpha_{M,j}^{j_1 \cdots j_M}; j = 1, ..., J_M\} THEN\ D\ is\{D_n, \beta_n^{j_1 \cdots j_M}; n = 1, ..., N\} \quad (1)$$
$$with\ \theta_{j_1 \cdots j_M}\ and\ \{\delta_i; i = 1, ..., M\}$$

where $\alpha_{i,j}^{j_1 \cdots j_M}$ and $\beta_n^{j_1 \cdots j_M}$ denote the belief degrees of referential value $A_{i,j}$ and consequent D_n in rule $R_{j_1 \cdots j_M}$. Moreover, the belief degrees in antecedent attribute satisfy $\alpha_{i,j_i}^{j_1 \cdots j_M} > \alpha_{i,j}^{j_1 \cdots j_M}$ $(j = 1, ..., J_i; j \neq j_i; i = 1, ..., M)$; $\theta_{j_1 \cdots j_M}$ denotes the weight of rule $R_{j_1 \cdots j_M}$; δ_i denotes the weight of attribute U_i.

In order to generate the rules shown in Eq. (1) for constructing Micro-EBRB, the following steps are performed:

Step 1: To generate belief distributions. Suppose $x_{t,i}$ is the tth $(t = 1, ..., T)$ input data of attribute U_i. A belief distribution $S(x_{t,i}) = \{(A_{i,j}, \alpha_{i,j}^t); j = 1, ..., J_i\}$ can be generated as follows:

$$a_{i,j}^t = \frac{u(A_{i,j+1}) - x_{t,i}}{u(A_{i,j+1}) - u(A_{i,j})}, a_{i,j+1}^t = 1 - a_{i,j}^t, \; if \; u(A_{i,j}) \le x_{t,i} \le u(A_{i,j+1}) \quad (2)$$

$$a_{i,s}^t = 0 \; for \; s = 1, ..., J_i \; and \; s \ne j, j+1 \quad (3)$$

where $u(A_{i,j})$ denotes the utility value of referential value $A_{i,j}$ in the ith attribute U_i; $a_{i,j}^t$ denotes the belief degree of referential value $A_{i,j}$ from data $x_{t,i}$.

Next, when y_t is assumed to be the jth class $D_j(j = 1, ..., N)$, the belief distribution $S(y_t) = \{(D_n, \beta_n^t); n = 1, ..., N\}$ is calculated as follows:

$$\beta_n^t = \left\{ \begin{array}{l} 1; if \; n = j \\ 0; otherwise \end{array} \right. \quad (4)$$

Step 2: To generate extended belief rules. All belief distributions generated from the tth input-output data pair $<x_{t,i}, y_t, i = 1, ..., M>$ is regarded as an initial extended belief rule $\bar{R}_t(t = 1, ..., T)$. All these rules should be mapped into a division domain according to the following map function:

$$\bar{R}_t \rightarrow D(A_{i,j_i}; i = 1, ..., M); j_i = \arg max_{j=1,...J_i}\{\alpha_{i,j}^t\} \quad (5)$$

where the map function means the collection of the rules with the maximum belief degree in the same referential values.

Consequently, for the division domain which has one rule at least, all rules in the same division domain are used to generate a new extended belief rule, in which the belief degrees of new rule $R_{j_1 \cdots j_M}$ are calculated as follows:

$$\alpha_{i,j}^{j_1 \cdots J_M} = \frac{\sum_{t=1}^{T_{j_1 \cdots j_M}} \alpha_{i,j}^t}{T_{j_1 \cdots j_M}}, \beta_n^{j_1 \cdots J_M} = \frac{\sum_{t=1}^{T_{j_1 \cdots j_M}} \beta_{i,j}^t}{T} \quad (6)$$

where $T_{j_1 \cdots j_M}$ is the number of rules in division domain $D(A_{i,j_i}; i = 1, ..., M)$.

Step 3: To calculate rule weights. Suppose that **R** denotes the set of all new extended belief rules. The rule weight of $R_k(R_k \in \mathbf{R})$ can be calculated by:

$$\theta_k = 1 - \frac{Incons(R_k)}{\sum_{R_j \in \mathbf{R}} Incons(R_j)} \quad (7)$$

where $Incons(R_k)$ denotes the inconsistency degree of R_k and it is calculated by the similarity of rule antecedent (SRA) and rule consequent (SRC) [3].

2.2   Micro-EBRB Inference Scheme

After constructing Micro-EBRB, the corresponding Micro-EBRBS can be used to classify given input data using Micro-EBRB inference scheme as follows:

Step 1: To calculate individual matching degrees. For a given input data $\mathbf{x} = (x_1, ..., x_M)$, each input $x_i(i = 1, ..., M)$ needs to be transformed into a belief distribution $S(x_i) = (A_{i,j}, \alpha_{i,j}); j = 1, ..., J_i$. Hence, the individual matching degree $S^{j_1 \cdots J_M}(x_i, U_i)$ between rule $R_{j_1 \cdots j_M}$ and data $\mathbf{x}$ for attribute U_i is calculated based on the similarity measure of belief distributions as follows:

$$S^{j_1 \cdots J_M}(x_i, U_i) = \begin{cases} 0, if \ d_i > 1 \\ 1 - d_i, otherwise \end{cases} d_i = \sqrt{\sum_{j=1}^{J_i}(\alpha_{i,j} - \alpha_{i,j}^{j_1 \cdots J_M})^2} \qquad (8)$$

where $\alpha_{i,j}^{j_1 \cdots J_M}$ is the belief degree of attribute U_i of rule $R_{j_1 \cdots j_M}$.

Step 2: To calculate activation weights. Based on the individual matching degrees shown in Eq. (8), the activation weight of rule $R_{j_1 \cdots j_M}$, denoted as $w_{j_1 \cdots j_M}$, is calculated by

$$w_{j_1 \cdots j_M} = \frac{\theta_{j_1 \cdots j_M} \prod_{i=1}^{M}(S^{j_1 \cdots J_M}(x_i, U_i))^{\bar{\delta}_i}}{\sum_{R_k}^{AR(\mathbf{x})} \theta_k \prod_{i=1}^{M}(S^k(x_i, U_i))^{\bar{\delta}_i}}, \bar{\delta}_i = \frac{\delta_i}{max_{t=1,...,M}\{\delta_t\}} \qquad (9)$$

where $\theta_{j_1 \cdots j_M}$ is the weight of rule $R_{j_1 \cdots j_M}$; δ_i is the weight of attribute U_i; $AR(\mathbf{x})$ is the rule set to classify data $\mathbf{x}$ and it constrains all rules of Micro-EBRB.

Step 3: To integrate activated rules. Suppose that all rules in $AR(\mathbf{x})$ are activated for classifying data $\mathbf{x}$ and they are further integrated using the analytical ER algorithm as follows:

$$\beta_n = \frac{\prod_{R_k}^{AR(\mathbf{x})}(w_k \beta_n^k + 1 - w_k) - \prod_{R_k}^{AR(\mathbf{x})}(1 - w_k)}{\sum_{i=1}^{N} \prod_{R_k}^{AR(\mathbf{x})}(w_k \beta_i^k + 1 - w_k) - N \prod_{R_k}^{AR(\mathbf{x})}(1 - w_k)} \qquad (10)$$

where β_n is the integrated belief degree. Hence, the output class of Micro-EBRBS is obtained as follows:

$$f(\mathbf{x}) = D_n, n = \arg max_{i=1,...,N}\{\beta_i\} \qquad (11)$$

3   Activation Factor to Improve Micro-EBRBS

In this section, the challenge of Micro-EBRBS is discussed to illustrate the purpose of this study, followed by using AF to improve Micro-EBRBS.

3.1   Challenge of Calculating Individual Matching Degrees

Consider that the calculation of individual matching degrees shown in Eq. (8) is related with the belief degree $\alpha_{i,j}$ generated from data x_i and the belief degree

$\alpha_{i,j}^{j_1 \cdots j_M}$ generated from data $x_{t,i}$ according to Sect. 2. For the sake of discussion, suppose that the domain of data x_i and $x_{t,i}$ is interval $[0, 1]$, and five utility values $\{u(A_{i,j}); j = 1, ..., 5\} = \{0.0, 0.25, 0.5, 0.75, 1.0\}$ are used to transform data x_i and $x_{t,i}$ into belief distributions. The corresponding individual matching degrees are shown in Fig. 1.

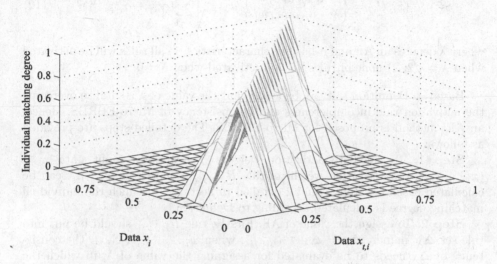

Fig. 1. An example of individual matching degrees

From Fig. 1, it is clear that there exists a distance between data x_i and $x_{t,i}$ to ensure that the individual matching degree is greater than 0. For example, in the case of $x_i = 0$, the individual matching degree is 0 if $x_{t,i}$ is smaller than 0.1768; otherwise the individual matching degree is greater than 0.

However, it is always hard to assign a certain distance for the data needed to be classified, e.g., x_i, and the data used to generate rules, e.g., $x_{t,i}$, leading to the activation rule incompleteness and inconsistency, in which the former occurs due to the situation that all the data used to generate rules are far from the data needed to be classified, namely Micro-EBRBS is unable to activate rules for producing an output class; the latter occurs when all the data used to generate rules not only are close to the data needed to be classified, but also contain conflicting information, namely Micro-EBRBS fails to produce a confident result because two or more rules with different consequents are activated.

3.2 Activation Factor-Based Inference Method

In order to overcome the activation rule incompleteness and inconsistency of Micro-EBRBS, a new definition regarding AF-based individual matching degree is provided as follows:

Definition 1 (AF-based individual matching degree): Suppose that the belief distribution of rule $R_{j_1 \cdots j_M}$ and data $\mathbf{x}$ in attribute U_i is $\{(A_{i,j}, \alpha_{i,j}^{j_1 \cdots j_M}); j = 1, ..., J_i\}$ and $\{(A_{i,j}, \alpha_{i,j}); j = 1, ..., J_i\}$, respectively. The new individual matching degree is calculated by

$$S_\lambda^{j_1 \cdots J_M}(x_i, U_i) = \begin{cases} 0, if \ d_i > \lambda \\ \lambda - d_i, otherwise \end{cases} d_i = \sqrt{\sum_{j=1}^{J_i} (\alpha_{i,j} - \alpha_{i,j}^{j_1 \cdots J_M})^2} \qquad (12)$$

where λ denotes an AF and it has two characteristics: 1) all rules will be activated when $\lambda = \sqrt{2}$; 2) none of rules will be activated when $\lambda = 0$.

Based on Definition 1, it can be found that the value of λ is vital to overcome the activation rule incompleteness and inconsistency of Micro-EBRBS. Hence, an AFI method is proposed for Micro-EBRBS. The detailed steps are provided as follows:

Step 1: To calculate activation weights. For a given input data $\mathbf{x} = (x_1, ..., x_M)$, the activation weight of rule $R_{j_1 \cdots j_M}$, denoted as $w_{j_1 \cdots j_M}$, can be calculated based on Steps 1 and 2 detailed in Sect. 2.2, in which the individual matching degree is calculated according to Definition 1.

Step 2: To assign the value of AF. Firstly, rule $R_{j_1 \cdots j_M}$ should be put into rule set $\mathbf{\Delta}_\lambda$, namely $\mathbf{\Delta}_\lambda = \mathbf{\Delta}_\lambda \cup R_{j_1 \cdots j_M}$, when $w_{j_1 \cdots j_M} > 0$; Next, the consistency of $\mathbf{\Delta}_\lambda$ needs to be evaluated for assigning the value of λ, in which the evaluation formula is shown as follows:

$$C(\mathbf{\Delta}_\lambda) = \frac{max_{n=1,...,N}\{C_n\}}{|\mathbf{\Delta}_\lambda|} \qquad (13)$$

where C_n is calculated by

$$C_n = |D_n; n = \arg max_{i=1,...,N}\{\beta_i^k\}; R_k \in \mathbf{\Delta}_\lambda| \qquad (14)$$

Step 3: To integrate activated rules. All activated rules should be integrated to produce an output class for classifying input data $\mathbf{x}$, in which the formulas of integrating rules and producing output class are shown in Eqs. (10) and (11).

4   Case Studies on Classification Problems

In this section, the introduction of classification datasets is provided firstly. Then, a comparative analysis is carried out on the purpose of system validation.

4.1   Classification Datasets

This section aims at introducing the classification datasets and experiment conditions used in the case study. Firstly, three classification datasets obtained from the KEEL dataset repository [5] and their detailed descriptions are shown in

Table 1. Description of three classification datasets

Datasets	#Data	#Attributes	#Classes
Ecoli	336	7	8
Glass	214	9	7
Thyroid	720	21	3

Table 1, in which these descriptions mainly include number of data (#Data), number of attributes (#Attributes), and number of classes (#Classes).

Additionally, 5-fold cross validation is considered in the experimental study. The attributes of each dataset are regarded as the antecedent attribute of Micro-EBRBS and all of them are supposed to have three referential values. The range of AF is set as interval $[0.7, \sqrt{2}]$.

4.2 Comparative Analysis

In order to validate the effectiveness of the AFI method in improving Micro-EBRBS, the result of each dataset is measured with accuracy and number of failed data, which indicates that the system could not retrieve any result due to lack of relevant rules activated. Table 2 shows the comparison results of Micro-EBRBS with or without AFI method. Table 3 shows the comparison results of Micro-EBRBS and some classical classifiers, including k-nearest neighbor (KNN), support vector machine (SVM), the fuzzy rule-based classification system proposed by Chi et al. [6] (Chi-FRBCS), and original EBRBS.

Table 2. Comparison of Micro-EBRBS with or without AFI method

Datasets	Accuracy		No. of failed data	
	Without AFI	With AFI	Without AFI	With AFI
Ecoli	73.81	77.08	3	0
Glass	65.89	67.29	5	0
Thyroid	88.19	92.08	29	0

As Table 2 illustrates, the AFI method not only can improve the accuracy of Micro-EBRBS, but also is able to reduce the number of failed data, i.e., the accuracy of Ecoli is increased from 73.81% to 77.08%, and the number of failed data is decreased from 29 to 0. The similar results can be also found in Glass and Thyroid. This is because the AFI method can adjust the value of AF to dynamically select activated rules, so that Micro-EBRBS is able to activate consistent rules for any given input data.

From Table 3, it can be found that Micro-EBRBS obtains the 2nd best accuracy in Ecoli and Glsss, respectively, and the 3rd best accuracy in Thyroid, and

Table 3. Comparison of Micro-EBRBS with other classifiers

Datasets	KNN	SVM	Chi-FRBCS	EBRBS	Micro-EBRBS
Ecoli	80.95(1)	42.56(5)	76.49(3.5)	76.49(3.5)	77.08(2)
Glass	49.53(5)	65.89(3)	50.00(4)	69.16(1)	67.29(2)
Thyroid	95.00(1)	92.50(2)	88.06(5)	88.19(4)	92.08(3)
Average rank	2.33	3.33	4.16	2.83	2.33

they are 77.08%, 67.29%, and 92.08%. Moreover, in the comparison of other classifiers using average rank, the result is Micro-EBRBS (2.33) = KNN (2.33) > EBRBS (2.83) > SVM (3.33) > Chi-FRBCS (4.16), indicating that the Micro-EBRBS can produce satisfactory results comparing to some machine learning-based classifiers and rule-based classification systems.

5   Conclusions

In this study, a new concept of AF was proposed to revise the calculation of individual matching degree and further overcome the activation rule incompleteness and data inconsistency of Micro-EBRBS. The results of experimental evaluations demonstrated that the proposed AFI method not only improved the accuracy of Micro-EBRBS, but also avoided the situation when none of the rules are activated for any given input data.

For future research, an offline method should be further studied to determine the value of AF for Micro-EBRBS, which would promote the application of Micro-EBRBS for various complex classification problems.

References

1. Barredo, A.A., Daz-Rodrguez, N., Del Ser, J., et al.: Explainable Artificial Intelligence (XAI): concepts, taxonomies, opportunities and challenges toward responsible AI. Inf. Fusion **58**, 82–115 (2020)
2. Yang, L.-H., Liu, J., Wang, Y.-M., et al.: A micro-extended belief rule-based system for big data multiclass classification problems. IEEE Trans. Syst. Man Cybern. Syst. **51**(1), 420–440 (2021)
3. Liu, J., Martínez, L., Calzada, A., et al.: A novel belief rule base representation, generation and its inference methodology. Knowl.-Based Syst. **53**, 129–141 (2013)
4. Calzada, A., Liu, J., Wang, H., et al.: A new dynamic rule activation method for extended belief rule-based systems. IEEE Trans. Knowl. Data Eng. **27**(4), 880–894 (2015)
5. Alcala-Fdez, J., Fernndez, A., Luengo, J., et al.: KEEL data-mining software tool-data set repository, integration of algorithms and experimental analysis framework. J. Mult.-Valued Log. Soft Comput. **17**, 255–287 (2011)
6. Chi, Z., Yan, H., Pham, T.: Fuzzy Algorithms with Applications to Image Processing and Pattern Recognition. World Scientific, Singapore (1996)

Orbit Classification for Prediction Based on Evidential Reasoning and Belief Rule Base

Chao Sun[1] , Xiaoxia Han[2], Wei He[1,2(✉)] , and Hailong Zhu[1]

[1] Harbin Normal University, Harbin 150025, China
[2] Rocket Force University of Engineering, Xi'an 710025, China

Abstract. At present, most of the modeling methods in orbit classification for prediction (OCP) are data-driven methods, these reasoning processes are not interpretable, and the modeling effect is not good under small samples. In this paper, a new interpretable small sample OCP method is proposed based on evidence reasoning (ER) and belief rule base (BRB). First, multiple indicators were integrated by ER iteration to reduce the parameters. Then the BRB model was constructed based on expert knowledge and quantitative data. Finally, the projection covariance matrix adaptation evolutionary strategy (P-CMA-ES) is used to optimize model parameters. A case study is constructed to verify the effectiveness of the proposed method.

Keywords: Belief rule base · Evidential reasoning · Orbit prediction

1 Introduction

Orbit prediction is the basis for research on aerospace science and technology and has important research significance both in theory and engineering [1].

In engineering practice, data samples are not always sufficient [2]. However, most of the research on OCP is mainly related to data-driven methods [3, 4], and its modeling process is uninterpretable. In this case, an interpretable method that can be modeled under small samples is need to be constructed. The belief rule base, developed based on the traditional IF-THEN rules and Dempster–Shafer (D–S) theory [5], can integrate expert knowledge and quantitative data to construct models, its reasoning process is interpretable [6–9]. Because of the addition of expert knowledge, the input information of the model is increased, which makes BRB becomes an ideal choice for modeling with small samples. At the same time, considering the multiple indicators and small samples, to reduce the modeling parameters, the indicators are first integrated by the evidential reasoning before inputting to the BRB [10–12]. Therefore, an OCP method based on ER and BRB is proposed.

The paper is organized as follows: Sect. 2 constructs an OCP model based on ER and BRB expert system, and its processes are presented. A case study is provided in Sect. 3. This paper is concluded in Sect. 4.

T. Denœux et al. (Eds.): BELIEF 2021, LNAI 12915, pp. 87–95, 2021.
https://doi.org/10.1007/978-3-030-88601-1_9

2 OCP Model Based on ER and BRB

In this section, an OCP model is established based on ER and BRB. The implementation processes of OCP are described in Sect. 2.1. An indicator integration model based on ER is constructed in Sect. 2.2. A BRB-based OCP model is constructed in Sect. 2.3.

2.1 Implementation Process of OCP

The implementation process of OCP is shown in Fig. 1 and is mainly composed of two parts. The first part is to obtain the combined indicators by ER iteration. The second part constructs the prediction model based on BRB and then uses the projection covariance matrix adaptation evolutionary strategy (P-CMA-ES) [7] algorithm to optimize the model parameters.

It is worth noting that the proposed method is not limited by the optimization method. Because the P-CMA-ES has a better effect on single-objective optimization under constrained conditions, it is selected as the optimization algorithm in this paper.

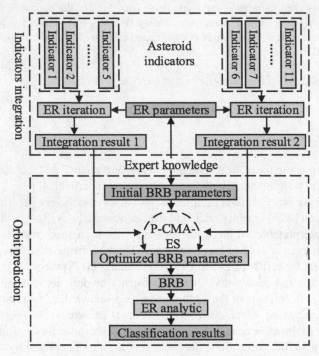

Fig. 1. The implementation process of OCP

To clearly describe the aggregation process, recursive ER is used in the integration process, and its modeling process is clear and interpretable. In the optimization process, analytical ER is used to train model parameters [13].

After integration, two combined indicators are obtained, then the rules in the BRB can be generated based on the indicators and their referential values. At the same time, because there are not many referential values in this paper, there will be no combinatorial explosion problem.

2.2 An ER Model for Indicators Integration

Evidential reasoning is developed based on the D-S and decision-making theory, it can solve the fusion problem of conflicting evidence by establishing a unified belief frame [14]. In this subsection, the recursive ER is used to integrate the indicators, and the processes are outlined as:

Step 1: The frame of discernment (FoD) in the ER model is $\{H_n, n = 1, \cdots, 4\}$. Let $\beta_{n,i}$ represents the belief degree that the sample is evaluated to the n_{th} grade in the i_{th} indicator. Then the belief distribution of the i_{th} indicator e_i can be described as:

$$S(e_i) = \{(H_n, \beta_{n,i}), n = 1, ..., N\}, i = 1, ..., L \tag{1}$$

where N denotes the grade number, L is the indicator number. $0 \le \beta_{n,i} \le 1 (n = 1, \cdots, N)$ and $\sum_{n=1}^{N} \beta_{n,i} \le 1$.

Step 2: Indicators fusion.

1) Calculate the basic probability masses.

$$m_{n,i} = \omega_i \beta_{n,i}$$

$$m_{H,i} = 1 - \omega_i \sum_{n=1}^{N} \beta_{n,i}$$

$$\overline{m}_{H,i} = 1 - \omega_i$$

$$\widetilde{m}_{H,i} = \omega_i (1 - \sum_{n=1}^{N} \beta_{n,i}) \tag{2}$$

where ω_i denotes the relative importance of the i_{th} indicator, $m_{n,i}$ is the basic probability mass of the i_{th} evaluated to the n_{th} grade. $m_{H,i}$ represents the basic probability mass that not assigned to any grade and is consists of two parts: $\overline{m}_{H,i}$ and $\widetilde{m}_{H,i}$. The former represents the importance of other indicators in the evaluation, the latter represents the incompleteness of the i_{th} indicator, in other words, if the distribution is complete, $\widetilde{m}_{H,i} = 0$.

2) Fusing indicators by the Dempster rule.

$$m_{n,I(i+1)} = K_{I(i+1)}[m_{n,I(i)}m_{n,i+1} + m_{H,I(i)}m_{n,i+1} + m_{n,I(i)}m_{H,i+1}]$$

$$m_{H,I(i)} = \widetilde{m}_{H,I(i)} + \overline{m}_{H,I(i)}$$

$$\widetilde{m}_{H,I(i+1)} = K_{I(i+1)}[\widetilde{m}_{H,I(i)}\widetilde{m}_{H,i+1} + \widetilde{m}_{H,I(i)}\overline{m}_{H,i+1} + \widetilde{m}_{H,I(i)}\overline{m}_{H,i+1}]$$

$$\overline{m}_{H,I(i+1)} = K_{I(i)}[\overline{m}_{H,I(i)}\overline{m}_{H,i+1}]$$

$$K_{I(i+1)} = [\frac{1}{1 - \sum_{t=1}^{N}\sum_{\substack{l=1 \\ l \ne t}}^{N} m_{t,I(i)}m_{l,i+1}}]^{-1} \tag{3}$$

where $m_{n,I(i+1)}$ and $m_{H,I(i)}$ are the integration results of the first $(k+1)$ indicators. The former denotes the combined belief degree, the latter denotes the residual belief degree.

3) Calculate the belief degree of the i_{th} indicator.

$$\beta_n = \frac{m_{n,I(L)}}{1 - \overline{m}_{H,I(L)}} \tag{4}$$

where β_n is the belief degree of the i_{th} indicator to the n_{th} grade.

4) The integration results are calculated using the utility function.

$$utility = \sum_{n=1}^{N} U(H_n)\beta_n \tag{5}$$

where $U(\cdot)$ denotes the utility function.

2.3 A BRB Model for Orbit Prediction

In the BRB-based classification model, there are multiple belief rules, and the k_{th} rule is profiled as:

$$R_k : \text{IF } x_1 \text{ is } A_1^k \wedge x_2 \text{ is } A_2^k \wedge \ldots \wedge x_M \text{ is } A_M^k$$
$$\text{Then } y \text{ is } \{(D_1, \beta_{1,k}), (D_2, \beta_{2,k}), \ldots, (D_N, \beta_{N,k})\}$$
$$\text{WITH rule weight } \theta_k$$
$$\text{AND attribute weight } \delta_1, \delta_2, \ldots, \delta_M \tag{6}$$

where R_k denotes k_{th} belief rule in BRB. $x_1, x_2, ..., x_M$ are the characteristics of the practical system. $A_1^k, A_2^k, ..., A_M^k$ denotes the reference points of the k_{th} belief rule of the system characteristics and they are determined by experts. $D_1, D_2, ..., D_N$ are the orbit classifications and $\beta_{1,k}, \beta_{2,k}, ..., \beta_{N,k}$ are their corresponding belief degrees. θ_k is the rule weight of the k_{th} belief rule. $\delta_1, \delta_2, ..., \delta_M$ are the indicator weights. L is the amount of the belief rules.

The orbit prediction is inferred by the analytic ER. The processes can be outlined as:

Step 1: The initial rule weights, attribute weights, and belief degrees of the BRB model are provided by the experts.

Step 2: Transform the input sample into belief distribution.

$$a_i^k = \begin{cases} \frac{A_i^{l+1} - x_i}{A_i^{l+1} - A_i^l}, & k = l, A_i^l \leq x_i \leq A_i^{l+1} \\ 1 - a_i^k, & k = l+1 \\ 0, & k = 1, \cdots, L, k \neq l, l+1 \end{cases} \tag{7}$$

where a_i^k denotes the k_{th} referential value, x_i denotes the i_{th} sample in the dataset.

Step 3: Calculate rule activation weight.

$$\omega_k = \frac{\theta_k \prod\limits_{i=1}^{M_k} (a_i^k)^{\overline{\delta_i}}}{\sum\limits_{l=1}^{L} \theta_l \prod\limits_{i=1}^{M_k} (a_i^l)^{\overline{\delta_i}}} \tag{8}$$

$$\overline{\delta_i} = \frac{\delta_i}{\max\limits_{i=1,\cdots,M_k} \{\delta_i\}} \tag{9}$$

where ω_k denotes the k_{th} rule's activation weight, $\overline{\delta_i}$ is the i_{th} attribute's normalized attribute weight.

Step 4: Combination of the activated rules.

$$\beta_n = \frac{\mu[\prod\limits_{k=1}^{L}(\omega_k\beta_{n,k}+1-\omega_k\sum\limits_{j=1}^{N}\beta_{j,k})-\prod\limits_{k=1}^{L}(1-\omega_k\sum\limits_{j=1}^{N}\beta_{j,k})]}{1-\mu[\prod\limits_{k=1}^{L}(1-\omega_k)]} \tag{10}$$

$$\mu = [\sum\limits_{n=1}^{N}\prod\limits_{k=1}^{L}(\omega_k\beta_{n,k}+1-\omega_k\sum\limits_{j=1}^{N}\beta_{j,k})-(N-1)\prod\limits_{k=1}^{L}(1-\omega_k\sum\limits_{j=1}^{N}\beta_{j,k})]^{-1} \tag{11}$$

where β_n denotes the belief degree of the i_{th} indicator integration level.

Step 5: Calculate the output by the utility function.

$$UTILITY = \sum\limits_{n=1}^{N}\mu(D_n)\beta_n \tag{12}$$

where $\mu(\cdot)$ denotes the utility function.

In this paper, the P-CMA-ES is used as the optimization algorithm of the BRB, and the optimization objective function is given by:

$$\min \ MSE(\theta_k, \beta_{n,k}, \delta_i)$$
$$st.$$
$$0 \le \theta_k \le 1, \ k = 1, \cdots, L,$$
$$0 \le \delta_i \le 1, \ i = 1, \cdots, M,$$
$$0 \le \beta_{n,k} \le 1, \ n = 1, ..., N, \ k = 1, \cdots, L,$$
$$\sum\limits_{n=1}^{N}\beta_{n,k} = 1, \ k = 1, \cdots, L \tag{13}$$

3 Case Study

In this section, to illustrate the effectiveness of the proposed method, a case study for the OCP is presented.

3.1 Problem Formulation

To classify the orbits of asteroids, the asteroid data from NASA are selected to be the experiment samples and the web link is https://www.kaggle.com/brsdincer/orbitc lassification. After selection, 81 samples were obtained. Each sample contains eleven indicators and one classification result. For indicator information, please refer to the following web link: https://www.kaggle.com/brsdincer/orbitclassification.

Each asteroid data has a corresponding classification, the FoD used in this paper is {AMO, APO, ATE, IEO} and the detailed information can be found in the following web link: https://pdssbn.astro.umd.edu/data_other/objclass.shtml.

3.2 Establishment of OCP Model Based on ER and BRB

The OCP model includes indicators integration and orbit prediction. The dataset of the asteroid is analyzed, then the indicators are divided into two groups. The first group has five indicators, which are related to distance. The second group has six indicators, which are not related to distance. When the integration model is established, different types of asteroid indicators are fused by the recursive ER algorithm, and then the combined indicators are obtained. Their referential values are shown in Table 1.

Table 1. The reference values of the combined indicators

Combined indicators	Reference values			
Combined indicator 1	0.1665	43.39564	101.3315	180.5768
Combined indicator 2	0.1856	44.50822	102.07603	181.2103

After integration, the BRB model is constructed, and its initial belief degrees are shown in Table 2.

Table 2. The initial belief degrees of BRB

No.	Rule weight	Belief degree {AMO APO ATE IEO}
1	1	{0.9 0.1 0 0}
2	1	{0.6 0.4 0 0}
3	1	{0.4 0.3 0.3 0}
4	1	{0.4 0.1 0.1 0.4}
5	1	{0.5 0.5 0 0}
6	1	{0.1 0.8 0.1 0}
7	1	{0 0.5 0.5 0}

(continued)

Table 2. (*continued*)

No.	Rule weight	Belief degree {AMO APO ATE IEO}
8	1	{0 0.3 0.4 0.3}
9	1	{0.3 0.3 0.3 0.1}
10	1	{0 0.4 0.6 0}
11	1	{0 0 0.9 0.1}
12	1	{0 0 0.7 0.3}
13	1	{0.1 0.2 0.3 0.5}
14	1	{0 0.4 0 0.6}
15	1	{0 0 0.3 0.7}
16	1	{0 0 0.2 0.8}

The BRB model is constructed by the above processes. Then the analytical ER is utilized to infer the model. Finally, the P-CMA-ES is used to optimize the parameters. The computational process and results can refer to in the paper [15].

3.3 Comparative Analysis

In this subsection, the BRB is compared with the fuzzy expert system [16], extreme learning machine (ELM) [17], and Random Forest [18]. Ten rounds of comparative experiments are carried out, with 50 experiments in each round. The training samples are 50% of the total samples, and the test samples are all samples. Take the average of the best 5 results in each round as the optimal result of each round. The accuracies of

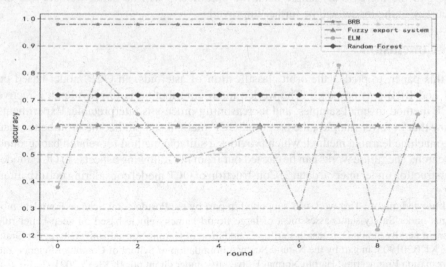

Fig. 2. Comparison of the four methods

10 rounds are shown in Fig. 2, and the average accuracies of them are shown in Table 3, the average Mean Square Errors (MSEs) are shown in Table 4.

Table 3. Average accuracy

	BRB	Fuzzy expert system	ELM	Random Forest
Accurate rate	98%	61%	54%	72%

Table 4. Mean MSEs

	BRB	Fuzzy expert system	ELM	Random Forest
MSE	0.0011	1.15	1.7347	1.1605

From the analysis above, the following conclusions can be drawn:

1) According to Tables 3 and 4. The proposed OCP model has the highest accuracy, which proves the effectiveness and superiority of the proposed OCP method.
2) The BRB modeling process has a clear causal relationship. Therefore, the BRB-based OCP method has better credibility than the method based on quantitative data.
3) Since BRB can be trained through samples, the OCP method based on BRB has better accuracy than the method based on qualitative knowledge.

With the increase of sample data, the parameters will be better trained, and the accuracy of the model will be further improved.

4   Conclusion

In this paper, to predict the orbit classification of asteroids, an OCP model based on ER and BRB was proposed. The work's main contribution is to provide an effective OCP method in small samples, and its reasoning process is interpretable. Experiments show that in fewer samples, the proposed OCP method can predict the orbit better than the machine learning method, which provides a suitable method for related background research. The future works can be carried out from the following aspects: 1) OCP model construction under interval data; 2) Construction of OCP model considering perturbation.

Acknowledgment. This work was supported in part by the Postdoctoral Science Foundation of China under Safety status assessment of large liquid launch vehicle based on deep belief rule base, in part by the Ph.D. research start-up Foundation of Harbin Normal University under Grant No. XKB201905, in part by the Natural Science Foundation of School of Computer Science and Information Engineering, Harbin Normal University, under Grant no. JKYKYZ202102.

References

1. Oh, H., Park, E., Lim, H.C., et al.: Orbit determination of high-earth-orbit satellites by satellite laser ranging. J. Astron. Space Sci. **34**(4), 271–279 (2017)
2. Yang, C.P., Sun, C., Su, J.L., He, W., Gao, Z.H.: A novel interval evidential reasoning approach to the physical and mechanical property assessment of particleboards. Math. Probl. Eng. https://doi.org/10.1155/2021/5581870.
3. Luo, F., Ren, H.L., Zhao, B.: Research on satellite orbit forecast based on neural network algorithm. Ship Sci. Technol. **42**(10), 146–151 (2020)
4. Peng, H., Bai, X.: Improving orbit prediction accuracy through supervised machine learning. Adv. Space Res. **61**(10), 2628–2646 (2018)
5. Shafer, G.: A Mathematical Theory of Evidence. Princeton University Press (1976)
6. Yang, J.B., Liu, J., Wang, J., Sii, H.S., Wang, H.W.: Belief rule base inference methodology using the evidential reasoning approach-RIMER. IEEE Trans. SMC - Part A: Syst. Hum. **36**(2), 266–285 (2006)
7. Zhou, Z.J., Hu, G.Y., Zhang, B.C., Hu, C.H., Zhou, Z.G., Qiao, P.L.: A model for hidden behavior prediction of complex systems based on belief rule base and power set. IEEE Trans. Syst. Man Cybern.: Syst. **48**(9), 1649–1655 (2018)
8. Kong, G., et al.: Combining principal component analysis and the evidential reasoning approach for healthcare quality assessment. Ann. Oper. Res. **271**(2), 679–699 (2018). https://doi.org/10.1007/s10479-018-2789-z
9. Li, G.L., et al.: A new safety assessment model for complex system based on the conditional generalized minimum variance and the belief rule base. Saf. Sci. **93**, 108–120 (2017)
10. Liu, Z.G., Liu, Y., Dezert, J., Cuzzolin, F.: Evidence combination based on credal belief redistribution for pattern classification. IEEE Trans. Fuzzy Syst. **28**(4), 618–631 (2020)
11. Huang, L., Liu, Z., Pan, Q., Dezert, J.: Evidential combination of augmented multi-source of information based on domain adaptation. Sci. China Inf. Sci. **63**(11), 1–14 (2020). https://doi.org/10.1007/s11432-020-3080-3
12. Liu, Z.G., Pan, Q., Dezert, J., Martin, A.: Combination of classifiers with optimal weight based on evidential reasoning. IEEE Trans. Fuzzy Syst. **26**(3), 1217–1230 (2018)
13. Feng, Z.C., Zhou, Z.J., Hu, C.H., Chang, L.L., Hu, G.Y., Zhao, F.J.: A new belief rule base model with attribute reliability. IEEE Trans. Fuzzy Syst. **27**(5), 903–916 (2019)
14. Yang, J.B., Singh, M.G.: An evidential reasoning approach for multiple-attribute decision making with uncertainty. IEEE Trans. Syst. Man Cybern.—Part A: Syst. Hum. 24(1), 1–18 (1994)
15. He, W., Yu, C.Q., Zhou, G.H., Zhou, Z.J., Hu, G.Y.: Fault prediction method for wireless sensor network based on evidential reasoning and belief-rule-base. IEEE Access **7**, 78930–78941 (2019)
16. Liao, T.W.: Classification of welding flaw types with fuzzy expert systems. Expert Syst. Appl. **25**(1), 101–111 (2003)
17. Yan, D., Chu, Y., Zhang, H., Liu, D.: Information discriminative extreme learning machine. Soft. Comput. **22**(2), 677–689 (2016). https://doi.org/10.1007/s00500-016-2372-y
18. Sun, J.Y., Zhong, G.Q., Huang, K.Z., Dong, J.Y.: Banzhaf random forests: cooperative game theory based random forests with consistency. Neural Netw. **106**, 20–29 (2018)

Imbalance Data Classification Based on Belief Function Theory

Jiawei Niu$^{(\boxtimes)}$ ⬚ and Zhunga Liu

School of Automation, Northwestern Polytechnical University, Xi'an, China
jiaweiniu@yeah.net

Abstract. Imbalance data is an important research for the classification and there are multiple techniques to deal with this problem. Each strategy has its particular advantage for solving imbalance data. To improve the classification performance, these strategies are combined in decision level via an appropriate way for taking fully advantages of the complementary information among different methods. Thus a new method is proposed as Evidence Redistributive Combination (ERC) for imbalance data. For query pattern, the classifier output produced by different techniques (i.e., undersampling, oversampling, hybridsampling) may have different reliabilities. So a cautious quality evaluation rule is created to estimate the credibility of each classification result based on the close neighborhoods. Then the revised classification results from different strategies are combined by Dempster's rule to reduce the ignorant information and to generate the final classification result. Multiple experiments are used to test the performance of the new ERC method, and it shows that ERC can efficiently improve the classification performance with respect to other related methods.

Keywords: Classification · Evidence theory · Imbalance data · Data sampling

1 Introduction

Traditional classification methods usually assume that each category in a dataset contains the same number of samples and the misclassification costs are equal. However, the data in the real world may have imbalanced distributions. A class with fewer instances is known as a positive class or a minority class, and a class with more examples is called a negative class or a majority class. The minority class is more important than the majority class in the real world, and the cost of misclassification is also higher. Nowadays, imbalanced classification is widely used in information security [1] and software prediction [2]. In such a way, the imbalanced data classification has attracted extensive interest from many researchers.

The imbalanced data classification methods are divided into three kinds: data preprocessing level [3], feature selection level [4], and classification methods improvement level [5]. In this work, we attempt to solve the problem at

© Springer Nature Switzerland AG 2021
T. Denœux et al. (Eds.): BELIEF 2021, LNAI 12915, pp. 96–104, 2021.
https://doi.org/10.1007/978-3-030-88601-1_10

data preprocessing level, which decreases the imbalance ratio of the dataset via creating minority data or deleting majority data. It focuses on undersampling [6], oversampling [7] and hybridsampling [8] methods to minimize the imbalance ratio by redistributing the data. In undersampling technique, it deletes the majority data to increase the classification accuracy of minority classes such as the Nearmiss [6] method. In the oversampling method, it creates the minority data by the Euclidean distance to balance the sample ratio such as Synthetic Minority Oversampling Techniques (SMOTE) [7]. The hybridsampling methods are linked with undersampling and oversampling techniques such as SmoteTomek [9] method.

These methods have their own advantages and drawbacks when they are utilized to deal with the imbalanced data classification. Oversampling method allows to generate minority data but it may cause the overfitting problems. Undersampling techniques remove majority data which may discard potentially important information. Hybridsampling algorithms are conducted with the connection of undersampling and oversampling methods. Each technique has its own particular benefits. To better improve the classification accuracy, we will propose a new method at decision level to combine these three algorithms via making full use of their complementary information.

Belief function theory provides an essential decision-level information fusion tool, and it is able to well combine the uncertain information. It has been already applied in data fusion and pattern classification fields [10,11]. In this paper, we want to propose a new method called Evidence Redistributive Combination (ERC) for imbalanced data. The output classification results generated by different methods (i.e., undersampling, oversampling, hybridsampling) may have different qualities/reliabilities. A reliability matrix through the neighborhood of the object is proposed to make a refined reliability evaluation. The classification outputs by different techniques will be cautiously revised utilizing the reliability matrix. Finally, the corrected classification results are combined by the evidence combination rule for making the final decision.

The remainder of this paper is organized as follows. Section 2 describes the proposed method in detail. The experimental applications are presented to test the performance of ERC in Sect. 3. Section 4 concludes this work.

2 A New Evidential Combination Method of Imbalance Data

The three imbalance data classification methods (undersampling, oversampling, hybridsampling) have their own benefits and drawbacks. To better improve the classification performance, these three methods are combined through an appropriate way for taking fully advantages of the complementary information among these methods. Belief function theory also called as Dempster's rule, which provides an efficient tool to combine the uncertain information at decision level. Thus, the evidence theory will be utilized here to combine these three techniques. A new method called Evidence Redistributive Combination (ERC) is proposed here to revise the classifier. We can obtain three pieces of classification results represented by evidence with three classifiers (i.e., undersampling, oversampling, hybridsampling), and we will combine these classification results under the framework of belief function efficiently.

The classification results of different data sampling method may have different reliabilities, and it may be harmful for the combination if the result with low reliability. So it is essential to evaluate the reliability of each classification output properly, and then revising the result based on the evaluation to improve the combination performance. We propose to estimate a refined reliability matrix to represent the qualities of each classification result. Such reliability matrix will be estimated based on the neighborhoods of objects in training dataset space, and it will show the possibility of the object misclassified to other classes. After that, the classification results are able to revised according to this matrix in a cautious way under belief function framework. The three corrected classification results are combined by belief function theory for predicting the class of object.

2.1   Evidential Combination of Classifier

In belief function theory, the mass function m, also called the basic belief assignment (BBA) is defined over the frame of discernment denoted by $\Omega = \{\omega_i, i = 1, 2, \ldots, c\}$, consisting of c exhaustive and exclusive hypotheses (classes) ω_i, $i \in 1, 2, \ldots, c$. The power-set 2^Ω is composed by all the subsets of Ω. A BBA is a mapping $m(.)$ from 2^Ω to $[0, 1]$ which satisfies $m(\phi) = 0$ and

$$\sum_{A \in 2^\Omega} m(A) = 1 \tag{1}$$

A is called a focal element of $m(.)$ which satisfy $m(A) > 0$. The BBA is called Bayesian BBA if the focal elements of BBA are all singleton classes. In this paper, we mainly assume that combining the classification results in form of BBAs.

Dempster's rule is usually utilized to combine the multiple classification results represented by BBA. DS rule for the combination of two BBA as $\mathbf{m} = \mathbf{m}_1 \oplus \mathbf{m}_2$ over 2^Ω is defined by $m(\phi) = 0$, and $\forall A \neq \phi \in 2^\Omega$ with the following equation

$$\mathbf{m} = \mathbf{m}_1 \oplus \mathbf{m}_2 = \begin{cases} \frac{\sum_{B \cap C = A} m_1(B) m_2(C)}{1 - K}, & \forall A \in 2^\Omega \setminus \{\phi\} \\ 0, & if A = \phi \end{cases} \tag{2}$$

where $K = \sum_{B, C \in 2^\Omega | B \cap C = \phi} m_1(B) m_2(C)$ is the total conjunctive conflicting masses. DS rule is associative, and the combination results are not influenced by the combination order for multiple BBA.

In reality, the classification result by different classification methods (i.e., undersampling, oversampling, hybridsampling) may have different reliabilities. It is essential to estimate the qualities of classification results and revised the results based on the evaluation before combination.

2.2   Evidence Quality Estimation

The classification results by different classifier may have different qualities. The undersampling method deletes the majority data which may change the distribution of data to affect the classification accuracy. The oversampling technique generates fake instance for minority class which may also has bad influence on classification result. The classification result of different methods can be seen as the evidence (BBA). The three methods (undersampling, oversampling, hybridsampling) are denoted by three classifiers as C_1, C_2, C_3 here. The object is classified over the frame of discernment $\Omega = \omega_1, \omega_2, \ldots, \omega_c$, and ω_i represents the class label. For each classifier $C_l, l \in 1, 2, 3$, the classification result for the training data x_i, $i \in 1, 2, \ldots, s$, is denoted by $\mathbf{p}_i = \{p_{i,1}, p_{i,2}, \ldots, p_{i,c}\}$, where $p_{i,j}$ represents the probability which x_i belongs to ω_j. The true classification result of training data is $t_i(\omega_j) = 1$ and $t_i(\omega_g) = 0$, $\omega_j \neq \omega_g$ when the true label of x_i is ω_j. Given a test pattern y, the classification result of y by different classifier can be shown as a BBA $\mathbf{m}_l$, $l \in 1, 2, 3$. The final label of y is calculated by the combination of these BBAs.

For each classifier $C_l, l \in 1, 2, 3$, it often shows close performance to close neighborhoods, and the close neighbors of object in dataset can be used to evaluate the quality of each classification result [12,13]. The classification results of training data are given by $\mathbf{p}_i$, and the true label of training data is also known. So the bias error of classifier can be computed by comparing the classifier output and the true label. Thus we can estimate the quality of the classification result of the y based on these neighbors.

How to select the suitable neighbors is an essential rule in reliability evaluation of each classification result. If we seek the close neighbors according to the attribute data, the selected neighbors seem near from the object, but the classification result of these neighbors as $\mathbf{p}_i$ may not close to the object of $\mathbf{m}_i$. These neighbors are not very useful to efficiently evaluate the quality of the classification result. The close neighbors are selected depending on both the attribute data and the classifier output (probability) at the same time. This ensures the selected neighbors with close attribute and the classification results to the object.

We select K nearest neighbors of the test data y using the attribute data and the classifier output. The selected attribute data are denoted by $x_1, x_2, \ldots, x_K$, and the classification results of the K nearest neighbors are given by $\mathbf{p}_1, \mathbf{p}_2, \ldots, \mathbf{p}_K$ with the corresponding ground truth are $\mathbf{t}_1, \mathbf{t}_2, \ldots, \mathbf{t}_K$. These two kinds of data can provide essential information for reliability evaluation. As there is a great difference between attribute data, the general linear normalization method is used to normalize the attribute data to [0,1].

Since these chosen neighbors are not totally same with the object, it can not be fully trusted during the reliability evaluation. The confidence factor mainly depends on the difference between the object and the selected neighbors, and both distance of attribute as well as the classifier output are considered to compute the difference. The majority of beliefs in the classification result will input the correction process when the confidence factor is high. Otherwise, a few beliefs will be redistributed.

The confidence factor α_l, $l \in 1, 2, 3$ is computed to the average distance between object and these neighbors.

$$\alpha_l = e^{-\beta_l d_l} \tag{3}$$

$$d_l = \frac{1}{2K} \left(\frac{\sum_{k=1}^{K} d_{yk}^A}{\bar{d}^A} + \frac{\sum_{k=1}^{K} d_{yk}^P}{\bar{d}^P} \right) \tag{4}$$

$$\bar{d}^A = \frac{1}{Ks} \sum_{i=1}^{s} \sum_{k=1}^{K} d_{xk}^A, \quad \bar{d}^P = \frac{1}{Ks} \sum_{i=1}^{s} \sum_{k=1}^{K} d_{xk}^P \tag{5}$$

β_l is a parameter distinct for each classifier C_l, which is used to adjust the influence of attribute distance and classifier output distance ratio on the confidence factor. d_l is the average distance between object and the K neighbors in regard to attribute as well as probability. $d_{yk}^A = \|y - x_k\|$ represents the Euclidean distance between the object and the K neighbors. $d_{yk}^P = \|\mathbf{m}_l - \mathbf{p}_k\|$ represents the Euclidean distance between the classifier output of the object and the K neighbors. $\bar{d}^A$ is the mean value of the average distance from each training data to its K neighbors. $\bar{d}^P$ represents the mean value of the average distance from the classification result for the training data to its K neighbors.

The beliefs are divided into two parts. One part will be redistributed in correction process on the basis of the reliability evaluation, and the other part will still preserved on each class as in original classification result.

$$m_{lr} = \alpha_l m_l, \quad m_{lo} = (1 - \alpha_l) m_l \tag{6}$$

2.3   Revision of Classification Result

The quality of the classification result of object will be evaluated in a refined way based on K neighbors, and then the classifier output will be revised according to the evaluation.

A reliability matrix Φ reflects the information about the misclassification error of the object, and the element ϕ_{ij} is the probability of the object classified to ω_i but the ground truth is ω_j. We are able to estimate the possibility of the object classified to ω_i when it truly belongs to ω_j with the aid of the true label t and the classification result $\mathbf{p}_i$. It is defined by the sum of the probabilities classified to ω_i with the true label ω_j.

$$w_{ji} = \sum_{t_{kj}=1} e^{-\tilde{d}_k} p_k (\omega_i) \tag{7}$$

where $\tilde{d}_k = \frac{1}{2} \left[\frac{d_{yk}^A}{\min_k d_{yk}^A} + \frac{d_{yk}^P}{\min_k d_{yk}^P} \right]$ is the relative distance between the object and the K neighbors.

The probability of the object classified to ω_i when it belongs to ω_j can be calculated by Bayesian rule.

$$\phi_{ij} = \frac{w_{ji}}{\sum_g w_{gi}} \qquad (8)$$

where $\sum_{j=1}^{c} \phi_{ij} = 1$, c is the number of the classes.

The classification result of the object can be revised by this matrix. The reliability matrix obtain the prior knowledge about the conditional probability of the object belonging to one class when it is classified to another class. We can get the belief of the object belonging to each class $\omega_j, j = 1, 2, \ldots, c$ as follows.

$$\tilde{m}_{lr}(\omega_j) = \sum_{i=1}^{c} \phi_{ij} m_{lr}(\omega_i) \qquad (9)$$

Thus we can obtain the evidence as

$$\begin{aligned}
\hat{m}_l(\omega_j) &= m_{lo}(\omega_j) + \tilde{m}_{lr}(\omega_j) \\
&= (1 - \alpha_l) m_l(\omega_j) + \sum_{i=1}^{c} \phi_{ij} m_{lr}(\omega_i)
\end{aligned} \qquad (10)$$

2.4 Parameter Optimization

In our proposed method, it includes a tuning parameter: β, which is used to determine the confidence factor α by (3). The optimal parameter is sought by minimizing an error criteria defined by the sum of distances between combined classifier result $\mathbf{m}^f$ and the true label $\mathbf{t}$. In MATLAB, the function *fmincon* is used to deal with this optimization problem.

$$\{\beta\} = \arg\min_{\beta} \sum_{i=1}^{s} \left\| \mathbf{m}_i^f - \mathbf{t}_i \right\|, \qquad \beta \in [0, 1] \qquad (11)$$

where $\|.\|$ is the Euclidean distance, and s is the number of the training dataset. $\mathbf{m}_i^f$ is the result of combining evidence concerning the ith training data, and $\mathbf{t}_i = [t_{i1}, t_{i2}, \ldots, t_{ic}]$. $t_{ij} = 1$ means the true label of x_i is ω_j.

3 Experimental Application

In this section, we will test the performance of our proposed ERC method with some benchmark datasets by comparing with other related imbalance data classification methods and information fusion method such as Smote, Nearmiss, SmoteTomek, and averaging fusion (AF). Some datasets are selected from UCI and KEEL dataset repository shown in Table 1.

As can be seen from Table 2–3, Evidence Redistributive Combination (ERC) method generally produces higher AUC values than a single data sampling method. This indicates that the complementary information among different techniques is very useful for improving classification performance. We can also find that the proposed ERC method typically yields the highest performance

Table 1. Imbalanced dataset description of the UCI and KEEL.

Data	Example	Attribute	Class	Majority examples	Minority examples	Imbalance ratio
Penbased	1100	16	10	80	75	1.06
Pima	768	8	2	400	215	1.87
Abalone	2560	8	3	1000	200	5
Page-blocks0	5472	10	2	3932	438	8.79
Yeast	1484	8	7	400	20	23.15
Yeast4	1484	8	2	1150	40	28.1
Yeast5	1484	8	2	971	29	32.73
Thyroid	720	21	3	533	13	36.94
Ecoli	336	7	8	100	7	71.5
Shuttle	2175	9	4	1200	9	853

comparing with the other combination methods. In ERC method, the reliability is evaluated based on the close neighbors in a refined way, and then the classifier output is cautiously revised to improve the quality. Moreover, the involved parameter in ERC is automatically optimized by minimizing an error criteria. Thus, ERC is able to produce the best classification performance in general.

Table 2. The AUC values using RF classifier

Datasets	Original	Smote	Nearmiss	Smotetomek	Voting	Average	ERC
Penbased	91.07	96.55	91.26	96.17	96.23	95.66	**99.14**
Pima	71.24	81.79	84.24	83.69	75.84	84.16	**84.34**
Abalone	66.96	76.48	70.32	76.65	66.51	77.05	**77.71**
Page-blocks0	94.48	89.69	98.93	98.76	94.79	98.00	**98.96**
Yeast	60.21	91.07	86.15	90.63	76.97	90.65	**91.16**
Yeast4	**88.96**	87.64	72.64	88.64	66.79	86.05	87.73
Yeast5	97.55	95.11	89.56	98.99	86.02	98.40	**99.41**
Thyroid	99.01	**99.37**	98.96	99.37	98.43	98.38	99.01
Ecoli	85.14	97.31	93.67	97.01	92.07	96.47	**97.75**
Shuttle	88.53	96.99	89.87	97.02	93.54	97.21	**99.99**
Average	84.32	91.2	87.56	92.69	86.52	92.98	**93.52**

Table 3. The AUC values using KNN classifier

Datasets	Original	Smote	Nearmiss	Smotetomek	Voting	Average	ERC
Pima	66.67	72.54	75.63	74.81	69.07	71.26	**75.84**
Penbased	96.61	99.15	98.63	99.15	98.11	99.16	**99.19**
Abalone	59.41	68.12	65.65	71.22	**71.47**	70.29	69.62
Page-blocks0	90.09	93.36	92.38	95.81	90.89	95.33	**97.12**
Yeast	80.53	84.42	84.56	87.42	76.78	**89.62**	89.61
Yeast4	62.93	66.06	79.54	81.86	78.46	82.15	**86.71**
Yeast5	79.46	86.02	96.84	92.51	86.02	98.83	**98.91**
Thyroid	86.11	93.75	62.18	91.38	80.21	93.96	**95.38**
Ecoli	81.19	93.31	93.35	93.29	90.38	95.09	**95.41**
Shuttle	89.37	97.78	86.64	98.58	96.69	98.99	**99.96**
Average	79.24	85.45	83.54	88.3	83.81	89.47	**90.78**

4   Conclusion

In this paper, we have proposed a new method for combination of classifiers to solve the imbalance data classification. Evidence Redistributive Combination (ERC) method is able to take advantage of essential complementary information among different data sampling techniques to improve classification performance. Multiple imbalanced datasets are used to validate the performance of the proposed method. The experimental results show that the ERC method is able to improve classification result comparing with other data sampling techniques and fusion methods.

Acknowledgements. This work was supported by the National Natural Science Foundation of China under Grants U20B2067, 61790552, 61790554.

References

1. Xiong, W., Gu, Q., Li, B., et al.: Collaborative web service QoS prediction via location-aware matrix factorization and unbalanced distribution. J. Internet Technol. **19**(4), 1063–1074 (2018)
2. Pouyanfar, S., Chen, S.C.: Automatic video event detection for imbalance data using enhanced ensemble deep learning. Int. J. Semant. Comput. **11**(01), 85–109 (2017)
3. Yang, X., Kuang, Q., Zhang, W., et al.: AMDO: an over-sampling technique for multi-class imbalanced problems. IEEE Trans. Knowl. Data Eng. **99**, 1 (2017)
4. Xiao, Y., Zhefu, W.U., Zhang, T., et al.: Feature selection based classification algorithm with imbalanced data. J. Integr. Technol. **5**(1), 68–74 (2016)
5. Dhar, S., Cherkassky, V.: Development and evaluation of cost-sensitive universum-SVM. IEEE Trans. Cybern. **45**(4), 806–818 (2017)
6. Chen, M., Xiaoyong, L.U.: Three random under-sampling based ensemble classifiers for Web spam detection. J. Comput. Appl. (2017)

7. Chawla, N.V., Bowyer, K.W., Hall, L.O., et al.: SMOTE: synthetic minority over-sampling technique (2011)
8. Seiffert, C., Khoshgoftaar, T.M., Hulse, J.V.: Hybrid sampling for imbalanced data. In: IEEE International Conference on Information Reuse & Integration. IEEE (2008)
9. Devi, D., Saroj, et al.: Redundancy-driven modified Tomek-link based undersampling: a solution to class imbalance. Pattern Recogn. Lett. (2017)
10. Zadeh, L.A.: Review of Shafer's: a mathematical theory of evidence. AI Mag. **5** (1984)
11. Skowron, A., Grzymala-Busse, J.: From rough sets theory to evidence theory. In: Advances in the Dempster-Shafer Theory of Evidence (1994)
12. Liu, Z., Pan, Q., Dezert, J., Han, J., He, Y.: Classifier fusion with contextual reliability evaluation. IEEE Trans. Cybern. **48**(5), 1605–1618 (2018)
13. Liu, Z.-G., Liu, Y., et al.: Evidence combination based on credal belief redistribution for pattern classification. IEEE Trans. Fuzzy Syst. **28**(4), 618–631 (2019)

A Classification Tree Method Based on Belief Entropy for Evidential Data

Kangkai Gao[1], Liyao Ma[2], and Yong Wang[1(✉)]

[1] University of Science and Technology of China, Hefei, China
gkk2010@mail.ustc.edu.cn, yongwang@ustc.edu.cn
[2] University of Jinan, Jinan, China
cse_maly@ujn.edu.cn

Abstract. Decision tree is widely applied in classification and recognition areas, but meanwhile it is hard to learn from evidential data with uncertainty. To solve this issue, we propose a decision tree method which can learn from uncertain data sets and guarantee a certain classification performance when handle problems with huge ignorance or uncertainty. This tree method selects attribute based on belief entropy, a kind of uncertainty measurement, which is calculated from the basic belief assignment. And especially the Evidential Expectation-Maximization algorithm is adopted to extract the distribution parameters from evidential likelihood to generate the basic belief assignment. The proposed method is an extension of decision tree based on belief entropy, which is supposed to handle problems with precise data. Some numerical experiments on Iris and Sonar data set are conducted and the experimental results suggested that the proposed tree method achieves good result on data with high-level uncertainty.

Keywords: Decision tree · Belief functions · Evidential data · Belief entropy · Evidential likelihood

1 Introduction

As one of the best-known classification methods, decision trees are widely used for their good learning capabilities and ease of understanding. However, traditional decision trees can only handle certain samples with precise data. The uncertain instances are usually ignored or removed by replacing them with precise instances when building decision trees, despite the fact that they may contain useful information [5], which may cause the lose of accuracy. We consider the situation that attributes are accurately observed, and their classes are imprecisely or incompletely observed, which generally happen when the observation is related to subjective opinion. An realistic example is the identification of outworn ancient stone inscriptions or scribbled handwritten numbers, in which the training set is built on experts' manual calibration of the fuzzy image samples, whose attributes, the image information, are certain, meanwhile the classes may be uncertain.

© Springer Nature Switzerland AG 2021
T. Denœux et al. (Eds.): BELIEF 2021, LNAI 12915, pp. 105–114, 2021.
https://doi.org/10.1007/978-3-030-88601-1_11

To solve the uncertain evidential data classification, some works based on probability theory have been carried out recent years. However various authors [1] have argued that probability cannot always adequately represent data uncertainty. Then the belief function theory is introduced into the tree method with the Evidential Expectation-Maximization (E^2M) algorithm [3,4] proposed by Denœux. Based on E^2M algorithm Sutton-Charani et al. [10] estimate tree parameters by maximizing evidential likelihood function. Ma et al. [7] proposed an active belief decision tree learning approach that can improve classification accuracy by querying while learning.

These above decision tree methods select best attributes according to the information entropy. Motivated by the idea of building decision tree on precise sample set by selecting attributes based on belief entropy [6], we extend it to evidential data set with uncertainty. The proposed method use E^2M algorithm to generate Basic Belief Assignment(BBA) from uncertain data labels, and choose best attribute by Deng entropy [2,11], which is calculated from BBAs of all samples corresponding to each attribute. The split strategy and stopping criterion are also adjusted accordingly. This method keeps fairly good accuracy even on sample set with high uncertainty level showed by experiments on UCI machine learning data set.

The sequel of the paper is organized as follows: Sect. 2 describes the generation of BBAs from sample set based normal mixture model and E^2M algorithm. Then we introduce the detailed procedure of the tree building in Sect. 3, which contains splitting strategy and stopping criterion. Some numerical experiments are conducted in Sect. 4. Concluded remarks are presented in Sect. 5.

2 Generation of BBAs

The proposed tree method primarily chooses split attribute based on belief entropy, which measures the uncertain level of instance in the form of BBA, the basic conception in belief function theory. So we firstly discuss the generation of BBA from each instance of the whole sample set in this section.

The purpose of a classification method is to build a model that maps an attribute vector $X = (x^1, ..., x^d)$ with D-dimensional attributes $(A^1, ..., A^D)$, to an output class $y \in \mathcal{C} = \{C_1, ..., C_K\}$ taking its value among k classes. The learning of the classification is based on a complete training set of precise data, denoted as

$$T = \begin{pmatrix} X_1, y_1 \\ \vdots \\ X_n, y_n \end{pmatrix} = \begin{pmatrix} x_1^1, ..., x_1^D, y_1 \\ \vdots \\ x_n^1, ..., x_n^D, y_n \end{pmatrix}.$$

In this paper, we consider the case that the output class labels of sample set contain uncertainty and model the uncertain output by mass function m_y : $2^{\mathcal{C}} \to [0, 1]$, such that $m_y(\varnothing) = 0$, and $\sum_{A \subseteq \mathcal{C}} m_y(A) = 1$, which is indeed the BBA. The subset A is called a *focal set* where $m_y(A) > 0$, and the $m_y(A)$ can be interpreted as the support degree of the evidence towards case that the true value is in set A.

One-to-one related to the mass function m_y, the belief function and plausibility function are defined as:

$$Bel_y(B) = \sum_{A \subseteq B} m_y(A), \tag{1}$$

$$Pl_y(B) = \sum_{A \cap B \neq \varnothing} m_y(A), \tag{2}$$

which respectively indicate the minimum and maximum belief degree of evidence towards set B. The function $pl_y(w) = Pl_y(\{w\})$ is called *contour function*.

When we model uncertain labels of evidential data with mass functions, the training set becomes

$$T = \begin{pmatrix} X_1, m_1 \\ \vdots \\ X_n, m_n \end{pmatrix} = \begin{pmatrix} x_1^1, ..., x_1^D, m_1 \\ \vdots \\ x_n^1, ..., x_n^D, m_n \end{pmatrix}.$$

2.1   Parameter Estimation of Normal Mixture Model

We denote the d-th attribute vector $X^d = \left(x_1^d, \cdots, x_n^d\right)^T, d \in \{1, ..., D\}$ by $W = (w_1, \cdots, w_n)^T$ in this chapter. The normal distribution is commonly encountered in practice, and is used widely in statistics as a simple model for complex issue. This is due to the normal distribution is very tractable analytically. When considering particular one attribute, we assume in this paper that this attribute value w of the sample is normal with mean μ_k and standard deviation σ independent on k:

$$w \sim \mathcal{N}\left(\mu_k, \sigma^2\right), k = 1, ..., K,$$

when this sample belongs to class C_k. It is actually a particular case of *linear discriminant analysis* [9] when the dimensionality of attribute is only one, in another word, a one-dimensional normal mixture model. Let π_k be the marginal probability when $Y = C_k$, and $\theta = (\mu_1, ..., \mu_K, \sigma, \pi_1, ..., \pi_K)$ the parameter vector. The complete-data likelihood is

$$L_c(\theta) = \prod_{i=1}^n p(w_i | Y_i = y_i) p(y_i) = \prod_{i=1}^n \prod_{k=1}^K \phi(w_i; \mu_k, \sigma)^{y_{ik}} \pi_k^{y_{ik}},$$

where the ϕ is normal distribution probability density, and y_{ik} is a binary indicator variable, such that $y_{ik} = 1$ if $y_i = k$ and $y_{ik} = 0$ if $y_i \neq k$.

when expended to evidential data, where we use contour function to describe the labels, we get the evidential likelihood

$$L(\theta) = \prod_{i=1}^n \sum_{k=1}^K pl_{ik} \phi(w_i; \mu_k, \sigma) \pi_k.$$

according the evidential EM algorithm. Then we compute the expectation of complete-data log likelihood

$$\ell_c(\theta) = \sum_{i=1}^{n} \sum_{k=1}^{K} y_{ik} \left[\log \phi\left(w_i; \mu_k, \sigma\right) + \log \pi_k \right]$$

with respect to the combined mass probability function

$$p_X\left(x \,\middle|\, pl; \theta^{(q)}\right) = \prod_{i=1}^{n} p\left(x_i \,\middle|\, pl_i; \theta^{(q)}\right)$$

in the E-step of the E^2M algorithm. We denote

$$p\left(x_i \,\middle|\, pl_i; \theta^{(q)}\right) = \zeta_{ik}^{(q)}, \, for \ k = (1, ..., K)$$

where

$$\zeta_{ik}^{(q)} = \frac{pl_{ik} \pi_k^{(q)} \phi\left(w_i; \mu_k^{(q)}, \sigma^{(q)}\right)}{\sum_{\ell} pl_{i\ell} \pi_{\ell}^{(q)} \phi\left(w_i; \mu_{\ell}^{(q)}, \sigma^{(q)}\right)}.$$

We get the to-be-maximized function

$$Q\left(\theta, \theta^{(q)}\right) = \sum_{i=1}^{n} \sum_{k=1}^{K} \zeta_{ik}^{(q)} \left[\log \phi\left(w_i; \mu_k, \sigma\right) + \log \pi_k \right]$$

whose formal is similar to the function computed in the EM algorithm on the normal mixture model [8]. Because of the similarity, the optimal parameter maximizing $Q\left(\theta, \theta^{(q)}\right)$ can be iteratively computed by

$$\pi_k^{(q+1)} = \frac{1}{n} \sum_{i=1}^{n} \zeta_{ik}^{(q)}, \mu_k^{(q+1)} = \frac{\sum_{i=1}^{n} \zeta_{ik}^{(q)} w_i}{\sum_{i=1}^{n} \zeta_{ik}^{(q)}},$$

$$\sigma^{(q+1)} = \sqrt{\frac{1}{n} \sum_{i=1}^{n} \sum_{k=1}^{K} \zeta_{ik}^{(q)} \left(w_i - \mu_k^{(q+1)}\right)^2}.$$

2.2   Determine BBA

We get K normal distributions on each attribute denoted as $\mathcal{N}_{kd}\left(\mu_{kd}, \sigma_d^2\right)$, which model data belong to the k-th class when concerning the d-th attribute only. The parameters of normal distribution corresponding to each class are calculated by the E^2M algorithm, during which all the samples participate, which means these parameters contains the general information of samples. Each mean of normal distribution indicates that the most possible value of this attribute if this sample belongs to corresponding class.

For the reason that we will measure the uncertainty each attribute contains in the attribute selection of decision tree building, we generate D, the number of attributes, BBAs to describe the uncertainty that one sample contains with the information of the whole training set.

It is denoted as $\phi_{kd}(x_i) = \phi(x_i^d; \mu_{kd}, \sigma_d)$ that the normal probability density of d-th attribute of sample x_i in the normal distribution $\mathcal{N}_{kd}$. According to the property of normal distribution, we assume that the probability that sample x_i belongs to class k is proportional to $\phi_{kd}(x_i)$ when consider attribute A^d only. So we propose a rule to assign mass function on subset of $\mathcal{C}$.

Firstly, normalize the $\phi_{kd}(x_i)$ with different class k such that

$$f_k = \phi_{kd}(x_i) / \sum_k \phi_{kd}(x_i).$$

Then rank f_k in decreasing order f_r' $(r = 1, ..., K)$, whose corresponding class is denoted as C_r' $(r = 1, ..., K)$. Assign f_r' to class set by following rule:

$$m(\{C_1'\}) = f_1'$$

$$m(\{C_1', C_2'\}) = f_2'$$

$$\cdots$$

$$m(\{C_1', ..., C_K'\}) = m(\theta) = f_K'.$$

By this rule, we obtain the BBA of x_i under the select attribute A^d, which we denote as m_i^d.

3   Tree Building Procedure

3.1   Attribute Selection and Splitting Strategy

Shannon entropy is often used to measure the information volume of a system or a process, and quantify the expected value of the information contained in a message under the probability theory. When extended to belief theory, there are a series of uncertainty entropy methods similar to Shannon entropy, are proposed to measure the uncertainty on BBA. We choose the Deng entropy [2]

$$E_{deng}(m) = -\sum_{A \subseteq \mathcal{C}} m(A) \log_2 \frac{m(A)}{2^{|A|} - 1}$$

to measure the uncertainty in this paper.

Based on the Deng entropy, the decision tree based on Deng Entropy can be constructed by attribute selection and split from the top to down. For the attribute A^d, we calculate the average entropy

$$E_{deng}(A^d) = \frac{1}{n} \sum_{i=1}^{n} E_{deng}(m_i^d),$$

where m_i^d denote the BBAs generated in Sect. 2. The less the entropy is, the less uncertainty the attribute contains. So on the split node of the decision tree, we choose the attribute $A^* = argmin_{A^d} E_{deng} \left(A^d \right)$ as the best attribute to split.

Compared to the precise situation, which can directly determine split intervals with maximum value and minimum value of the attribute data which exactly belong to a same class, the split on the best attribute with imprecise labels can not be solved directly by adopting the same strategy. We generate the split intervals from the parameter of the K normal distributions modeled from attribute.

According to the property of normal distribution, it is reasonable to declare that the closer the distance between attribute of a instance and the mean μ, the more likely this instance sampled from this distribution. Based on this opinion, we use the intervals with mid-value μ as the split interval, which is denoted as $I = [\mu - \varepsilon, \mu + \varepsilon]$. The width of the interval is adjusted flexibly to control how quickly the data are split to the branches, in another word, how deep the tree is on different sample set. We choose $\varepsilon = \sigma$ in this paper.

It is obvious that there is conflict between the intervals such that $I_a \cap I_b \neq \varnothing, \exists I_a, I_b \in \{I_1, ..., I_K\}$ if the width are not narrow enough when consider K normal distributes. So we assign an instance into k-th branch with class label A^k if the attribute is contained by the I_k only, which is equivalent of cutting the intervals into what do not intersect with each other.

3.2 Stopping Criterion

In the process of building the decision tree, we iteratively proceed the attribute selection and training sample splitting. There exist two issues during this process: firstly, the overfitting when the amount of non-split samples is too small; and secondly, failure to assign any sample into branches on a tree node caused by the gathering too much of non-split samples. The first issue causes more error and the second causes endless loop during building process.

To handle these issues, the stopping criterion is set such that stopping when the percentage of non-split samples is below a well-setting number or no sample is assigned into branches on a tree node. The non-split samples are assigned into corresponding class according to their distance to means of each class.

The decision tree procedure is summarized by the Algorithm 1.

4 Experiments

In this section we present experiments on several UCI data sets and compare the classification accuracy of the proposed method with traditional decision trees and belief entropy trees using only precise instances.

We chose two data sets from UCI repository, Iris and Sonar, the former contains 150 instances with 4 attributes and 3 classes, and the latter contains 208 instances with 60 attributes and 2 classes. 5-fold cross-validations on the two data sets are performed to validate the proposed methodology.

Algorithm 1. belief decision tree building

Input: A evidential data sample set T with n instances, attribute $\mathcal{A} = \{A_1, ..., A_D\}$ and class $\mathcal{C} = \{C_1, ..., C_K\}$, stop threshold α, split interval width ε

Output: tree node set$\{N_1, ..., N_{depth}\}$, corresponding split attribute $\{A_1^*, ..., A_{depth}^*\}$, and corresponding split interval set sequence $\{S_1^*, ..., S_{depth}^*\}$ in which each split interval set is $S^* = \{I_1, ..., I_K\}$

1:   $i = 1$
2: **for** $|T| > \alpha n$ **do**
3:     calculate normal distribute parameter matrix $\mathcal{N}_{kd} (k = 1, ..., K, d = 1, ..., D)$;
4:     determine split attribute $A_i^* = argmin_{A \in \mathcal{A}} E_{deng}(A)$;
5:     calculate the split interval $S_i^* = \{I_k^*\} = \{[\mu_k^* - \varepsilon^*, \mu_k^* + \varepsilon^*]\} (k = 1, ..., K)$;
6:     **for** $j < |T|$ **do**
7:       **if** $\exists k, x_j^* \in I_k^*$ and $x_j^* \notin I_q^*$ for any $q \neq k$ **then**
8:         delete the instance x_j from T;
9:       **end if**
10:    **end for**
11:    **if** there is no instance is deleted **then**
12:      **break**;
13:    **end if**
14:    build the node N_k with split attribute A_i^* and split interval S_i^*;
15:    $i = i + 1$;
16: **end for**
17: build the bottom node N_{depth} with the non-split samples T;

Denote the true label of a instance by C_i^*, and give its uncertain observation m_{y_i}. Due to the characters of belief function, we can simulate several situations from precise data:

- a *precise* observation is such that $pl_{y_i}(C_i^*) = 1$, and $pl_{y_i}(C_j) = 0, \forall C_j \neq C_i^*$;
- a *vacuous* observation is such that $pl_{y_i}(C_j) = 1, \forall C_j \in \mathcal{C}$;
- an *imprecise* observation is such that $pl_{y_i}(C_j) = 1$ if $C_j = C_i^*$ or $C_j \in \mathcal{C}_{rm}$, and $pl_{y_i}(C_j) = 0$ otherwise, where $\mathcal{C}_{rm}$ is a set of randomly selected labels;
- an *uncertain* observation is such that $pl_{y_i}(C_i^*) = 1$, and $pl_{y_i}(C_j) = r_j, \forall C_j \neq C_i^*$, where r_j are sampled independently from uniform distribution $\mathcal{U}([0, 1])$.

We repeated all experiments ten times and computed an average classification accuracy since the random generation of uncertain data. Three tree building techniques were compared:

- Tree 1(C4.5), which uses only precise data to build the tree;
- Tree 2(belief entropy tree with precise data), which chooses split attribute by belief entropy but generates BBAs with precise data only;
- Tree 3(belief entropy tree with uncertain data), which chooses split attribute by belief entropy and the BBAs are generated from uncertain data.

We tested the three tree building techniques on data sets in several situations: with vacuousness level $V \in [0, 1]$; with imprecision level $I \in [0, 1]$; and with uncertainty level $U \in [0, 1]$. The V, I and U controls the chance of an instance to become vacuous, imprecise or uncertain. For each instance, a number V_i was randomly generated on [0,1] and it will be replaced by a vacuous one if $V_i < V$. Similar operation was taken to generate imprecision and uncertainty.

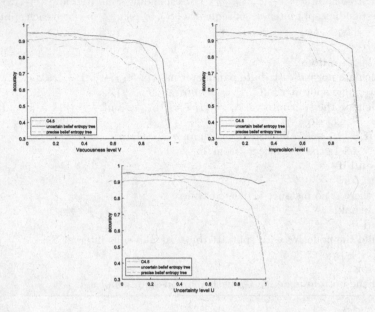

Fig. 1. Classification accuracies on Iris data

The Fig. 1 and Fig. 2 showed the classification accuracies of trees in different situations. We learn from the Fig. 1 that belief entropy tree with uncertain data roundly performed better than the one with precise data only and classical C4.5 tree when tested on Iris data set. The Fig. 2 shows that, on Sonar data, the proposed method got slightly weaker but almost same classification accuracy as classical C4.5 tree method when the V, I and U are low. When processing data with high vacuousness, Imprecision and uncertainty level, specifically, when $V > 0.6$, $I > 0.6$ or $U > 0.7$, the belief entropy trees performed obviously better than C4.5 tree. Particularly, the belief entropy tree with uncertain data and the one with precise data achieved similar performance on different vacuousness, Imprecision level and on uncertainty level below 0.7, but the former performed much better on higher uncertainty level.

We model the value of each attribute by one-dimensional normal mixture model, which is widely adapted in clustering problems. For this reason, we can achieve pretty good classification accuracy when handling data with huge uncertainty. However, this method makes a request to the normality of attributes' distribution, which makes it not always a suitable method for every data set and results in the different performance between Iris and Sonar data set.

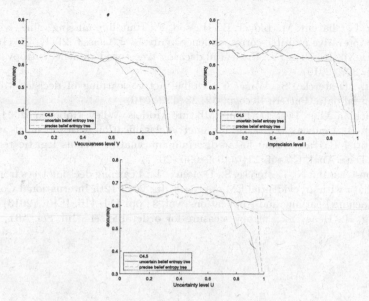

Fig. 2. Classification accuracies on Sonar data

5   Conclusion

A new classification tree method based on belief entropy with uncertain data modeled by belief function theory is proposed in this paper. It generates BBAs by E^2M algorithm and selects attributes according to belief entropy. As the experimental results show, the proposed method is robust to different sorts of uncertainty especially to high-level uncertainty. It is proved that the classification tree based belief entropy with evidential data have a potentially broad field of application.

In future works, we will focus on the details of the tree method such as the tree node building after reaching stopping criterion. And ensemble methods, such as bagging and boosting, will be considered to improve the average performance of the tree, in which the output of single tree will keep in the form of belief function.

References

1. Couso Blanco, I., Sánchez Ramos, L.: Harnessing the information contained in low-quality data sources. Int. J. Approximate Reasoning 1485–1486 (2014)
2. Deng, Y.: Deng entropy. Chaos Solitons Fractals **91**, 549–553 (2016)
3. Denœux, T.: Maximum likelihood from evidential data: an extension of the EM algorithm. In: Borgelt, C., et al. (eds.) Combining Soft Computing and Statistical Methods in Data Analysis. AINSC, vol. 77, pp. 181–188. Springer, Heidelberg (2010). https://doi.org/10.1007/978-3-642-14746-3_23
4. Denoeux, T.: Maximum likelihood estimation from uncertain data in the belief function framework. IEEE Trans. Knowl. Data Eng. **25**(1), 119–130 (2011)

5. Josse, J., Chavent, M., Liquet, B., Husson, F.: Handling missing values with regularized iterative multiple correspondence analysis. J. Classif. **29**(1), 91–116 (2012)
6. Li, M., Xu, H., Deng, Y.: Evidential decision tree based on belief entropy. Entropy **21**(9), 897 (2019)
7. Ma, L., Destercke, S., Wang, Y.: Online active learning of decision trees with evidential data. Pattern Recogn. **52**, 33–45 (2016)
8. McLachlan, G.J., Peel, D.: Finite Mixture Models. Wiley, Hoboken (2004)
9. Quost, B., Denoeux, T., Li, S.: Parametric classification with soft labels using the evidential EM algorithm: linear discriminant analysis versus logistic regression. Adv. Data Anal. Classif. **11**(4), 659–690 (2017)
10. Sutton-Charani, N., Destercke, S., Denœux, T.: Learning decision trees from uncertain data with an evidential EM approach. In: 2013 12th International Conference on Machine Learning and Applications, vol. 1, pp. 111–116. IEEE (2013)
11. Zhang, H., Deng, Y.: Entropy measure for orderable sets. Inf. Sci. **561**, 141–151 (2021)

Statistical Inference and Learning

Entropy-Based Learning of Compositional Models from Data

Radim Jiroušek[1,2], Václav Kratochvíl[1,2]([✉]), and Prakash P. Shenoy[3]

[1] The Czech Academy of Sciences, Institute of Information Theory and Automation,
Prague, Czech Republic
{radim,velorex}@utia.cas.cz
[2] Faculty of Management, Prague University of Economics and Business,
Jindřichův Hradec, Czech Republic
[3] University of Kansas School of Business, Lawrence, KS 66045, USA
pshenoy@ku.edu
https://pshenoy.ku.edu/

Abstract. We investigate learning of belief function compositional models from data using information content and mutual information based on two different definitions of entropy proposed by Jiroušek and Shenoy in 2018 and 2020, respectively. The data consists of 2,310 randomly generated basic assignments of 26 binary variables from a pairwise consistent and decomposable compositional model. We describe results achieved by three simple greedy algorithms for constructing compositional models from the randomly generated low-dimensional basic assignments.

Keywords: Compositional models · Entropy of Dempster-Shafer belief functions · Decomposable entropy of Dempster-Shafer belief functions · Mutual information · Information content

1 Introduction

Probabilistic compositional models were first proposed in [3] for discrete variables. It has since been generalized for many other uncertainty calculi [7]. In this paper, we are concerned with compositional models for Dempster-Shafer (DS) belief functions [5].

In the probabilistic framework, one strategy for learning models from data is to use information-theoretic concepts such as information content or mutual information based on the concept of Shannon's entropy [13]. In this paper, we investigate the use of two measures of entropy of belief functions defined by Jiroušek and Shenoy in 2018 [8] and 2020 [9]. The 2018 definition does not satisfy the subadditivity property, whereas the 2020 definition is the only one that is decomposable in the sense that $H(m_X \oplus m_{Y|X}) = H(m_X) + H(m_{Y|X})$. Here, m_X is a basic assignment for some variable X, $m_{Y|X}$ is a conditional

Supported by the Czech Science Foundation – Grant No. 19-06569S (to the first two authors), and by the Harper Professorship (to the third author).

T. Denœux et al. (Eds.): BELIEF 2021, LNAI 12915, pp. 117–126, 2021.
https://doi.org/10.1007/978-3-030-88601-1_12

basic assignment for $Y|X$ such that its marginal for X is vacuous, $\oplus$ denotes Dempster's combination rule, and $H(m)$ denotes entropy of basic assignment m.

Unfortunately, in contrast to probabilistic model learning, in the framework of belief function, we have to cope with several additional problems arising from the fact that we cannot support the respective procedures by belief function information theory. Not having an analog to probabilistic Kullback-Leibler divergence [12], we have problems even with determining, which of two different models is a better approximation of a given multidimensional belief function.

To study the applicability of the above-mentioned entropies, we concentrate only on a part of a complete model learning procedure. As we will see below, to define a joint compositional model, one starts with a set of low-dimensional marginal belief functions and then compose them in some order. In the computational experiments, we will randomly generate sets of pairwise consistent basic assignments, and compare three different algorithms seeking their best ordering. The first algorithm is based on decomposable entropy where we learn a compositional model that minimizes mutual information. The second is based on maximizing information content using the 2018 Jiroušek-Shenoy's definition of entropy that has two components—Dubois-Prade's entropy [2] of a basic assignment and Shannon's entropy of plausibility transform of a basic assignment [1]. The third is a modification of the second definition where the plausibility transform is replaced by the pignistic transform [15]. Not having a general tool allowing us to compare the results, we randomly generate only the situations when the optimality of a solution can be easily recognized. It occurs, as we will see below, when the learned model is decomposable. Our results indicate that the second and third algorithms are more effective than the first one in learning decomposable compositional models.

2   Preliminaries

Consider a finite set of binary variables $\mathcal{W} = \{S, T, U, \ldots\}$. A *basic assignment* for variables $\mathcal{V} \subseteq \mathcal{W}$ is a mapping $m_{\mathcal{V}} : 2^{\Omega_{\mathcal{V}}} \to [0, 1]$, such that $\sum_{\mathbf{a} \in 2^{\Omega_{\mathcal{V}}}} m_{\mathcal{V}}(\mathbf{a}) = 1$ and $m_{\mathcal{V}}(\emptyset) = 0$, where $\Omega_{\mathcal{V}} = \{0, 1\} \times \{0, 1\} \times \ldots \times \{0, 1\}$ is a $|\mathcal{V}|$-dimensional Cartesian product of values of the variables in $\mathcal{V}$. When the set of variables is evident from the context, or, if the set of variables is irrelevant, we omit the index $\mathcal{V}$. We say that $\mathbf{a} \subseteq \Omega$ is said to be a *focal element* of m if $m(\mathbf{a}) > 0$.

For basic assignment $m_{\mathcal{V}}$, we often consider its *marginal* basic assignment for $\mathcal{U} \subseteq \mathcal{V}$, denoted by $m_{\mathcal{V}}^{\downarrow \mathcal{U}}$. An analogous notation is used also for *projections*: for $a \in \Omega_{\mathcal{V}}$, let $a^{\downarrow \mathcal{U}}$ denote the element of $\Omega_{\mathcal{U}}$ that is obtained from a by omitting the values of variables from $\mathcal{V} \setminus \mathcal{U}$, i.e., for $\mathbf{a} \subseteq \Omega_{\mathcal{V}}$, $\mathbf{a}^{\downarrow \mathcal{U}} = \{a^{\downarrow \mathcal{U}} : a \in \mathbf{a}\}$. The marginal of basic assignment $m_{\mathcal{V}}$ for $\mathcal{U} \subseteq \mathcal{V}$ is defined as follows: $m_{\mathcal{V}}^{\downarrow \mathcal{U}}(\mathbf{b}) = \sum_{\mathbf{a} \subseteq \Omega_{\mathcal{V}} : \mathbf{a}^{\downarrow \mathcal{U}} = \mathbf{b}} m_{\mathcal{V}}(\mathbf{a})$ for all for all $\mathbf{b} \subseteq \Omega_{\mathcal{U}}$.

A basic assignment m can be described by equivalent functions such as *belief function*, *plausibility function*, or *commonality function*. The latter two are defined as follows:

$$Pl_m(\mathbf{a}) = \sum_{\mathbf{b} \subseteq \Omega:\, \mathbf{b} \cap \mathbf{a} \neq \emptyset} m(\mathbf{b}), \qquad Q_m(\mathbf{a}) = \sum_{\mathbf{b} \subseteq \Omega:\, \mathbf{b} \supseteq \mathbf{a}} m(\mathbf{b}).$$

When normalizing the plausibility function on singletons, one gets a probability mass function on Ω called a *plausibility transform* of basic assignment m [1]. Another popular probabilistic representation of a belief function is the so-called *pignistic transform* advocated by Philippe Smets [15] (though, as argued in [1], it is inconsistent with Dempster's combination rule). Let λ_m and π_m denote these two transforms, respectively, as follows. Suppose $a \in \Omega$. Then,

$$\lambda_m(a) = \frac{Pl_m(\{a\})}{\sum_{b \in \Omega} Pl_m(\{b\})}, \quad \text{and} \quad \pi_m(a) = \sum_{\mathbf{b} \subseteq \Omega:\, a \in \mathbf{b}} \frac{m(\mathbf{b})}{|\mathbf{b}|}.$$

3 Compositional Models

To construct multidimensional models from low-dimensional building blocks, we need a binary operator combining two low-dimensional (marginal) basic assignments into one (joint) basic assignment. One such binary operator $\triangleright$ is called a *composition operator* if it satisfies the following four axioms.

A1 *(Domain):* $m_{\mathcal{U}_1} \triangleright m_{\mathcal{U}_2}$ is a basic assignment for variables $\mathcal{U}_1 \cup \mathcal{U}_2$.

A2 *(Composition preserves first marginal):* $(m_{\mathcal{U}_1} \triangleright m_{\mathcal{U}_2})^{\downarrow \mathcal{U}_1} = m_{\mathcal{U}_1}$.

A3 *(Commutativity under consistency):* If $m_{\mathcal{U}_1}$ and $m_{\mathcal{U}_2}$ are consistent, i.e., $m_{\mathcal{U}_1}^{\downarrow \mathcal{U}_1 \cap \mathcal{U}_2} = m_{\mathcal{U}_2}^{\downarrow \mathcal{U}_1 \cap \mathcal{U}_2}$, then $m_{\mathcal{U}_1} \triangleright m_{\mathcal{U}_2} = m_{\mathcal{U}_2} \triangleright m_{\mathcal{U}_1}$.

A4 *(Associativity under special condition):* If $\mathcal{U}_1 \supset (\mathcal{U}_2 \cap \mathcal{U}_3)$, or, $\mathcal{U}_2 \supset (\mathcal{U}_1 \cap \mathcal{U}_3)$ then, $(m_{\mathcal{U}_1} \triangleright m_{\mathcal{U}_2}) \triangleright m_{\mathcal{U}_3} = m_{\mathcal{U}_1} \triangleright (m_{\mathcal{U}_2} \triangleright m_{\mathcal{U}_3})$.

For two operators satisfying these axioms see [5]. These operators account for the common information in two marginal basic assignments when there is overlap in the domain of the marginals.

By a *compositional model*, we mean a basic assignment $m_1 \triangleright \cdots \triangleright m_n$ obtained by multiple applications of the composition operator. Since the composition operator is generally neither associative nor commutative, if not specified otherwise by parentheses, the operators are always performed from left to right, i.e.,

$$m_1 \triangleright m_2 \triangleright m_3 \triangleright \ldots \triangleright m_n = (\ldots((m_1 \triangleright m_2) \triangleright m_3) \triangleright \ldots \triangleright m_{n-1}) \triangleright m_n.$$

Thus, for a given operator of composition, a (joint) compositional model is uniquely defined by an ordered sequence of low-dimensional (marginal) belief functions. In this paper, we consider only a part of the complete model learning process. Namely, given a set of low-dimensional marginal belief functions, what sequence should we use to construct the joint. To specify this step properly, consider a (finite) system $\mathbb{W}$ of small subsets of the considered variables $\mathcal{W}$. The vague assumption that $\mathcal{U} \in \mathbb{W}$ is small is made to avoid the computational problems connected with computations with the corresponding basic assignments. Thus, we assume that for each $\mathcal{U} \in \mathbb{W}$ we have (or we can easily get) a basic assignment $m_{\mathcal{U}}$ and that this basic assignment, as well as the corresponding commonality function, can be effectively represented in computer memory.

Given system $\mathbb{W}$, we study finding a sequence of sets $\{\mathcal{U}_i\}_{i=1,\ldots,n}$ from $\mathbb{W}$ such that the model $m_{\mathcal{U}_1} \triangleright m_{\mathcal{U}_2} \triangleright \cdots \triangleright m_{\mathcal{U}_n}$ represents as much of the relations among the variables as possible. As discussed in Sect. 1, we do not have a general tool for comparing two models. Therefore, we will consider a specific situations in which one can recognize an optimal solution regardless of the composition operator used.

To describe the necessary theoretical results consider the following notation. Let m_i denote $m_{\mathcal{U}_i}$. Thus, we speak about a compositional model $m_1 \triangleright m_2 \triangleright \ldots \triangleright m_n$, which is a $|\mathcal{U}_1 \cup \ldots \cup \mathcal{U}_n|$-dimensional basic assignment, in which basic assignment m_i is defined for variables $\mathcal{U}_i$. It is said to be *perfect* if all m_i's are marginals of the model. Recall that pairwise consistency of m_i's is a necessary but not sufficient condition for perfectness of model $m_1 \triangleright \ldots \triangleright m_n$. A perfect model reflects all the information contained in the low-dimensional basic assignments from which it is composed. So, it is not surprising that the optimal solution of a model learning algorithm is, if it exists, a perfect model. Quite often we can take advantage of the fact that such a solution is not defined by a unique sequence of low-dimensional basic assignments. In [4,7], the following two propositions are proved.

Proposition 1 (on perfect models). *Consider a perfect model $m_1 \triangleright \ldots \triangleright m_n$, and a permutation of its indices $i_1, \ldots, i_n$ such that $m_{i_1} \triangleright \ldots \triangleright m_{i_n}$ is also perfect. Then $m_1 \triangleright \ldots \triangleright m_n = m_{i_1} \triangleright \ldots \triangleright m_{i_n}$.*

Compositional model $m_1 \triangleright m_2 \triangleright \ldots \triangleright m_n$ is said to be *decomposable* if the sequence of sets $\mathcal{U}_1, \mathcal{U}_2, \ldots, \mathcal{U}_n$ satisfies the so-called *running intersection property*: $\forall i = 3, \ldots, n \ \exists j < i : \mathcal{U}_i \cap (\mathcal{U}_1 \cup \ldots \cup \mathcal{U}_{i-1}) \subseteq \mathcal{U}_j$.

Proposition 2 (on consistent decomposable models). *Decomposable model $m_1 \triangleright \ldots \triangleright m_n$ is perfect if and only if basic assignments $m_1, \ldots, m_n$ are pairwise consistent, i.e., $\forall \{i,j\} \subset \{1,2,\ldots,n\}: \ m_i^{\downarrow \mathcal{U}_i \cap \mathcal{U}_j} = m_j^{\downarrow \mathcal{U}_i \cap \mathcal{U}_j}$.*

4 Entropy and Information Content

The goal of this paper is to study the learning of compositional models from data using entropy and related information quantities. Probabilistic model learning algorithms are often based on characteristics of information theory. They may maximize the information content of the probability distribution $P(\mathcal{U})$ defined as follows: (H_s denotes the classical Shannon's entropy)

$$IC(P(\mathcal{U})) = \sum_{X \in \mathcal{U}} H_s(P(X)) - H_s(P(\mathcal{U}))$$

$$= \sum_{a \in \Omega_{\mathcal{U}} : P(a) > 0} P(a) \log \left(\frac{P(a)}{\prod_{X \in \mathcal{U}} P(a^{\downarrow X})} \right).$$

Alternatively, model construction may be based on mutual information defined as follows: ($\mathcal{U}$ and $\mathcal{V}$ are disjoint)

$$MI(P(\mathcal{U} \parallel \mathcal{V})) = H_s(P(\mathcal{U})) + H_s(P(\mathcal{V})) - H_s(P(\mathcal{U} \cup \mathcal{V}))$$
$$= \sum_{a \in \Omega_{\mathcal{U} \cup \mathcal{V}}:P(a)>0} P(a) \log \left(\frac{P(a)}{P(a^{\downarrow \mathcal{U}}) \cdot P(a^{\downarrow \mathcal{V}})} \right).$$

Notice that both information content and mutual information are non-negative. The information content IC measures the strength of dependence among the variables. All variables are independent (which is much stronger requirement than the pairwise independence of variables) under probability distribution P if and only if $IC(P) = 0$. Therefore, a model learning algorithm maximizing $IC(P)$ looks for a distribution that represents as much knowledge as possible. Thus, the goal is to find a model maximizing the information content, which is, due to its definition, equivalent to minimizing Shannon entropy within the class of models with the same one-dimensional marginals.

In this paper, we investigate learning compositional models in the framework of belief functions with the help of similar information-theoretic characteristics of basic assignments. We consider two definitions of entropy introduced in [8] and [9]. The former paper proposes

$$H_A(m) = \sum_{a \subseteq \Omega} m(\mathbf{a}) \log(|\mathbf{a}|) + H_s(\lambda_m),$$

where the first part of this expression is the Dubois-Prade entropy [2], and the second part is the Shannon entropy of the plausibility transform of m. This entropy is computationally inexpensive, and, as argued in [8], it is among the few that are consistent with the semantics of Dempster-Shafer theory of evidence. Its disadvantage is that it is not subadditive, and therefore the derived information-theoretic characteristics $IC_A(m_{\mathcal{U}}) = \sum_{X \in \mathcal{U}} H_A(m_{\mathcal{U}}^{\downarrow X}) - H_A(m_{\mathcal{U}})$, and $MI_A(m(\mathcal{U} \| \mathcal{V})) = H_A(m^{\downarrow \mathcal{U}}) + H_A(m^{\downarrow \mathcal{V}}) - H_A(m^{\downarrow \mathcal{U} \cup \mathcal{V}})$ need not be positive. Unfortunately, this manifests itself quite often even in very simple situations, and therefore we also study its approximation defined by $H_P(m) = \sum_{\mathbf{a} \subseteq \Omega} m(\mathbf{a}) \log(|\mathbf{a}|) + H_s(\pi_m)$ based on the pignistic transform π_m [10]. Though this entropy has also been shown to be not subadditive [11], in our computational experiments (described in Sect. 6), we encountered that the information content based on this entropy $IC_P(m_{\mathcal{U}}) = \sum_{X \in \mathcal{U}} H_P(m_{\mathcal{U}}^{\downarrow X}) - H_P(m_{\mathcal{U}})$, or the corresponding mutual information $MI_P(m(\mathcal{U} \parallel \mathcal{V})) = H_P(m^{\downarrow \mathcal{U}}) + H_P(m^{\downarrow \mathcal{V}}) - H_P(m^{\downarrow \mathcal{U} \cup \mathcal{V}})$ was rarely negative[1].

The other entropy considered in this paper is the decomposable entropy introduced in [9]. It is defined as follows:

$$H_S(m) = \sum_{\mathbf{a} \subseteq \Omega} (-1)^{|\mathbf{a}|} Q_m(\mathbf{a}) \log(Q_m(\mathbf{a})). \tag{1}$$

---

[1] In our experiments, MI_A was negative in about 12% of situations, whilst MI_P was negative only in 0.1% of cases.

It is defined using the commonality function of basic assignment m, and therefore the conversion of m to Q_m is required. In general, this function is not always non-negative. However, its merit is that it is the only definition of belief function entropy that satisfies an additivity property in the sense that $H_S(m_X \oplus m_{Y|X}) = H_S(m_X) + H_S(m_{Y|X})$ (here, m_X is a basic assignment for X, $m_{Y|X}$ is a conditional basic assignment for Y given X such that its marginal for X is vacuous, and $\oplus$ denotes Dempster's combination rule). Such a property characterizes Shannon's entropy for probability mass functions, and is often used in machine learning when constructing probabilistic models from data. To use this property when computing the entropy for compositional models, the conditional entropy is defined as follows ($\mathcal{U}$ and $\mathcal{V}$ are disjoint sets of variables, for which m is defined):

$$H_S(m_{\mathcal{U}|\mathcal{V}}) = \sum_{\mathbf{a} \subseteq \Omega_{\mathcal{U} \cup \mathcal{V}}} (-1)^{|\mathbf{a}|} Q_{m \downarrow \mathcal{U} \cup \mathcal{V}}(\mathbf{a}) \log(Q_{m_{\mathcal{U}|\mathcal{V}}}(\mathbf{a})), \qquad (2)$$

where $Q_{m_{\mathcal{U}|\mathcal{V}}}(\mathbf{a}) = Q_{m \downarrow \mathcal{U} \cup \mathcal{V}}(\mathbf{a}) / Q_{m \downarrow \mathcal{V}}(\mathbf{a}^{\downarrow \mathcal{V}})$ for all $\mathbf{a} \subseteq \Omega_{\mathcal{U} \cup \mathcal{V}}$ (note that for $\mathcal{V} = \emptyset$, $H_S(m_{\mathcal{U}|\mathcal{V}}) = H_S(m_{\mathcal{U}})$). Thus, we see that this entropy can be computed for compositional models of large dimensions if the composition operator satisfies the following axiom:

A5 *(Conditional independence):* For basic assignment $m_{\mathcal{U}_1} \triangleright m_{\mathcal{U}_2}$, variables $\mathcal{U}_1 \setminus \mathcal{U}_2$ and $\mathcal{U}_2 \setminus \mathcal{U}_1$ are conditionally independent given variables $\mathcal{U}_1 \cap \mathcal{U}_2$.

Axiom A5 implicitly defines conditional independence for sets of variables in the DS theory. This definition is consistent with the definition of conditional independence in valuation-based systems [14].

Using the notation from Sect. 3, let $\hat{\mathcal{U}}_j$ denote $\mathcal{U}_1 \cup \ldots \cup \mathcal{U}_{j-1}$. We get for such compositional models:

$$H_S(m_1 \triangleright \ldots \triangleright m_n) = H_S(m_1) + \sum_{j=2}^{n} H_S\big(m_j(\mathcal{U}_j \setminus \hat{\mathcal{U}}_j | \mathcal{U}_j \cap \hat{\mathcal{U}}_j)\big). \qquad (3)$$

5   Algorithms

Based on an analogy with probabilistic model learning processes, we may either look for a model with the smallest possible entropy or equivalently, a model maximizing the corresponding informational content. Therefore, we consider the following simple heuristic algorithm to minimize Eq. (3).

Min-entropy Greedy Algorithm.

1. Define $\mathcal{U}_1 := \arg\max_{\mathcal{U} \in \mathbb{W}}(IC_S(m_{\mathcal{U}}))$, $\hat{\mathcal{U}} = \mathcal{U}_1$, and n:=1.
2. Until $\big(\hat{\mathcal{U}} = \mathcal{W}\big)$

   find $\mathcal{U}_{n+1} := \arg\min_{\mathcal{U} \in \overline{\mathbb{W}}}(H_S(m_{\mathcal{U}}(\mathcal{U} \setminus \hat{\mathcal{U}} | \mathcal{U} \cap \hat{\mathcal{U}})))$,

   where $\overline{\mathbb{W}} = \big\{\mathcal{U} \in \mathbb{W} : \mathcal{U} \setminus \hat{\mathcal{U}} \neq \emptyset\big\}$,

   and redefine $\hat{\mathcal{U}} := \hat{\mathcal{U}} \cup \mathcal{U}_{n+1}$, $n := n+1$.

This algorithm cannot be used for entropy other than H_S. If all basic assignments are sufficiently small (the current version of our code cannot compute H_S entropy for basic assignments of dimensions larger than four), the algorithm is very efficient. Note that the algorithm (as well as the one from bellow) ends when all variables from W are covered by specified sequence $\mathcal{U}_1, \ldots, \mathcal{U}_n$. If there are some sets left, then adding respective basic assignments to the compositional model would not make any change because of Axiom A2 from Sect. 3.

An alternative model learning algorithm is based on the computation of information content using entropies H_A and H_M.

Max-information Greedy Algorithm.

1. Define $\mathcal{U}_1 := \arg\max_{\mathcal{U} \in \mathbb{W}}(IC_A(m_\mathcal{U}))$, $\hat{\mathcal{U}} = \mathcal{U}_1$, and n:=1.
2. Until $\left(\hat{\mathcal{U}} = W\right)$

    find $\mathcal{U}_{n+1} := \arg\max_{\mathcal{U} \in \overline{\mathbb{W}}} \left(MI_A(m_\mathcal{U}(\mathcal{U} \setminus \hat{\mathcal{U}} \parallel \mathcal{U} \cap \hat{\mathcal{U}})) \right)$,

    where $\overline{\mathbb{W}} = \left\{ \mathcal{U} \in \mathbb{W} : \mathcal{U} \setminus \hat{\mathcal{U}} \neq \emptyset \right\}$,

    and redefine $\hat{\mathcal{U}} := \hat{\mathcal{U}} \cup \mathcal{U}_{n+1}$, $n := n + 1$.

Similar to the case of min-entropy greedy algorithm, the efficiency of this algorithm follows from the fact that all the necessary computations are realized with basic assignments $m_\mathcal{U}$, $\mathcal{U} \in \mathbb{W}$. The algorithm does not compute any information-theoretic quantity of a complete model. Naturally, in general, this greedy algorithm doesn't find an optimal model either.

6 Results of Experiments

In this section, we briefly describe results achieved when applying the algorithms described in Sect. 5 to randomly generated systems of low-dimensional basic assignments. When constructing several compositional models from a system of low-dimensional assignments, we do not have a criterion enabling us to say, which of them is the best. The only characteristic we can compute for the multidimensional compositional models is their H_S entropy. Unfortunately, as it can be shown by examples, neither this characteristic guarantees that it achieves the lowest value for the optimal model. Thus, as the main criterion for the comparison of the considered approaches we consider how often they find decomposable models. We know that if it exists, then it is optimal.

In our computational experiments, we considered 26 binary variables. Randomly generated systems of basic assignments were such that

- the dimension of any basic assignment was not greater than 4,
- the basic assignment in a system were pairwise consistent,
- the basic assignments could be ordered so that the sets of variables met the running intersection property.

According to these rules, we generated 2,130 systems of basic assignments. Each system was generated by the following procedure: First, an ordered covering $\mathcal{U}_1, \ldots \mathcal{U}_n$ of all 26 binary variables satisfying the running intersection property was generated. This systems of sets was used for sequential generation of corresponding basic assignments (defined over respective variables) as follows.

1. $m_{\mathcal{U}_1}$ is randomly generated
2. for $i \in 2 \ldots n$
    find $j < i : \mathcal{U}_i \cap (\mathcal{U}_1 \cup \ldots \cup \mathcal{U}_{i-1}) \subseteq \mathcal{U}_j$
    $m_{\mathcal{U}_i}$ is randomly generated
    $m_{\mathcal{U}_i} = (m_{\mathcal{U}_j} \triangleright m_{\mathcal{U}_i})^{\downarrow \mathcal{U}_i}$

By *randomly generated* we mean the following. Randomly set the number of focal elements, randomly generate focal elements, and randomly generate respective mass assignments. This procedure guarantees pairwise consistency of respective basic assignments [6]. Using the ibelief package [16], we generated random belief functions of four types with respect to their focal elements: random, random with Ω guaranteed, quasibayesian, and nested with Ω guaranteed. However, it appeared that the type does not have any significant impact on the result of the experiment and therefore the type is not reported bellow. All calculations were performed in R language using our experimental routines based on relational databases.

To each generated system of basic assignments, we applied the min-entropy greedy algorithm, and two versions of the max-information greedy algorithm using entropies H_A and H_P. As mentioned earlier, the main criterion to evaluate the results was how often the algorithms found decomposable models. Even though all the generated systems could be ordered to meet the running intersection property, we could not expect that this goal would be always achieved. As an extreme situation consider a system of basic assignments consisting of basic assignments for independent variables. Then all models describe a multidimensional basic assignment of independent variables regardless of the ordering of low-dimensional assignments in a model. The existence of only one basic assignment meeting an improper conditional independence may prevent the construction of a decomposable model. Since we control only systems of variables and not the values of basic assignments (these were left to the random generator), we could not expect that there would be a chance that a model learning process would find decomposable models for all generated data. Actually, in 295 cases (from 2,130 generated) none of the three algorithms found a decomposable model.

7   Conclusions

A summary of results from the experiments is shown in Fig. 1, where the numbers indicate the number of successes of the respective algorithms. One can see that in 524 cases all three algorithms found decomposable models. The min-entropy greedy algorithm with H_S entropy found $524 + 43 + 132 + 18 = 717/2{,}130 = 0.337\%$ decomposable models, the max-information greedy algorithm using H_A

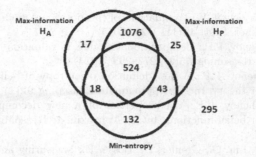

Fig. 1. A Venn diagram indicating the number of successes of the three algorithms.

entropy found $17 + 1076 + 524 + 18 = 1{,}635/2{,}130 = 0.768\%$ decomposable models, and the max-information greedy algorithm using H_P entropy found $1076 + 25 + 43 + 524 = 1{,}668/2{,}130 = 0.783\%$ decomposable models. Thus, we conclude that the min-entropy greedy process with H_S entropy is not as efficient as max-information greedy process with either H_A or H_M entropies for learning decomposable compositional models.

Notice also that the max-information greedy algorithm does not depend very much on the entropy used. They both succeed for about 0.77% of randomly generated systems of basic assignments. When using H_P, it found a decomposable model only in 33 more cases (about 1.5%) than when using H_A.

The computations required by the min-entropy greedy algorithm required about 35 times more time than that of the max-information greedy algorithm. This is because the computations of H_S require transformation of a basic assignment into a commonality function. If the data were given in a form of commonality functions, the difference would not be so striking (but the space complexity would noticeably increase).

References

1. Cobb, B.R., Shenoy, P.P.: On the plausibility transformation method for translating belief function models to probability models. Int. J. Approximate Reasoning **41**(3), 314–340 (2006)
2. Dubois, D., Prade, H.: Properties of measures of information in evidence and possibility theories. Fuzzy Sets Syst. **24**(2), 161–182 (1987)
3. Jiroušek, R.: Composition of probability measures on finite spaces. In: Geiger, D., Shenoy, P.P. (eds.) Proceedings of the Thirteenth Conference on Uncertainty in Artificial Intelligence (UAI 1997), pp. 274–281. Morgan Kaufmann (1997)
4. Jiroušek, R.: Foundations of compositional model theory. Int. J. Gener. Syst. **40**(6), 623–678 (2011)
5. Jiroušek, R.: On two composition operators in Dempster-Shafer theory. In: Augustin, T., Doria, S., Miranda, E., Quaeghebeur, E. (eds.) Proceedings of the 9th International Symposium on Imprecise Probability: Theories and Applications (ISIPTA 2015), pp. 157–165. Society for Imprecise Probability: Theories and Applications (2015)

6. Jiroušek, R., Kratochvíl, V.: Foundations of compositional models: structural properties. Int. J. Gener. Syst. **44**(1), 2–25 (2015)
7. Jiroušek, R., Shenoy, P.P.: Compositional models in valuation-based systems. Int. J. Approximate Reasoning **55**(1), 277–293 (2014)
8. Jiroušek, R., Shenoy, P.P.: A new definition of entropy of belief functions in the Dempster-Shafer theory. Int. J. Approximate Reasoning **92**(1), 49–65 (2018)
9. Jiroušek, R., Shenoy, P.P.: On properties of a new decomposable entropy of Dempster-Shafer belief functions. Int. J. Approximate Reasoning **119**(4), 260–279 (2020)
10. Jousselme, A.L., Liu, C., Grenier, D., Bossé, E.: Measuring ambiguity in the evidence theory. IEEE Trans. Syst. Man Cybern. Part A Syst. Hum. **36**(5), 890–903 (2006)
11. Klir, G.J., Lewis III, H.W.: Remarks on "Measuring ambiguity in the evidence theory". IEEE Trans. Syst. Man Cybern. Part A: Syst. Hum. **38**(4), 995–999 (2008)
12. Kullback, S., Leibler, R.A.: On information and sufficiency. Ann. Math. Stat. **22**, 76–86 (1951)
13. Shannon, C.E.: A mathematical theory of communication. Bell Syst. Tech. J. **27**(379–423), 623–656 (1948)
14. Shenoy, P.P.: Conditional independence in valuation-based systems. Int. J. Approximate Reasoning **10**(3), 203–234 (1994)
15. Smets, P.: Constructing the pignistic probability function in a context of uncertainty. In: Henrion, M., Shachter, R., Kanal, L.N., Lemmer, J.F. (eds.) Uncertainty in Artificial Intelligence, vol. 5, pp. 29–40. Elsevier (1990)
16. Zhou, K., Martin, A.: ibelief: belief function implementation (2021). https://CRAN.R-project.org/package=ibelief, R package version 1.3.1

Approximately Valid and Model-Free Possibilistic Inference

Leonardo Cella[✉] and Ryan Martin

Department of Statistics, North Carolina State University, Raleigh, NC 27695, USA
{lolivei,rgmarti3}@ncsu.edu

Abstract. Existing frameworks for probabilistic inference assume the inferential target is a feature the posited statistical model's parameters. In this paper, we develop a new version of the so-called generalized inferential model framework for possibilistic inference on unknowns that are well-defined independent of a statistical model. We provide a bootstrap-based implementation and establish approximate validity.

Keywords: Belief · Bootstrap · Inferential model · M-estimation · Plausibility · Quantile regression

1 Introduction

In statistics, it is common to work under an assumed statistical model. There are a number of advantages to a model-based approach, including (a) the inferential targets—the model parameters—are well-defined, and (b) a model is necessary in order to apply existing *probabilistic inference* frameworks, including Bayes and generalized Bayes (de Finetti 1990; Walley 1991), fiducial (Fisher 1935), generalized fiducial (Hannig et al. 2016), structural inference (Fraser 1968), Dempster–Shafer theory (Dempster 1967, 1968, 2008; Shafer 1976), and inferential models (IMs, Martin and Liu 2015), which are model-based.

In machine learning, however, the inferential targets are rarely model parameters. Instead, they are features that characterize a "best" action, the one that would minimize a suitable risk function. A risk minimizer is a "real-world" quantity, a functional of the true distribution, not something whose existence relies on the correctness of a posited statistical model. And if the quantity of interest is well-defined independent of a model, then the use of a model-based approach for inference on that quantity creates a risk of model misspecification bias. For example, consider estimation of a qth population quantile, θ, where $q \in (0,1)$. One approach might be to specify an exponential model, with scale parameter λ, and estimate the quantile as $\hat{\theta} = -\hat{\lambda}^{-1} \log(1 - q)$, where $\hat{\lambda}$ is, say, the maximum likelihood estimator. This would be a reliable estimator if the underlying distribution were at least approximately exponential, but not so otherwise.

R. Martin—Partially supported by the U.S. National Science Foundation, DMS–1811802.

T. Denœux et al. (Eds.): BELIEF 2021, LNAI 12915, pp. 127–136, 2021.
https://doi.org/10.1007/978-3-030-88601-1_13

The sample quantile, on the other hand, is a direct, model-free estimate of the quantity of interest and, therefore, would be reliable more generally.

In light of the potential biases model specification can create, when the quantity of interest is not determined by parameters of a statistical model, a model-free approach would be desirable. In this paper we develop a new approach, a further generalization of the so-called *generalized IMs* in Martin (2015, 2018). The motivation behind the original generalized IM was to avoid giving a complete specification of the data-generating process, which generally simplifies the IM construction without sacrificing its desirable validity property. In this paper, we push this idea further by developing a generalized IM that requires no model specifications. While the original generalized IM construction proceeds without a complete description of the data-generating process, its implementation still requires a statistical model. Here we side-step this requirement by leveraging the powerful *bootstrap* procedure (e.g., Efron 1979) that provides a model-free approximation of the relevant sampling distributions.

Following some brief background about generalized IMs in Sect. 2, the main results are presented in Sect. 3. There we lay out the model-free generalized IM framework, with its bootstrap-based implementation, and offer some theoretical support in the form of an asymptotically approximate validity property. An brief illustration in the context of quantile regression is presented in Sect. 4, with some concluding remarks in Sect. 5.

2   Generalized IMs

Assume that data $Z^n = (Z_1, \dots, Z_n)$ are independent and identically distributed (iid) with distribution P_θ, and inference on the parameter $\theta \in \Theta$ is desired. Note that the individual Z_i's could be response–predictor variable pairs (X_i, Y_i) as in Sect. 4.1. The IM approach quantifies uncertainty about θ using belief and plausibility functions, denoted by $(\underline{\varPi}_{Z^n}, \overline{\varPi}_{Z^n})$. These are interpreted as data-dependent degrees of belief/plausibility in assertions about the unknown θ. These measures are meaningful thanks to their so-called *validity* property, which states that, for any $A \subset \Theta$, as a function of $Z^n \overset{\text{iid}}{\sim} P_\theta$, the random variable $\overline{\varPi}_{Z^n}(A)$ tends not to be small when $\theta \in A$, i.e., when A is a true assertion. More precisely, validity means that the IM' plausibility function satisfies

$$\sup_{\theta \in A} P_\theta\{\overline{\varPi}_{Z^n}(A) \leq \alpha\} \leq \alpha, \quad \text{for all } \alpha \in [0, 1], \text{ all } A \subseteq \Theta. \tag{1}$$

Given the duality between belief and plausibility functions, i.e., $\underline{\varPi}_{Z^n}(A) = 1 - \overline{\varPi}_{Z^n}(A^c)$, and the fact that (1) is required to hold for all A, an equivalent definition of validity can be stated in terms of the belief function.

The original IM developments in Martin and Liu (2013) utilized a description of how the data Z^n are generated to construct the valid belief and plausibility functions. More specifically, P_θ is written in terms of an association between data Z^n, parameter θ and a set of unobservable auxiliary variables U^n through

$$Z^n = a(\theta, U^n), \tag{2}$$

where $U^n \stackrel{\text{iid}}{\sim} P_U$ and P_U is known. The IM construction proceeds with the specification of a random set $\mathcal{S}$ to predict the unobserved value of U^n. Easy to arrange properties of this user-specified random set ensure that the prediction of the auxiliary variable is done in a reliable/calibrated way, which turns out to be fundamental for the IM validity. Lastly, this random set gets fused with the observed data and the statistical model, originating a new random set on Θ, the parameter space. The distribution of this random set, as a function of the random $\mathcal{S}$ for fixed z^n, defines the IM's belief and plausibility functions.

In this formulation, it is important to write the association in (2) in terms of data summaries that match the dimension of the parameter of interest. Doing so reduces the dimension of the auxiliary variable too, which makes its prediction by a (lower-dimensional) random set easier and more efficient. Unfortunately, this dimension reduction step is not always easy to do. The generalized IM approach in Martin (2018) bypasses this potential obstacle through the specification of a real-valued function T that associates Z^n, θ, and an unobservable auxiliary variable. For example, consider the log-relative likelihood

$$T_{z^n}(\theta) = \log L_{z^n}(\theta) - \log L_{z^n}(\hat{\theta}_{z^n}), \tag{3}$$

where $L_{z^n}(\theta)$ is the likelihood function and $\hat{\theta}_{z^n}$ the corresponding maximum likelihood estimator. Let $F_\theta(t) = P_\theta\{T_{Z^n}(\theta) \leq t\}$ the distribution function of $T_{Z^n}(\theta)$ under P_θ. Then the generalized association takes the form

$$T_{Z^n}(\theta) = F_\theta^{-1}(U), \quad U \sim \mathsf{Unif}(0,1). \tag{4}$$

Since θ values having large likelihood are most "plausible", the recommended random set in this case is

$$\mathcal{S} = [0, \tilde{U}], \quad \tilde{U} \sim \mathsf{Unif}(0,1). \tag{5}$$

This random set gets pushed forward, through the generalized association (4) at the observed $Z^n = z^n$, to a corresponding random set on the θ-space:

$$\Theta_{z^n}(\mathcal{S}) = \{\vartheta : F_\vartheta(T_{z^n}(\vartheta)) \ni \mathcal{S}\} = \{\vartheta : F_\vartheta(T_{z^n}(\vartheta)) \leq \tilde{U}\}.$$

Since $\mathcal{S}$ is nested, so is $\Theta_{z^n}(\mathcal{S})$, so the IM output has the special form of a *consonant* belief/plausibility function or, equivalently, necessity/possibility measure (Dubois and Prade 1988), and, therefore, can be fully characterized by its plausibility contour function

$$\pi_{z^n}(\vartheta) = \mathsf{P}_{\mathcal{S}}\{\Theta_{T_{y^n}}(\mathcal{S}) \ni \vartheta\} = 1 - F_\vartheta(T_y(\vartheta)), \quad \vartheta \in \Theta.$$

That is, $\overline{\Pi}_{z^n}(A) = \sup_{\vartheta \in A} \pi_{z^n}(\vartheta)$ and $\underline{\Pi}_{z^n}(A) = 1 - \overline{\Pi}_{z^n}(A^c)$. Validity, in the sense of (1), follows from the easily-verified fact that $\pi_{Z^n}(\theta) \sim \mathsf{Unif}(0,1)$ when $Z^n \stackrel{\text{iid}}{\sim} P_\theta$; see Martin (2015, 2018) for details.

Computation of the generalized IM can be difficult because the distribution function F_θ is needed, at least approximately, for all θ in a dense grid. See Martin (2020, 2021) for details. For specific inferential tasks, at least, simplification would be possible (e.g., Syring and Martin 2021).

3 Generalized IMs Without a Model

3.1 Construction

As described above, the benefit of the generalized IM is that it does not require a full specification of the data-generating process, which in turn simplifies the entire construction. However, implementing that generalized IM still depends on a statistical model in several ways, including: the relative likelihood and the distribution function F_θ defined in (4) are features of the model. Here we describe a new approach that avoids the model dependencies.

Here we assume that data Z^n are iid but with an entirely unspecified distribution P. The quantity of interest is some specified feature $\theta = \theta(P)$ of the distribution P, taking values in a space Θ. Our focus here will be on features defined as minimizers of a *risk function*. Start with a loss function $\ell_\vartheta(z)$ that measures the discrepancy between a single data point z and a generic feature value ϑ; see Sect. 4 for a practical example. Define the risk $R(\cdot)$ function

$$R(\vartheta) \equiv R_P(\vartheta) := \int \ell_\vartheta(z)\, P(dz), \quad \vartheta \in \Theta,$$

the P-expected loss. Then define the quantity of interest $\theta = \theta(P)$ as

$$\theta = \arg\min_{\vartheta \in \Theta} R(\vartheta). \tag{6}$$

Of course, the risk-minimizer framework does not cover all cases, but we leave these further generalizations for future work. Our present goal is valid, model-free, (imprecise) probabilistic uncertainty quantification about θ.

Naturally, θ can be estimated by replacing the risk function R, that depends on the unknown P, with its empirical version

$$R_{Z^n}(\vartheta) = \frac{1}{n}\sum_{i=1}^n \ell_\vartheta(Z_i), \quad \vartheta \in \Theta.$$

The corresponding estimate of θ is

$$\hat{\theta}_{Z^n} := \arg\min_{\vartheta \in \Theta} R_{Z^n}(\vartheta).$$

This is the so-called *M-estimator*. Analogous to the log-relative likelihood in (3), define the empirical risk difference

$$T_{Z^n}(\vartheta) = n\{R_{Z^n}(\vartheta) - R_{Z^n}(\hat{\theta}_{Z^n})\}, \quad \vartheta \in \Theta. \tag{7}$$

This is exactly the log-relative likelihood if $\ell_\vartheta(z) = -\log p_\vartheta(z)$ and p_ϑ is the density of a parametric model for the true distribution. The multiplier "n" in (7) could be removed, but we keep it here for the direct connection to the log-relative likelihood. Of course, $T_{Z^n}(\theta)$ is a random variable as a function of Z^n, and it has a distribution function $F(t) = P\{T_{Z^n}(\theta) \leq t\}$ where $Z^n \overset{\text{iid}}{\sim} P$. The choice to only use the true θ, and not generic ϑ, is deliberate.

If the distribution function F was known, then we could proceed exactly as in Sect. 2 above. That is, the generalized association would be

$$T_{Z^n}(\theta) = F^{-1}(U), \quad U \sim \mathsf{Unif}(0,1),$$

and, for observed data $Z^n = z^n$, using the same random set $\mathcal{S}$ as in (5), we end up with a plausibility contour

$$\pi_{z^n}(\vartheta) = 1 - F\big(T_{z^n}(\vartheta)\big), \quad \vartheta \in \Theta, \tag{8}$$

which, as before, defines a consonant plausibility function/possibility measure that can be used for inference on θ. And just like above, the validity of the corresponding generalized IM would follow from

$$Z^n \overset{\text{iid}}{\sim} P \implies \pi_{Z^n}(\theta) \text{ is stochastically no smaller than } \mathsf{Unif}(0,1). \tag{9}$$

Things above are much more difficult than they appear. Indeed, the "if the distribution function F was known" statement never holds, so we stumble immediately when trying to carry out the above construction. But we can pick ourselves up by our bootstraps by making use the powerful *bootstrap* procedure developed in the seminal work by Efron (1979). The basic idea behind the bootstrap is that iid samples from the empirical distribution of the observed data z^n should closely resemble iid samples from P. Our proposal is to approximate the unknown distribution F using this bootstrap strategy.

The bootstrap requires an extra level of randomization and here is a convenient way to describe that. Let $\xi = (\xi_1, \ldots, \xi_n)$ denote a random vector having a multinomial distribution with size n and bin probabilities $(n^{-1}, \ldots, n^{-1})$. Then the bootstrap version of the empirical risk function is given by

$$R_{z^n}^{\xi}(\vartheta) = \frac{1}{n} \sum_{i=1}^{n} \xi_i \, \ell_\vartheta(z_i), \quad \vartheta \in \Theta,$$

and $\hat{\theta}_{z^n}^{\xi}$ is the corresponding bootstrap version of the empirical risk minimizer. This leads naturally to the bootstrap version of the empirical risk difference, i.e.,

$$T_{z^n}^{\xi}(\vartheta) = n\{R_{z^n}^{\xi}(\vartheta) - R_{z^n}^{\xi}(\hat{\theta}_{z^n}^{\xi})\}, \quad \vartheta \in \Theta.$$

Finally, the bootstrap approximation to the distribution function F is given by

$$\widehat{F}^{\text{boot}}(t) = Q_n\{T_{z^n}^{\xi}(\hat{\theta}_{z^n}) \le t\}, \quad t \in \mathbb{R},$$

where Q_n is the aforementioned multinomial distribution for ξ. To be clear, the data z^n is fixed in the above calculation, the probability is with respect to the distribution of ξ. Note also that where the definition of F involved the true θ, the definition of $\widehat{F}^{\text{boot}}$ involves the risk minimizer from the original data z^n. This leads to a bootstrap-based plausibility contour

$$\pi_{z^n}^{\text{boot}}(\vartheta) = 1 - \widehat{F}^{\text{boot}}\big(T_{z^n}(\vartheta)\big), \quad \vartheta \in \Theta, \tag{10}$$

which can be readily converted into a consonant belief/plausibility function for inference on θ, via the rule $\overline{\Pi}_{z^n}^{\text{boot}}(A) = \sup_{\vartheta \in A} \pi_{z^n}^{\text{boot}}(\vartheta)$, $A \subset \Theta$.

Of course, in practice it is not possible to evaluate even this bootstrap-based plausibility contour exactly, because there are too many distict ξ's to enumerate. Instead, we can do a simple Monte Carlo approximation,

$$\hat{\pi}_{z^n}^{\text{boot}}(\cdot) = \frac{1}{B} \sum_{b=1}^{B} 1\{T_{z^n}^{\xi(b)}(\hat{\theta}_{z^n}) > T_{z^n}(\cdot)\}, \quad \xi(b) \sim Q_n, \quad b = 1, \dots, B, \tag{11}$$

where B is a user-specified bootstrap sample size.

3.2 Theoretical Properties

Recall that if F, the distribution of (7), were available and used in the IM construction, validity would follow immediately from (9). However, in the present context where F is unknown, we have recommended the bootstrap-based plausibility contour $\pi_{z^n}^{\text{boot}}$ in (10). Technically, we also approximate this by a Monte Carlo average (11), but since for large enough B, which is in our control, we clearly have $\hat{\pi}_{z^n}^{\text{boot}} \approx \pi_{z^n}^{\text{boot}}$, it suffices to work with the latter in our theoretical investigation. Dependence on the bootstrap also suggests that the best we can hope for is asymptotically approximate validity.

Suppose that the risk minimization problem is sufficiently smooth that it can be restated as a root finding problem, i.e.,

$$\theta = \arg\min_{\vartheta} R(\vartheta) \iff \dot{R}(\theta) = 0, \tag{12}$$

where the dot denotes differentiation. Assume that the loss function $\ell_\vartheta(z)$ is such that $\dot{R}(\vartheta) = \int \dot{\ell}_\vartheta \, dP$, i.e., that the order of differentiation and integration can be interchanged. Then the M-estimation problem can be recast as Z-estimation, and the asymptotic theory for $\hat{\theta}_{Z^n}$ and $\hat{\theta}_{Z^n}^\xi$ can be studied using the general results in, e.g., Kosorok (2008, Ch. 10.3). His *Z-estimator master theorem* provides sufficient conditions for our Assumption A1 below. Then smoothness of the loss (Assumption A2) allows us to show that $T_{Z^n}(\theta)$ and $T_{Z^n}^\xi(\hat{\theta}_{Z^n})$ have the same limiting distribution, hence validity in a sense similar to (1).

Theorem 1. *Assume that the loss function is such that M-estimation is equivalent to Z-estimation in the sense of (12). In addition, assume*

1. *$n^{1/2}(\hat{\theta}_{Z^n} - \theta)$ and $n^{1/2}(\hat{\theta}_{Z^n}^\xi - \hat{\theta}_{Y^n})$ have the same limiting distribution;*
2. *there exists a function f on the Z-space, with $Pf < \infty$, such that $\vartheta \mapsto \dot{\ell}_\vartheta(z)$ is $f(z)$-Lipschitz in a neighborhood of θ.*

Then $P\{\overline{\Pi}_{Z^n}^{\text{boot}}(A) \leq \alpha\} \to \alpha$ as $n \to \infty$, for all $\alpha \in [0,1]$ and all $A \ni \theta(P)$.

A direct consequence of Theorem 1 is that the set

$$\{\vartheta : \hat{\pi}_{y^n}^{\text{boot}}(\vartheta) > \alpha\} \tag{13}$$

constitutes an approximately valid $100(1 - \alpha)\%$ plausibility region.

The detailed proof of Theorem 1 is too lengthy to provide here, but the idea is simple so we provide just a brief sketch. Define $H_n(t)$ and H_n^{boot}, the survival functions of $T_{Z^n}(\theta)$ and $T_{Z^n}^{\xi}(\hat{\theta}_{Z^n})$, respectively, and observe that $\pi_{Z^n}(\theta) = H_n(T_{Z^n}(\theta))$ and $\pi_{z^n}^{\text{boot}}(\theta) = H_n^{\text{boot}}(T_{Z^n}(\theta))$. Then we have

$$\pi_{Z^n}^{\text{boot}}(\theta) = \pi_{Z^n}(\theta) + \Delta_n,$$

where $|\Delta_n| \le \sup_t |H_n^{\text{boot}}(t) - H_n(t)|$. Since $\pi_{Z^n}(\theta)$ is exactly valid, the claim follows from Slutsky's theorem if we can show that $\Delta_n \to 0$ in probability. We proceed by using Taylor approximations of $T_{Z^n}(\theta)$ and $T_{Z^n}^{\xi}(\hat{\theta}_{Z^n})$, with Assumption 2 to control the remainder, and Assumption 1 to show that both $T_{Z^n}(\theta)$ and $T_{Z^n}^{\xi}(\hat{\theta}_{Z^n})$ have the same limiting distribution. This implies H_n and H_n^{boot} merge together as $n \to \infty$, and hence that Δ_n vanishes in probability.

4 Examples

4.1 Quantile Regression

Consider a typical regression context where a response variable Y is coupled a covariate $X \in \mathbb{R}^p$. The goal is to estimate the quantile for the conditional distribution of Y, given X. This problem is known as *quantile regression*, an extension of linear regression useful in cases where the linear regression assumptions are violated. Fix a probability $\tau \in (0,1)$ and let $Q_{Y|X}(\tau)$ denote the τth conditional quantile. Then the quantile regression "model" says

$$Q_{Y|X=x}(\tau) = x^\top \theta,$$

where $\theta = \theta_\tau \in \mathbb{R}^p$ is the vector of regression coefficients of interest. The goal is inference on θ. Koenker and Bassett (1978) show that θ is a risk-minimizer with respect to the loss function $\ell_\theta(x,y) = |y - x^\top \theta| - (2\tau - 1)x^\top \theta$.

Let $X_i \overset{\text{iid}}{\sim} \text{Unif}(0,4)$, $i = 1, \dots, n$, with $n = 200$, and let $Y_i = \mu(X_i) + \varepsilon(X_i)$, where $\mu(x) = \theta_0 + \theta_1 x$, and $\varepsilon(x) \sim \mathsf{N}\big(0, (0.1 + 0.1x)^2\big)$. Suppose the interest is θ for $\tau = 0.75$. Figure 1 (left) displays the data and the estimated quantile regression line corresponding to the empirical risk minimizer $\hat{\theta}_{z^n}$. A 95% plausibility region for θ is obtained from (13) and the plot shows the corresponding marginal plausibility band for μ. Approximate validity of the plausibility bands is implied by the approximate validity of the generalized IM. To check this claim empirically, we simulate 1000 data sets according the above scheme and calculated $\hat{\pi}_{Z^n}^{\text{boot}}(\theta)$, in each replication. Figure 1 (right) shows the empirical distribution of these values over replications and it is clear this closely matches a uniform distribution, confirming Theorem 1. The same plots for $\tau = 0.25$ and $\tau = 0.5$ are included and all suggest the uniform approximation is accurate.

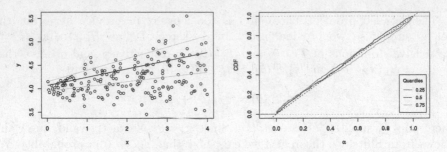

Fig. 1. Left: plot of the data, the fitted third quartile regression line and the 95% plausibility band, based on $B = 500$. Right: empirical distribution function of $\hat{\pi}_{Z^n}^{boot}(\theta)$ based on 1000 replications; this is show for $\tau \in \{0.25, 0.5, 0.75\}$.

4.2   Multivariate Median

In univariate analysis, it is well known that the median is a more robust measure of the distribution's center than the mean. This is also the case in multivariate analysis. However, replacing the multivariate mean by a multivariate median is not so straightforward; since multivariate data do not have a natural ordering, there are various different was of defining the multivariate median or, more generally, multivariate quantiles (Oja 2013).

The most common version of a multivariate median is the *spatial median* (e.g., Brown 1983). For two-dimensional data $y = (y_1, y_2)$, the spatial median is defined as the point in the plane minimizing the sum of absolute distances to y. That is, the two-dimensional spatial median is an M-estimator associated with the loss function $\ell_\theta(y) = \|y - \theta\|_2 - \|y\|_2$, where $\|\cdot\|_2$ is the usual ℓ_2-norm.

For a quick illustration, Fig. 2(a) shows the data pairs $y_i = (y_{i1}, y_{i2})$, for $i = 1, \ldots, n = 200$, which are samples from bivariate normal with mean $\theta = (1, 1)^\top$, unit variances, and correlation 0.7. In Fig. 2(b), the plausibility contour in (11) is shown, based on the loss function above and $B = 500$. The shaded area in Fig. 2(c) represents the 95% plausibility region for θ derived by (13) and, in red, the classic 95% confidence ellipse based on the asymptotic normality. Figure 2(d) shows the resulting empirical distribution of the simulation study where the above scenario is repeated 1,000 times and, for each data set, $\hat{\pi}_{Y^n}^{boot}(\theta)$ is evaluated. Approximate validity is once again verified.

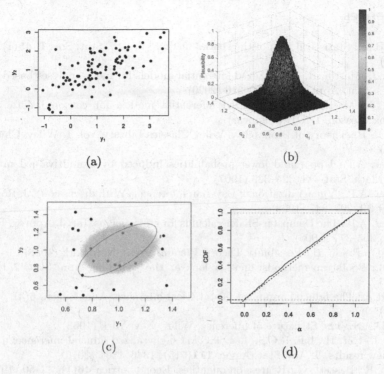

Fig. 2. Panel (a): Scatter plot of the data. Panel (b): Plausibility contour in Equation (11), based on $B = 500$. Panel (c): Classic 95% confidence ellipse based on the asymptotic normality in red and 95% plausibility region shaded in grey. Panel (d): Empirical distribution function of $\hat{\pi}_{Z^n}^{boot}(\theta)$ based on 1000 replications.

5   Conclusion

Here we focused on data-driven uncertainty quantification for unknowns that are defined outside the context of a statistical model. We presented a new generalized IM that not only avoids the explicit description of the data generating process, but does not require a model at all—only a loss function is needed to define the inferential target. We showed that this construction leads to approximately valid uncertainty quantification in the sense of Theorem 1. This provides guarantees beyond those from classical confidence regions. The IM's validity property applies to belief assignments to all assertions about the unknown.

Applications in cases beyond the simple, low-dimensional problems above will be reported elsewhere. Of course, larger dimension creates computational challenges, so calculating the plausibility contour on a grid may not be practically feasible. In such cases, powerful techniques like *stochastic gradient descent* are expected to be useful.

References

Brown, B.M.: Statistical uses of the spatial median. J. Roy. Stat. Soc. B **45**(1), 25–30 (1983)

Cahoon, J., Martin, R.: Generalized inferential models for meta-analyses based on few studies. Stat. Appl. **18**(2), 299–316 (2020)

Cahoon, J., Martin, R.: Generalized inferential models for censored data. Int. J. Approx. Reason. (2021, to appear)

de Finetti, B.: Theory of Probability. Wiley Classics Library, vol. 1. Wiley, Chichester (1990)

Dempster, A.P.: Upper and lower probabilities induced by a multivalued mapping. Ann. Math. Stat. **38**, 325–339 (1967)

Dempster, A.P.: A generalization of Bayesian inference. (With discussion). J. Roy. Stat. Soc. B **30**, 205–247 (1968)

Dempster, A.P.: The Dempster-Shafer calculus for statisticians. Int. J. Approx. Reason. **48**(2), 365–377 (2008)

Dubois, D., Prade, H.: Possibility Theory. Plenum Press, New York (1988)

Efron, B.: Bootstrap methods: another look at the jackknife. Ann. Stat. **7**(1), 1–26 (1979)

Fisher, R.A.: The fiducial argument in statistical inference. Ann. Eugen. **6**(4), 391–398 (1935)

Fraser, D.A.S.: The Structure of Inference. Wiley, New York (1968)

Hannig, J., Iyer, H., Lai, R.C.S., Lee, T.C.M.: Generalized fiducial inference: a review and new results. J. Am. Stat. Assoc. **111**(515), 1346–1361 (2016)

Koenker, R., Bassett, G.: Regression quantiles. Econometrica **46**(1), 33–50 (1978)

Kosorok, M.: Introduction to Empirical Processes and Semiparametric Inference. Springer Series in Statistics, Springer, New York (2008). https://doi.org/10.1007/978-0-387-74978-5

Martin, R.: Plausibility functions and exact frequentist inference. J. Am. Stat. Assoc. **110**(512), 1552–1561 (2015)

Martin, R.: On an inferential model construction using generalized associations. J. Stat. Plann. Inference **195**, 105–115 (2018)

Martin, R., Liu, C.: Inferential models: a framework for prior-free posterior probabilistic inference. J. Am. Stat. Assoc. **108**, 301–313 (2013)

Martin, R., Liu, C.: Inferential Models: Reasoning with Uncertainty. Monographs in Statistics and Applied Probability Series. Chapman & Hall/CRC Press (2015)

Oja, H.: Multivariate median. In: Fried, R., Kuhnt, S. (eds.) Robustness and Complex Data Structures, pp. 3–15. Springer, Heidelberg (2013). https://doi.org/10.1007/978-3-642-35494-6_1

Shafer, G.: A Mathematical Theory of Evidence. Princeton University Press, Princeton (1976)

Syring, N., Martin, R.: Stochastic optimization for numerical evaluation of imprecise probabilities. arXiv:2103.02659 (2021)

Walley, P.: Statistical Reasoning with Imprecise Probabilities. Chapman & Hall/CRC Monographs on Statistics & Applied Probability. Taylor & Francis (1991)

Towards a Theory of Valid Inferential Models with Partial Prior Information

Ryan Martin[✉]

Department of Statistics, North Carolina State University, Raleigh, NC 27695, USA
rgmarti3@ncsu.edu

Abstract. Inferential models (IMs) are used to quantify uncertainty in statistical inference problems, and *validity* is a crucial property that ensures the IM's reliability. Previous work has focused on validity in the special case where no prior information is available. Here I allow for prior information in the form of a non-trivial credal set, define a notion of validity and investigate its implications.

Keywords: Belief function · Coherence · Consonance · Generalized bayes · Possibility measure · Statistical inference

1 Introduction

In statistical inference, there are two dominant schools of thought: *Bayesian* and *frequentist*. The most significant difference between the two is that the former quantifies uncertainty about unknowns in a formal way, using the classical/ordinary/precise probability theory, while the latter does so in a less formal way, focusing on procedures—hypothesis tests, confidence sets, and other decision rules—that have appropriate control on their error rates. Numerous attempts, with different motivations, have been made to reconcile the two frameworks, including fiducial inference (e.g., Fisher 1935) and Dempster's extension (e.g., Dempster 1968), structural inference (Fraser 1968), generalized fiducial (Hannig et al. 2016), and confidence distributions (Schweder and Hjort 2016). Modern developments in this area are largely focused on the construction of data-dependent (precise) probability distributions from which procedures having frequentist error rate control properties (at least approximately) can be derived.

A different thread of work has focused on the development of data-dependent, imprecise probabilities that have a certain calibration or validity property designed to ensure that inferences drawn based on the magnitudes of the (lower and upper) probabilities would be reliable in a frequentist sense. Although ideas along similar lines appeared earlier in Balch (2012), to my knowledge, the first formal definition of *validity* and construction of an imprecise probability that achieves it was given in Martin and Liu (2013) and, later, in Martin and Liu

---

R. Martin—Partially supported by the U.S. National Science Foundation, DMS–1811802.

T. Denœux et al. (Eds.): BELIEF 2021, LNAI 12915, pp. 137–146, 2021.
https://doi.org/10.1007/978-3-030-88601-1_14

(2015); see, also, Martin (2019). Their construction of a valid *inferential model* (IM) makes use of random sets and, therefore, the imprecise probabilities take the form of (consonant) belief functions (Shafer 1976). These and other efforts to construct calibrated belief functions are surveyed in Denoeux and Li (2018).

In the spirit of de Finetti (1937), the focus in the imprecise probability literature is largely on the behavioral interpretation of the lower and upper probabilities; see Walley (1991) and Troffaes and de Cooman (2014). In particular, what minimal conditions on the mathematical structure of those lower and upper probabilities, treated as bounds on the prices an agent sets for gambles, are needed to protect him from sure loss? Since these *coherence* properties concern the internal reliability of the lower and upper probabilities, while the aforementioned validity property concerns a sort of external reliability, it makes sense to investigate the connections between the two.

After some brief background in Sect. 2, I give a definition of validity that is more general than those presented in the references above, and investigate its consequences in Sect. 3. In particular, I allow for available prior information in the form of a credal set—a collection of prior distributions—and present a definition of validity in such cases; previous work focus on the case where the credal set contains all possible priors. The motivation behind this extension is two-fold. First, the introduction of prior information brings the formulation closer to the subjective approach of de Finetti and Walley, where it's natural to consider behavioral implications, and I show in Proposition 2 that an agent adopting a pricing scheme based on lower and upper probabilities derived from a valid IM avoids sure loss. Second, in modern statistical problems involving high-dimensional unknowns, it's often believed that there's an underlying low-dimensional structure. These beliefs can be quantified using a set of prior distributions, so it's important to understand how the notion of validity might extend to such cases. I show that generalized Bayes is valid in the sense I defined. I also claim that a variation of Dempster's generalization of Bayesian inference would be valid too, but a precise statement and proof will be presented elsewhere.

However, it's important to emphasize that an IM being valid does not necessarily make it "good." For example, the IM could be inefficient in the sense that validity is achieved in a trivial way and the inferences drawn are not practically useful. In certain cases, especially those where little or no reliable prior information is available, there are other constructions—including one from Walley (2002) and one I refer to as "p-value + consonance"—that are more efficient without sacrificing efficiency. In cases where reliable prior information is available, efficiency can be gained by taking this into account. An open question is how to incorporate prior information so that both validity and this gain in efficiency is realized; see Cella and Martin (2019) for some first thoughts.

Finally, I present a notion of strong validity, which allows for a practically relevant uniformity over assertions, and I show that the approach advocated for in Martin (2019) and elsewhere achieves this stronger notion of validity and is also efficient, at least in the case of a vacuous prior.

For the sake of space, many details have been omitted. The full-length version (Martin 2021b), still in progress, contains proofs and more.

2   Problem Setup

Let Y denote observable data taking values in a sample space $\mathbb{Y}$; note that the sample space is general, so the data could be a vector, a matrix, etc. Next, consider a statistical model, $\mathscr{P} = \{P_\theta : \theta \in \Theta\}$, a family of probability distributions on $\mathbb{Y}$ indexed by the parameter space Θ, which too is general. The goal is to quantify uncertainty about θ based on the observed data $Y = y$.

Prior information about θ might available in the form of a (closed and convex) credal set $\mathscr{Q}$ of prior distributions Q for θ. The "size" of $\mathscr{Q}$ controls the prior's precision, with $\mathscr{Q} = \{Q\}$ being the most precise and $\mathscr{Q} = \{$all probability distributions$\}$ being the least. These two extreme $\mathscr{Q}$'s are special: the former is classical Bayes while the latter matches the frequentist setup.

By uncertainty quantification, here I mean a data-dependent (precise or imprecise) probability distribution defined on a collection $\mathcal{A}$ of subsets of Θ. I will associate a subset $A \in \mathcal{A}$ with an assertion about the unknown, i.e., both A and "$\theta \in A$" will be called an *assertion*. Since the goal is to have something like a posterior distribution for θ, here I'll take $\mathcal{A}$ to be the Borel σ-algebra on Θ.

Following Martin (2019), define an *inferential model* (IM) as a mapping from data y, model $\mathscr{P}$, and prior information $\mathscr{Q}$ to a pair of lower and upper probabilities $(\underline{\Pi}_y, \overline{\Pi}_y)$ defined on $\mathcal{A}$. I'll interpret $\underline{\Pi}_y(A)$ and $\overline{\Pi}_y(A)$ as the y-dependent belief in and plausibility of the assertion A, respectively. It will be assumed throughout that $y \mapsto \overline{\Pi}_y(A)$ is Borel measurable for all $A \in \mathcal{A}$.

In the imprecise probability literature, it is common to give the lower and upper probabilities a behavioral interpretation. Imagine a situation where, after data $Y = y$ has been observed, the value of θ will be revealed and any gambles made on the truthfulness/falsity of assertions could be settled. Then the (subjective/personal) behavioral interpretation of my (data-dependent) lower and upper probabilities are

$$\underline{\Pi}_y(A) = \text{my maximum buying price for } 1(\theta \in A)$$
$$\overline{\Pi}_y(A) = \text{my minimum selling price for } 1(\theta \in A).$$

Here and in what follows, $1(E)$ denotes the indicator function of the event E. This behavioral interpretation, and one's clear desire to avoid being made a sure loser imposes certain constraints on the mathematical structure of the lower and upper probabilities. However, as I mentioned in Sect. 1, these mathematical constraints do not provide any assurance that the lower and upper probabilities are reliable in a statistical sense.

3   Statistical Properties

3.1   Validity

As discussed above, motivated by the behavioral interpretation, the imprecise probability literature mainly focuses on coherence. For data analysts on the front lines, the ones crunching the numbers behind real-world decisions, this kind of internal rationality is important. From the perspective of a statistician who is developing methods for front-line data analysts to use off-the-shelf, there are other considerations. The only reason someone might use my method for their analysis is that they believe it's reliable, that it "works" in some specific sense. This goes beyond the internal rationality of coherence—lots of things that are coherent won't "work"—and this external rationality is what I call *validity*. A formal definition, more general than those in Martin and Liu (2013; 2015) and Martin (2019; 2021a), and its immediate consequences are below. These results extend the ideas developed by Cella and Martin (2020) in the context of prediction to cover the statistical inference problem.

First some additional notation. For the distribution P_θ of Y and a prior Q for θ, let P_Q denote the corresponding marginal distribution for Y and Q_y the corresponding conditional distribution of θ, given $Y = y$. Next, for a $Q \in \mathcal{Q}$, let M_Q denote the joint distribution of (Y, θ) under the corresponding Bayes model. Similarly, let $M_{\mathcal{Q}}$ denote the image of $\mathcal{Q}$ under $Q \mapsto M_Q$, and $\underline{M}_{\mathcal{Q}}$ and $\overline{M}_{\mathcal{Q}}$ as the lower and upper envelopes, respectively, corresponding to the assertion-wise infimum and supremum of M_Q over $Q \in \mathcal{Q}$. So, if E is any (appropriately measurable) joint event about (Y, θ), then the upper probability $\overline{M}_{\mathcal{Q}}(E)$ can be expressed more concretely as

$$\overline{M}_{\mathcal{Q}}(E) = \sup_{Q \in \mathcal{Q}} \iint 1\{(y, \theta) \in E\} \, P_\theta(dy) \, Q(d\theta)$$

$$= \sup_{Q \in \mathcal{Q}} \iint 1\{(y, \theta) \in E\} \, Q_y(d\theta) \, P_Q(dy).$$

Similarly, there is a corresponding lower probability, $\underline{M}_{\mathcal{Q}}$ that simply replaces the supremum above with an infimum, but this will not be used here.

Definition 1. *An IM $(\underline{\Pi}_Y, \overline{\Pi}_Y)$ is valid, relative to $(\mathscr{P}, \mathcal{Q})$, if either (and, hence, both) of the following equivalent conditions holds:*

$$\overline{M}_{\mathcal{Q}}\{\overline{\Pi}_Y(A) \leq \alpha, \, \theta \in A\} \leq \alpha, \quad \text{for all}(\alpha, A) \in [0,1] \times \mathcal{A}, \qquad (1)$$

$$\overline{M}_{\mathcal{Q}}\{\underline{\Pi}_Y(A) > 1 - \alpha, \, \theta \notin A\} \leq \alpha, \quad \text{for all}(\alpha, A) \in [0,1] \times \mathcal{A}. \qquad (2)$$

The equivalence of (1) and (2) follows from the duality $\overline{\Pi}_Y(A) = 1 - \underline{\Pi}_Y(A^c)$ and the "for all A" part of the conditions. The intuition behind this notion of validity is as follows. In applications, the data analyst will use the magnitudes of the IM's lower and upper probabilities to decide if the data support various assertions about θ. Of course, large values of $\underline{\Pi}_Y(A)$ support the truthfulness of A and small values $\overline{\Pi}_Y(A)$ support the truthfulness of A^c. So the events

$$\{(y, \theta) : \overline{\Pi}_y(A) \leq \alpha, \, \theta \in A\} \quad \text{and} \quad \{(y, \theta) : \underline{\Pi}_y(A) > 1 - \alpha, \, \theta \notin A\}$$

are situations when an erroneous conclusion may be made—or gamble may be lost—and the validity property controls the probability of these undesirable events, thus making the IM's uncertainty quantification reliable.

That this is a generalization of the valid inference framework presented in, say, Martin (2021a Definition 2), can be seen by considering the case where $\mathscr{Q}$ is the set of all probability distributions on Θ. In that case, validity in the sense of (1) reduces to

$$\sup_{\theta \in A} P_\theta \{\overline{\Pi}_Y(A) \leq \alpha\} \leq \alpha, \quad \text{for all}(\alpha, A) \in [0,1] \times \mathcal{A},$$

which is precisely the definition of validity in Martin (2021a).

A very basic requirement is that validity ought to imply that statistical procedures derived from the IM have certain error rate control guarantees. Proposition 1 below makes this precise.

Proposition 1. *Let* $(\underline{\Pi}_Y, \overline{\Pi}_Y)$ *be a valid IM in the sense of Definition 1. Then the following error rate control properties hold.*

1. *A hypothesis testing rule that says reject* "$\theta \in A$"*iff* $\overline{\Pi}_Y(A) \leq \alpha$ *satisfies*

$$\overline{M}_{\mathscr{Q}}\{\text{test rejects and}\theta \in A\} \leq \alpha.$$

2. *The set* $C_\alpha(y) = \bigcap\{A \in \mathcal{A} : \underline{\Pi}_y(A) > 1 - \alpha\}$ *satisfies*

$$\overline{M}_{\mathscr{Q}}\{C_\alpha(Y) \not\ni \theta\} \leq \alpha. \tag{3}$$

For some intuition about these results, consider two important (extreme) special cases corresponding to the traditional frequentist and Bayes approaches. For the frequentist case, where $\mathscr{Q}$ is all possible distributions, (3) immediately reduces to the familiar non-coverage probability bound, $\sup_\theta P_\theta\{C_\alpha(Y) \not\ni \theta\} \leq \alpha$, which is satisfied if C_α is a $100(1-\alpha)\%$ confidence region in the traditional sense. Next, for the purely Bayes case, where $\mathscr{Q}$ is a singleton $\{Q\}$, $M_{\mathscr{Q}}$ corresponds to a specific joint distribution of (Y, θ) and (3) is the condition automatically satisfied when C_α is the $100(1-\alpha)\%$ posterior credible region.

Validity not only has implications for the operating characteristics of procedures derived from the IM, it also has behavioral implications. Proposition 2 below can be interpreted as saying that *validity implies no sure loss*. Avoiding sure loss is related to the aforementioned coherence properties (e.g., Walley 1991, Sect. 6.5.2), establishing a new perspective on validity compared to what had been discussed in previous works. This helps solidify the intuition that a procedure which is externally reliable shouldn't be internally irrational. The results below focus on the upper probability $\overline{\Pi}_y$; there are analogous properties expressed in terms of the corresponding lower probability $\underline{\Pi}_y$.

Proposition 2. *If an IM* $(\underline{\Pi}_Y, \overline{\Pi}_Y)$ *satisfies*

$$\sup_y \overline{\Pi}_y(A) < \underline{Q}(A) := \inf_{Q \in \mathscr{Q}} Q(A) \quad \text{for some } A, \tag{4}$$

then it's not valid, relative to $(\mathscr{P}, \mathscr{Q})$*, in the sense of Definition 1.*

A closer look at the validity property (1) reveals a relatively simple sufficient condition, namely, *dominance*. Indeed, by the iterated expectation formula,

$$\overline{M}_{\mathscr{Q}}\{\overline{\Pi}_Y(A) \leq \alpha,\, \theta \in A\} = \sup_{Q \in \mathscr{Q}} \int 1\{\overline{\Pi}_y(A) \leq \alpha\}\, Q_y(A)\, P_Q(dy), \qquad (5)$$

so if $\overline{\Pi}_y(A) \geq Q_y(A)$ for all y, all $A \in \mathcal{A}$, and all $Q \in \mathscr{Q}$, then it follows immediately that (1) holds. But bounding an integral doesn't require uniform bounds on the integrand, it's enough for the above dominance to hold in an average sense. The following proposition makes this precise.

Proposition 3. *If the IM $(\underline{\Pi}_Y, \overline{\Pi}_Y)$ satisfies the following dominance property,*

$$\sup_{Q \in \mathscr{Q}} \int \frac{Q_y(A)}{\overline{\Pi}_y(A)}\, P_Q(dy) \leq 1, \quad \textit{for all } A, \qquad (6)$$

then it's valid, relative to $(\mathscr{P}, \mathscr{Q})$, in the sense of Definition 1.

An immediate consequence of Proposition 3 and the preceding discussion, if $\overline{\Pi}_y$ is the upper envelope in the generalized Bayes rule (e.g., Walley 1991, Sect. 6.4), that is, if

$$\overline{\Pi}_y(A) = \overline{Q}_y(A) := \sup_{Q \in \mathscr{Q}} Q(A \mid y), \qquad (7)$$

then (6) holds and, therefore, so does validity in the sense of Definition 1. So the conservatism built in to the generalized Bayes rule, motivated by subjective coherence properties, is sufficient to achieve validity as well.

This also sheds light on what kinds of IM (likely) are not valid in the sense of Definition 1. For example, consider an approach like that described in Dempster (2008), where independent random sets/belief functions for θ—one based on prior information, the other based on data and statistical model—are combined, via Dempster's rule, to produce an IM $(\underline{\Pi}_Y, \overline{\Pi}_y)$. The probability intervals $[\underline{\Pi}_Y(A), \overline{\Pi}_Y(A)]$ obtained by Dempster's rule tend to be narrower than those corresponding to the generalized Bayes lower and upper envelopes (e.g., Kyburg 1987, Theorems A.3 and A.6). So, while I don't yet have a concrete counter-example at this time, the above sufficient condition generally doesn't hold, hence validity is questionable.

It's important to emphasize that dominance in the sense of (6) above is a sufficient but not necessary condition for validity. Indeed, there are other IM constructions besides the generalized Bayes lower/upper envelopes that are valid. One such construction is discussed below. Another is the combination of the prior-free IM for θ constructed in Martin and Liu (2015) with a prior belief function for θ via Dempster's rule; there's insufficient space to describe this here, so I'll present the result in a follow-up paper.

It's also worth emphasizing that validity, on its own, doesn't make the IM "good"—it may happen that (6) is achieved in a trivial way, which is not practically useful. For example, if the credal set $\mathscr{Q}$ is large, then the upper envelope

(7) in the generalized Bayes rule could be close to 1, for all/many A's, and then the inference would not be informative. So, beyond validity, it is necessary to consider the IM's *efficiency*.

3.2   Efficiency

As pointed out above, it's easy to see that the generalized Bayes solution is valid but perhaps in an inefficient, even trivial way. So, in a certain sense, the strong coherence properties satisfied by the generalized Bayes solution come at the cost of statistical efficiency. Since that formulation using lower and upper envelopes is only a sufficient condition for validity, there is an opportunity to find a more efficient solution, which is the focus of this subsection.

Towards finding a more efficient solution, let's consider a different strategy. If it can be shown that the IM's upper probability satisfies

$$\sup_\theta P_\theta\{\overline{\Pi}_Y(\{\theta\}) \leq \alpha\} \leq \alpha, \quad \text{for all } \alpha \in [0, 1], \tag{8}$$

then validity in the sense of Definition 1 holds by monotonicity:

$$\overline{\Pi}_Y(\{\theta\}) \leq \alpha \implies \overline{\Pi}_Y(A) \leq \alpha \text{ for all } A \ni \theta. \tag{9}$$

The condition in (8) is (roughly) what Walley (2002) calls the *fundamental frequentist principle*, or *FFP*; Walley's version says "$\alpha \subset [0, \bar{\alpha}]$," for $\bar{\alpha} \leq 1$. He then constructs an IM based on generalized Bayes applied to a special but broad credal set of the form $\mathscr{Q}_W = (1 - \varepsilon)Q_0 + \varepsilon \mathscr{Q}_{\text{all}}$, where Q_0 is a fixed prior distribution on Θ, $\mathscr{Q}_{\text{all}}$ is the set of all priors on Θ, and $\varepsilon \in (0, \frac{1}{2})$. Walley shows that the IM with upper probability $\overline{\Pi}_y = \overline{Q}_y$, with supremum over the special $\mathscr{Q}_W$, satisfies FFP which, for all practical purposes, implies validity in the sense of Definition 1 for all $\mathscr{Q}$, not just $\mathscr{Q}_W$. Most importantly, Walley's solution is far more efficient than, e.g., using generalized Bayes directly on $\mathscr{Q}_{\text{all}}$. However, as Walley notes, this solution is still inefficient in the sense that its plausibility intervals tend to be wider than classical confidence intervals.

A second option, more in line with the approach in Martin and Liu (2015) and Liu and Martin (2020), is as follows. Suppose one can find a function $\pi_y : \Theta \to [0, 1]$ with the property that the random variable $\pi_Y(\theta)$ satisfies the stochastic inequality in (8). This is precisely the property that typical p-values satisfy, so these functions are quite common. If that function also satisfies $\sup_\theta \pi_y(\theta) = 1$ for all y, then I can construct an IM whose upper probability is given by

$$\overline{\Pi}_y(A) = \sup_{\theta \in A} \pi_y(\theta), \quad A \in \mathcal{A}.$$

Under this construction, $\overline{\Pi}_y$ is a consonant plausibility function (Shafer 1976) or, equivalently, a possibility measure (Dubois and Prade 1988; Hose and Hanss 2021), and π_y is its corresponding *plausibility contour*. I'll refer to this below as the "p-value + consonance" IM construction. It's easy to show that, like Walley's above, this IM is valid in the sense of Definition 1 for any $\mathscr{Q}$. However, this app-roach is generally more efficient than Walley's. For example, in a normal mean problem, Walley's plausibility interval has width of the order $(\log n)^{1/2} n^{-1/2}$, whereas the p-value + consonance intervals like in Martin and Liu (2015) have width of the order $n^{-1/2}$, just like classical confidence intervals.

A subtle point is that the meaning of "efficient" varies by the context. For example, when θ is relatively low-dimensional, the IM based on the construc-tion in Martin and Liu (2015) is guaranteed to be valid and would generally be efficient. However, if θ is high-dimensional, then the same IM would tend to be inefficient. This is a sort of "curse of dimensionality"—increasing the dimension θ tends to inflate the plausibility function. More efficient solutions are possible when, as is typical in high-dimensional inference problems, there is an underly-ing low-dimensional structure. Combining this assumed low-dimensional struc-ture/prior information with the data in an appropriate way would lead to a valid IM with improved efficiency compared to the no-prior IM. An open question is how to quantify and then incorporate that structural information so that both validity and efficiency are achieved? This will be answered elsewhere.

3.3 Strong Validity

While the validity condition in Definition 1 seems strong in the sense that it requires the inequalities (1–2) to hold for all assertions A, there is another sense in which it is too weak. In a gambling scenario, the agent will advertise his buying and selling prices based on his specified IM $(\underline{\Pi}_Y, \overline{\Pi}_Y)$, depending on data Y, and his opponents can decide what, if any, transactions they'd like to make. If the opponents also have access to data Y, then surely they will use that information to make a strategic choice of A in order to beat the agent. If the opponents can use data-dependent assertions, then it's not enough to consider the assertion-wise guarantees provided by Definition 1—some kind of uniformity in A is required. This scenario is not so far-fetched. Imagine a statistician who's developing a method for the applied data analyst to use. If the statistician can prove that his method satisfies (1–2), then his method is reliable for any fixed A. But what if the data analyst peeks at the data for guidance about relevant assertions? Without some uniformity, validity cannot be ensured in such cases. With this in mind, consider the following stronger notion of validity.

Definition 2. *An IM* $(\underline{\Pi}_Y, \overline{\Pi}_Y)$ *is strongly valid, relative to* $(\mathscr{P}, \mathscr{Q})$, *if*

$$\overline{M}_{\mathscr{Q}}\{\overline{\Pi}_Y(A) \leq \alpha \text{ for some} A \ni \theta\} \leq \alpha, \quad \text{for all} \alpha \in [0,1] \qquad (10)$$

$$\overline{M}_{\mathscr{Q}}\{\underline{\Pi}_Y(A) > 1 - \alpha \text{ for some } A \not\ni \theta\} \leq \alpha, \quad \text{for all} \alpha \in [0,1]. \qquad (11)$$

Both Walley's and the p-value + consonance IM construction above achieve validity quite easily, arguably too easily. Perhaps this stronger notion of validity, with uniformity in A, is "just right." Indeed, it is not difficult to show that

$$\overline{\Pi}_y(A) \leq \alpha \text{ for some } A \ni \theta \iff \overline{\Pi}_y(\{\theta\}) \leq \alpha. \qquad (12)$$

If the IM satisfies (8), which is akin to Walley's FFP, then strong validity follows.

Proposition 4. *If the IM's upper probability $\overline{\Pi}_Y$ satisfies (8), then the strong validity property in Definition 2 holds for any $\mathcal{Q}$.*

The fact that the p-value + consonance construction presented here achieves strong validity and is generally more efficient than Walley's clever neighborhood model construction suggests that the former might be the "right" type of construction, and that IMs having this consonant structure are fundamental for statistical inference. I hope to present verification of these latter claims in follow-up work.

4   Conclusion

In this paper, I've investigated some more general version of the validity property first put forward in Martin and Liu (2015). An overarching goal of this and other ongoing work is to better understand the spectrum between the classical Bayesian setup with a single precise prior and the frequentist setup whose "prior" is completely imprecise/vacuous indexed by the level of imprecision, i.e., the size/complexity of $\mathcal{Q}$. Previous work had focused primarily on the latter frequentist setup and this paper gives a definition of validity that could be applied across a range of different precision levels in $\mathcal{Q}$.

The conclusion I draw from Proposition 4 above is that the p-value + consonance construction can be used to achieve (strong) validity for every $\mathcal{Q}$ and that, in a certain yet-to-be-formalized sense, is the "best" in the frequentist setup with a vacuous $\mathcal{Q}$; see, also, Martin (2021a). But this doesn't directly address the question of how to use genuine prior information in a non-extreme $\mathcal{Q}$ in a way that's both valid and efficient. Proposition 3 provides some minimal guidance, in particular, it says that generalized Bayes is a valid IM, but more investigation into its statistical efficiency is needed.

Acknowledgments. Thanks to Jasper De Bock, Leonardo Cella, Harry Crane, and Matthias Troffaes for helpful discussions, and to three anonymous conference program committee members for their feedback on a previous version.

References

Balch, M.S.: Mathematical foundations for a theory of confidence structures. Internat. J. Approx. Reason. **53**(7), 1003–1019 (2012)

Cella, L., Martin, R.: Incorporating expert opinion in an inferential model while maintaining validity. In: De Bock, J., de Campos, C.P., de Cooman, G., Quaeghebeur, E., and Wheeler, G. (eds.) Proceedings of the Eleventh International Symposium on Imprecise Probabilities: Theories and Applications. Proceedings of Machine Learning Research, vol. 103, pp. 68–77. PMLR, Thagaste, Ghent (2019)

Cella, L., Martin, R.: Validity, consonant plausibility measures, and conformal prediction. arXiv:2001.09225 (2020). https://doi.org/10.1016/j.ijar.2021.07.013

de Finetti, B.: La prévision: ses lois logiques, ses sources subjectives. Ann. Inst. H. Poincaré **7**(1), 1–68 (1937)

Dempster, A.P.: A generalization of Bayesian inference. (With discussion). J. Roy. Statist. Soc. Ser. B **30**, 205–247 (1968)

Dempster, A.P.: The Dempster-Shafer calculus for statisticians. Int. J. Approx. Reason. **48**(2), 365–377 (2008)

Denœux, T., Li, S.: Frequency-calibrated belief functions: review and new insights. Int. J. Approx. Reason. **92**, 232–254 (2018)

Dubois, D., Prade, H.: Possibility Theory. Plenum Press, New York (1988)

Fisher, R.A.: The fiducial argument in statistical inference. Ann. Eugen. **6**, 391–398 (1935)

Fraser, D.A.S.: The Structure of Inference. Wiley, New York (1968)

Hannig, J., Iyer, H., Lai, R.C.S., Lee, T.C.M.: Generalized fiducial inference: a review and new results. J. Am. Stat. Assoc. **111**(515), 1346–1361 (2016)

Hose, D., Hanss, M.: A universal approach to imprecise probabilities in possibility theory. Int. J. Approx. Reason. **133**, 133–158 (2021)

Kyburg, H.E., Jr.: Bayesian and non-Bayesian evidential updating. Artif. Intell. **31**(3), 271–293 (1987)

Liu, C., Martin, R.: Inferential models and possibility measures. arXiv:2008.06874 (2020)

Martin, R.: False confidence, non-additive beliefs, and valid statistical inference. Int. J. Approx. Reason. **113**, 39–73 (2019)

Martin, R.: An imprecise-probabilistic characterization of frequentist statistical inference (2021a). https://researchers.one/articles/21.01.00002

Martin, R.: Towards a theory of valid inferential models with partial prior information (2021b). https://researchers.one/articles/21.05.00001

Martin, R., Liu, C.: Inferential models: a framework for prior-free posterior probabilistic inference. J. Am. Stat. Assoc. **108**(501), 301–313 (2013)

Martin, R., Liu, C.: Inferential Models: Reasoning with Uncertainty. Monographs on Statistics and Applied Probability, vol. 147. CRC Press, Boca Raton (2015)

Schweder, T., Hjort, N.L.: Confidence, Likelihood, Probability. Cambridge Series in Statistical and Probabilistic Mathematics, vol. 41. Cambridge University Press, New York (2016)

Shafer, G.: A Mathematical Theory of Evidence. Princeton University Press, Princeton (1976)

Troffaes, M.C.M., de Cooman, G.: Lower Previsions. Wiley Series in Probability and Statistics, Wiley, Chichester (2014)

Walley, P.: Statistical Reasoning with Imprecise Probabilities. Monographs on Statistics and Applied Probability, vol. 42. Chapman & Hall Ltd., London (1991)

Walley, P.: Reconciling frequentist properties with the likelihood principle. J. Stat. Plann. Inference **105**(1), 35–65 (2002)

Ensemble Learning Based on Evidential Reasoning Rule with a New Weight Calculation Method

Cong Xu[1] , Zhi-Jie Zhou[2], Wei He[1,2(✉)] , Hailong Zhu[1], and Yan-Zi Gao[3]

[1] Harbin Normal University, Harbin 150025, China
[2] Rocket Force University of Engineering, Xi'an 710025, China
[3] University of Science and Technology Beijing, Beijing 100083, China

Abstract. As a derivative of DS (Dempster-Shafer) theory, ER (Evidential Reasoning) rule can be used as a combination strategy in ensemble learning to dig the classifier information. However, when ER rule are used to integrate classifiers, it is sometimes difficult to assign weight to classifier by traditional ER rule. In view of the above problems, ER rule are improved by using combination weighting instead of expert knowledge weighting in traditional ER rule in this paper, so as to reduce the loss of information and set a more reasonable weight. Firstly, the subjective weight and objective weight are combined to get the combination weight. Then the value range of weight in ER rule is studied, and the regularization of weight is discussed. Finally, the validity of the proposed weight setting method is verified through the classification of the English Bay weather image data set.

Keywords: Ensemble learning · ER rule · Weight · Combination weighting method

1 Introduction

As a common machine learning method, Ensemble learning usually consists of two parts: the generation of classifiers and the combination of classifiers. As a combination strategy, ER rule [1] can combine classifiers by means of decision reasoning and get better integration results. However, when the classifier in ensemble learning is generated by the black box model or the data set is too complex, it is difficult for the traditional ER rule to set the appropriate weight for the classifier.

In traditional ER rule, the weight of evidence is given by expert knowledge, such as Tang [2] and Zhou [3] in the setting of weight are given by experts with actual industrial process, but the human factors of this subjective way is too strong. In some non-industrial fields, it is also difficult for experts to give proper weight. Recently, some scholars proposed to use objective weighting method [4, 5] set weight for evidence. But this method is highly dependent on samples and may lead to wrong judgments. Other scholars set the weight of evidence through mathematical optimization [6]. Mathematical optimization can give more precise weight to the evidence, but it needs a large number of evidence samples for training. However, in ensemble learning, the number of classifiers

© Springer Nature Switzerland AG 2021
T. Denœux et al. (Eds.): BELIEF 2021, LNAI 12915, pp. 147–156, 2021.
https://doi.org/10.1007/978-3-030-88601-1_15

is generally less. Based on the above problems, the combination weighting method is used in this paper to assign the weight of each classifier integrated by ER rule. The combination weighting method has been widely applied in many fields, such as fault analysis [7], electric energy evaluation [8], ecological evaluation [9]. This method can minimize the loss of information, reflect both subjective and objective information, and give more reasonable weight [10–12].

In this paper, the weight of evidence based on ER rule in ensemble learning is studied. The method of combination weight is proposed to set weight for classifier. In addition, this paper studies the weight constraints in ER rule based on accuracy.

2 Problem Description

As a part of the integration learning process, the combination strategy has certain influence on the integration result. A good combination strategy can dig the information of classifier deeply and improve the accuracy of integration model [13]. Based on its advantages in uncertainty and reasoning, ER rule can make the integrated model get better results compared with strategies such as the voting method.

The main problems solved in this paper include the following two:
Problem 1: When using ER rule for ensemble learning, appropriate methods should be adopted to set the weight of evidence. In ensemble learning, Sometimes classifiers may be generated by black box patterns, so it is difficult for experts to give accurate weights. and the number of classifiers is generally small, so it is difficult to set weights by mathematical optimization. Therefore, this paper combines the subjective assignment method and the objective assignment method to establish a combined assignment model for weight.

Problem 2: At present, there is no clear range for the sum of weights of evidence in ER rule. In ER rule, the range of a single weight is $0 \leq \omega_k \leq 1$. There is no requirement for the range of the sum of the weights. But the sum of the weights of all the evidence is sometimes greater than 1, sometimes equal to 1, without a clear definition [14]. Therefore, whether the weight of evidence should be regularized has always been a controversial issue among scholars. In view of this, this paper takes the accuracy rate as the evaluation index and studies whether the weight needs regularization through experiments.

3 The Integration Process of Combination Weighting ER Rule

In this paper, the ER rule improved on Dempster's rule in DS theory [15] is used as the combination strategy in ensemble learning. The classifier is regarded as evidence, and the classifier is integrated by ER rule. The integration process is shown in Fig. 1.

3.1 The Setting of Reliability of Evidence

The reliability of evidence is an inherent characteristic of evidence, which reflects the ability of the evidence to provide a correct assessment or solution for the hypothesis [15].

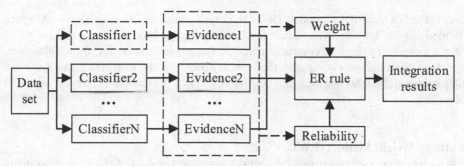

Fig. 1. Integration process of ER rule

In ensemble learning, the probability that each classifier correctly classifies each sample in the sample set is its ability to correctly evaluate. Therefore, by means of mathematical statistics, the accuracy of each classifier for the correct classification in the data set is its reliability r_k in ER rule.

3.2 The Setting of Weight of Evidence

The weight of evidence is usually set by subjective or objective weighting methods. In this paper, several representative methods are combined to give evidential weight. The weighting process of evidence is shown in Fig. 2.

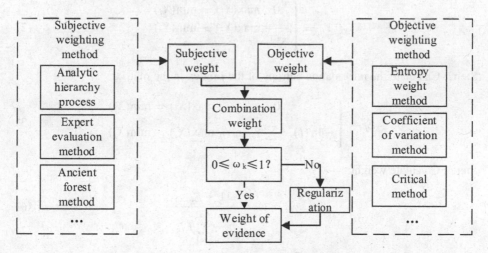

Fig. 2. Evidential weight assignment process

Analytic Hierarchy Process Method (AHP)
AHP is based on the experience of decision-makers, combines quantitative analysis and qualitative analysis, judges the relative importance of each measurement target, and gives the weight of the decision scheme.

Step1: Build a hierarchical model. The goal, criterion and object of decision-making are divided into the highest level, middle level and the lowest level.

Step2: Construct a pairwise comparison matrix. Based on the accuracy of each classifier, the classifier is pairwise compared, the scale is given, and the matrix is formed.

Step3: Hierarchy total ordering and consistency check. Calculate the weight $\overrightarrow{\omega_k}$ of all classifier.

Entropy Weight Method (EW)

EW is based on the information content contained in the data of the evaluation index system, and determines the weight of the index by calculating the information entropy. The calculation steps are as follows:

Step1: An evaluation matrix Y_{ij} is constructed by means of standardization. If there are I evaluation samples, K classifier, and T categories, then there are KT evaluation indexes, $X_1, X_2, \cdots, X_{KT}$. Where $X_j = \{X_{1j}, X_{2j}, \cdots X_{Ij}\}$, x_{ij} is the value of the $j(j = 1, \cdots, KT)$ th index of sample i $(i = 1, \cdots, I)$ th. The indicators used in this paper are all positive indicators, and the value y_{ij} of the evaluation indicator is obtained through standardization:

$$y_{ij} = \begin{cases} 0 & \max(X_j) = \min(X_j) \\ \frac{x_{ij} - \min(X_j)}{\max(X_j) - \min(X_j)} & \max(X_j) > \min(X_j) \end{cases} \tag{1}$$

$$Y_{ij} = \begin{cases} 0 & \max(X_j) = \min(X_j) \\ \frac{y_{ij}}{\sum\limits_{i}^{I} y_{ij}} & \max(X_j) > \min(X_j) \end{cases} \tag{2}$$

Step2: Calculate the information entropy of the j index in the matrix:

$$E_j = \begin{cases} 0 & \max(X_j) = \min(X_j) \\ -\ln(I)^{-1} \sum\limits_{i}^{I} Y_{ij} \ln Y_{ij} & \max(X_j) > \min(X_j) \end{cases} \tag{3}$$

Step3: Calculate weight:

$$\omega_{kt} = \omega_j = \frac{1 - E_j}{n - \sum\limits_{j=1}^{KT} E_j} \tag{4}$$

Coefficient of Variation Method (COV)

COV is based on the degree of variation between the current value and the target value of each index in the index system, and the index weight is determined by calculating the degree of variation of each index in the system. The calculation steps are as follows:

Step1: Same as Step1 in entropy weight method.
Step2: The coefficient of variation of index j th in the matrix is calculated as follows:

$$v_j = \frac{\sqrt{\frac{\sum_{i=1}^{I} (y_{ij} - \frac{1}{I} \sum_{i=1}^{I} y_{ij})^2}{I-1}}}{\frac{1}{I} \sum_{i=1}^{I} y_{ij}} \tag{5}$$

Step3: Calculate weight:

$$\omega_{kt} = \omega_j = \frac{v_j}{\sum_{j=1}^{KT} v_j} \tag{6}$$

CRITIC Method

CRITIC is based on the comparative strength and conflict of each index in the evaluation index system, and determines the index weight through the objective attribute of the data itself. The calculation steps are as follows:

Step1: Same as Step1 in entropy weight method.
Step2: The information content of item j th in the matrix is calculated as follows:

$$C_j = \sqrt{\frac{\sum_{i=1}^{I} (y_{ij} - \frac{1}{I} \sum_{i=1}^{I} y_{ij})^2}{I-1}} \sum_{i=1}^{KT} (1 - r_{ij}) \tag{7}$$

Step3: Calculate weight:

$$\omega_{kt} = \omega_j = \frac{C_j}{\sum_{j=1}^{KT} C_j} \tag{8}$$

Combination

AHP takes into account the knowledge and intention of the decision-maker and makes full use of the decision-maker's experience, but it is subjective and arbitrary. EW, COV, CRITIC method and other methods can mine data information, which are relatively objective, but they are too dependent on data, and sometimes there may be phenomena that do not conform to the actual situation. Based on the advantages and disadvantages of the above methods, this paper combines the weight result $\overrightarrow{\omega_k}$ determined by the subjective assignment method with the weight result $\overleftarrow{\omega_k}$ determined by the objective assignment method to obtain the weight ω_k of each classifier in ER rule integration [16–18]. The combination mode is shown in Eq. (9):

$$\omega_k = \overrightarrow{\omega_k} + \overleftarrow{\omega_k} \text{ or } \omega_k = \overrightarrow{\omega_k} * \overleftarrow{\omega_k} \tag{9}$$

3.3 Evidential Reasoning Process

It is assumed that each classifier in the integration process is an independent evidence with a total of K evidence. The evidential reasoning process is as follows:

Step1: The category is taken as the evaluation level, and the probability of the classifier's judgment on the sample category is taken as its belief corresponding to the evaluation level. The reliability distribution of each evidence can be expressed as $e_k = \{(\theta_n, p_{n,k}), n = 1, \cdots, N; (\Theta, p_{\Theta,k})\}$.

Step2: The reliability r_k and weight ω_k of evidence have been determined by Sects. 3.1 and 3.2 respectively. The weighted belief distribution of article k th evidence with reliability is $m_k = \{(\theta, m_{\theta,k}), \forall \theta \subseteq \Theta; (P(\Theta), m_{P(\Theta),k})\}$.

Step3: The evidential fusion process is as follows:

$$\widehat{m}_{\theta,e(b)} = [(1 - r_b)m_{\theta,e(b-1)} + m_{P(\Theta),e(b-1)}m_{\theta,b}] + \sum_{A\cap B=\theta} m_{A,e(b-1)}m_{B,b}, \forall \theta \subseteq \Theta \tag{10}$$

$$\widehat{m}_{P(\Theta),e(b)} = (1 - r_b)m_{P(\Theta),e(b-1)} \tag{11}$$

$$m_{\theta,e(b)} = \begin{cases} 0, & \theta = \varnothing \\ \dfrac{\widehat{m}_{\theta,e(b)}}{\sum_{A\subseteq\Theta}\widehat{m}_{A,e(b)}+\widehat{m}_{P(\Theta),e(b)}}, & \theta \subseteq \Theta, \theta \neq \varnothing \end{cases} \tag{12}$$

belief after fusion:

$$P_{\theta,e(b)} = \begin{cases} 0, & \theta = \varnothing \\ \dfrac{\widehat{m}_{\theta,e(b)}}{\sum_{A\subseteq\Theta}\widehat{m}_{A,e(b)}}, & \theta \subseteq \Theta, \theta \neq \varnothing \end{cases} \tag{13}$$

Step4: Let the utility of assessment level θ_n be $u(\theta_n)$, and the expected utility u of the final model be:

$$u = \sum_{1}^{N} u(\theta_n)P_{\theta,e(b)} \tag{14}$$

Compare the expected utility u with the utility $u(\theta_n)$ of the evaluation level. If $u = u(\theta_n)$, then the result of integration using ER rule is the n th category.

4 Experimental Analysis

The data set used in this experiment was a Kat Kam webcam image of the English Bay and Burrard Street Bridge in Vancouver, British Columbia, Canada, under different weather conditions. Including 484 clear pictures, 816 cloudy pictures, 648 rain pictures, 90 snow pictures, a total of 2038 pictures. The specific experimental steps are as follows:

Step1: DenSent201(D), GoogleNet(InceptionV3,G), Vgg16(V), AlexNet(A) and ResNet50(R), five currently common deep learning algorithms are used as classifiers. Model parameters are shown in Table 1. 60% of the samples in the dataset are taken as the training set and all the samples are taken as the test set to make classification prediction. During ensemble learning, three deep learning algorithms are randomly selected from the five deep learning algorithms for integration. Five classifier combinations of DGA, GAV, DVR, DGR and GVR were generated. A matrix of 2038×12 is formed by taking the number of samples as rows and the combination of categories and deep learning algorithms as columns. The value in i th row and kt th column represents the probability that the classifier predicts that the sample i th will belong to case t th.

Table 1. The parameters of each model

	Epochs	Learning rate	Batch_size	Activation	Padding
D	100	0.01	20	ReLu	Y
G	100	0.0001	32	ReLu	Y
V	100	0.01	20	ReLu	Y
A	100	0.01	30	ReLu	Y
R	100	0.001	32	ReLu	Y

Step2: The accuracy of the classifier for the sample prediction is calculated as the reliability of the classifier. The probability matrix is used as the original evaluation matrix, and the subjective weight and objective weight are calculated by the AHP, EW, COV and CRITIC method respectively. The combined weight of the classifier is obtained by the combination weighting method of multiplication or addition. To reduce the influence of diversity on classifier integration. In the comparative experiment, the same group of classifier combinations were used.

Step3: In this paper, the classifier is taken as evidence, the category is taken as the evaluation level of evidence, and the probability of prediction by different classifier is taken as the belief distribution. The classifier was integrated with evidential reasoning algorithm, and the expected utility value was compared with the evaluation level to obtain the judgment of the image category by ER rule. The accuracy rate after ER rule integration was obtained by mathematical statistics. This is shown in Table 2, Table 3 and Table 4.

As shown in Table 2, different methods to determine weight have their advantages and disadvantages in integration. For example, when the three classifier of DGR are integrated, the weight determined by AHP is the best; when the three classifier of DGA are integrated, the weight determined by COV is the best; when the three classifier of GAV, DVR and GVR are integrated, the weight determined by the CRITIC is the best. So it is difficult to judge which method is more excellent in determining the weight.

Table 2. Comparison of integration accuracy of ER rule with different weight determination methods

	DGA	GAV	DVR	DGR	GVR
EW	0.7620	0.7448	0.8189	0.8184	0.8131
COV	**0.7674**	0.7444	0.8155	0.8229	0.8131
CRITIC	0.7655	**0.7463**	**0.8219**	0.8229	**0.8135**
AHP	0.7591	0.7399	0.8165	**0.8268**	0.8069

Table 3. Comparison of ER rule integration accuracy after combination weight without regularization

	DGA	GAV	DVR	DGR	GVR
EW+AHP	0.7625	0.7458	0.8209	0.8229	–
COV+AHP	**0.7689**	**0.7483**	**0.8229**	0.8263	**0.8155**
CRITIC+AHP	0.7659	0.7458	0.8224	**0.8268**	0.8140
EW*AHP	0.7645	0.7365	0.8077	0.8155	0.8081
COV*AHP	0.7571	0.7380	0.8081	0.8175	0.8042
CRITIC*AHP	0.7610	0.7370	0.8081	0.8189	0.8047

According to the comparison between Table 2 and Table 3, the weight determined by the combined weighting method makes the accuracy rate of ER rule integration greatly improved compared with the objective weighting method or subjective weighting method alone. The method of weight combination by addition is better than the method of weight combination by multiplication. The method of weight combination by addition of COV and AHP achieves the highest accuracy in the four classifier combinations of DGA, GAV, DVR and GVR, which is slightly weaker than that of the method of weight combination by addition of CRITIC and AHP in DGR. It is worth noting that in the GVR combination, the weight of the ResNet model given by EW is 0.5253, and the weight of the ResNet model given by AHP is 0.5396. The sum of the two exceeds the range of evidential weight in ER rule, and it is impossible to calculate the accuracy after integration. Therefore, —is used instead of accuracy.

The comparison between Table 3 and Table 4 shows that the accuracy of weight regularization in the case of addition combination is usually lower than that in the case of unregulated weight. In the case of multiplication combination, the accuracy of weight regularization is close to that without weight regularization. Sometimes it is higher than that without weight regularization. In Table 4, COV and AHP determining weight by addition are still the best. This may be caused by the size of the weight. The weight value of evidence given by objective weighting method and subjective weighting method is between 0 and 1. So in the additive combination, weight value will increases, in the multiplicative combination, weight value will decrease. The larger weight value may

Table 4. Comparison of ER rule integration accuracy after combination weight in the case of weight regularization

	DGA	GAV	DVR	DGR	GVR
EW+AHP	0.7596	0.7444	0.8184	0.8209	0.8121
COV+AHP	**0.7635**	**0.7458**	**0.8194**	**0.8238**	**0.8126**
CRITIC+AHP	0.7596	0.7424	**0.8194**	**0.8238**	0.8121
EW*AHP	0.7615	0.7390	0.8111	0.8189	0.8057
COV*AHP	0.7596	0.7380	0.8140	0.8189	0.8096
CRITIC*AHP	0.7596	0.7375	0.8145	0.8189	0.8096

more effectively respond to the relative importance between different classifiers and provide the basis for the evidential reasoning process. Therefore, when the combination weighting method is used to assign values to the evidence in ER rule, it is recommended to use a larger weight combination when the weight of each evidence is guaranteed to be greater than or equal to 0 and less than or equal to 1.

5 Summary and Prospect

1. When ER rule is applied to ensemble learning as a combination strategy, a more reasonable weight can be set through the combination weighting method, which makes the final model obtain higher accuracy. Except in the field of ensemble learning, whether the combined weighting method is applicable to ER rule in other fields has yet to be verified.
2. Among the common methods of combination weighting, the combination effect of coefficient of variation method and analytic hierarchy process is the best, and the combination of subjective weight and objective weight by addition is more effective than multiplication. This paper only studies the several classical combination of objective weighting methods and subjective weighting methods. There are more combination methods worth studying, and more combination methods of combination weighting method can be used, such as deviation maximization, game theory and other methods.
3. Under the premise that the weight of each evidence in ER rule is greater than or equal to 0 or less than or equal to 1. The larger the weight value of evidence is, the better effect of the final model integration will be. Therefore, regularization of the weight of evidence is generally not recommended.

Acknowledgments. In this paper, Yan-Zi Gao and Xu Cong have the same contribution. This work was supported in part by the Postdoctoral Science Foundation of China under Safety status assessment of large liquid launch vehicle based on deep belief rule base, in part by the Ph.D. research start-up Foundation of Harbin Normal University under Grant No. XKB201905, in part by the Natural Science Foundation of School of Computer Science and Information Engineering, Harbin Normal University, under Grant no. JKYKYZ202102.

References

1. Yang, J.-B., Xu, D.-L.: Evidential reasoning rule for evidential combination. Artif. Intell. **205**(1), 1–29 (2013)
2. Tang, S.-W., Zhou, Z.-J.: Perturbation analysis of evidential reasoning rule. IEEE Trans. Syst. Man Cybern.: Syst. 1–16 (2019)
3. Zhou, Z.-J., Liu, T.-Y.: A fault detection method based on data reliability and interval evidential reasoning. Acta Autom. Sinica **46**(12), 2628–2637 (2020)
4. Huang, L., Liu, Z., Pan, Q., Dezert, J.: Evidential combination of augmented multi-source of information based on domain adaptation. Sci. China Inf. Sci. **63**(11), 1–14 (2020). https://doi.org/10.1007/s11432-020-3080-3
5. Liu, Z., Quan, P., Dezert, J., et al.: Combination of classifiers with optimal weight based on evidential reasoning. IEEE Trans. Fuzzy Syst. **PP**(99), 1 (2017)
6. Xu, X.-B., Jin, Z., Xu, D.-L., et al.: Information fusion fault diagnosis method based on evidence reasoning rules. Control Theory Appl. **32**(9), 1170–1182 (2015)
7. Liu, H.-C., You, J., You, X.-Y., et al.: A novel approach for failure mode and effects analysis using combination weighting and fuzzy VIKOR method. Appl. Soft Comput. **28**(C), 579–588 (2015)
8. Li, L.-J., Yao, J.-G.: Application of combinatorial weighting method in fuzzy comprehensive evaluation of power quality. Power Syst. Autom. (4), 56–60 (2007)
9. Cui, M.-Z., Yang, F.-H.: Evaluation of cultivated land ecological security in Harbin based on combination weighting method. Res. Soil Water Convers. (06), 188–191+196 (2012)
10. Yu, J.X., Tan, Z.D.: Research on quantificational performance evaluation based on combination weighting and TOPSIS. Syst. Eng.-Theory Pract. **25**(11), 46–50 (2005)
11. Nabavi-Kerizi S.H., Abadi, M., Kabir, E.: A PSO-based weighting method for linear combination of neural networks. Comput. Electr. Eng. **36**(5), 886–894 (2010)
12. Li, L.-J., Yao, J.-G., Long, L.-B., et al.: Application of combinatorial weighting method in fuzzy comprehensive evaluation of power quality. Power Syst. Autom **4**, 56–60 (2007)
13. Dietterich, T.G.: Ensemble methods in machine learning. In: Kittler, J., Roli, F. (eds.) MCS 2000. LNCS, vol. 1857, pp. 1–15. Springer, Heidelberg (2000). https://doi.org/10.1007/3-540-45014-9_1
14. Zhou, Z.-J.: Belief Rule Base Expert System and Complex System Modeling. Science Press (2011)
15. Zhou, Z.-J., Tang, S.-W., Hu, C.-H.: Evidential reasoning theory and its application [J/OL]. Acta Automat. Sinica. https://doi.org/10.16383/j.aas.c190676
16. Bai, X.-C.: Statistics Comprehensive Evaluation Method and Application. China Statistics Press (2013)
17. Zeng, X.-B.: New exploration of combinatorial weighting method. Forecasting **05**, 70–73 (1997)
18. Gang, L., Li, J.-P.: Study on the combination of subjective and objective weights and its rationality. Manag. Rev. **029**(012), 17–26, 61 (2017)

Deep Learning

Evidential Segmentation of 3D PET/CT Images

Ling Huang[1,2]([✉]), Su Ruan[2], Pierre Decazes[3], and Thierry Denœux[1,4]

[1] Université de technologie de Compiègne, CNRS, Heudiasyc, Compiègne, France
ling.huang@utc.fr
[2] University of Rouen Normandy, Quantif, LITIS, Rouen, France
[3] CHB Hospital, Rouen, France
[4] Institut universitaire de France, Paris, France

Abstract. Positron Emission Tomography (PET) and Computed Tomography (CT) are two modalities widely used in medical image analysis. Accurately detecting and segmenting lymphomas from these two imaging modalities are critical tasks for cancer staging and radiotherapy planning. However, this task is still challenging due to the complexity of PET/CT images, and the computation cost to process 3D data. In this paper, a segmentation method based on belief functions is proposed to segment lymphomas in 3D PET/CT images. The architecture is composed of a feature extraction module and an evidential segmentation (ES) module. The ES module outputs not only segmentation results (binary maps indicating the presence or absence of lymphoma in each voxel) but also uncertainty maps quantifying the classification uncertainty. The whole model is optimized by minimizing Dice and uncertainty loss functions to increase segmentation accuracy. The method was evaluated on a database of 173 patients with diffuse large b-cell lymphoma. Quantitative and qualitative results show that our method outperforms the state-of-the-art methods.

Keywords: Lymphoma segmentation · 3D PET/CT · Belief functions · Dempster-Shafer theory · Uncertainty quantification · Deep learning

1 Introduction

Positron Emission Tomography - Computed Tomography (PET/CT) scanning is an effective imaging tool for lymphoma segmentation with application to clinical diagnosis and radiotherapy planning. The standardized uptake value (SUV) for PET images is widely used to locate and segment lymphomas thanks to its high sensitivity and specificity to the metabolic activity of tumor. CT images

This work was supported by the China Scholarship Council (grant 201808331005). It was carried out in the framework of the Labex MS2T (Reference ANR-11-IDEX-0004-02).

are usually used jointly with PET images because of their anatomical feature representation capability.

Although a lot of progress has been made in computer-aided lymphoma segmentation, the segmentation of whole-body lymphomas is still challenging. (Figure 1 shows an example of lymphoma patient. There is great variation in intensity distribution, shape, type and number of lymphomas). The methods can be classified into three main categories: SUV-threshold-based [5], region-growing-based [4] and Convolutional Neural Network (CNN)-based [7] methods. For PET images, it is common to segment lymphomas with a set of fixed SUV thresholds. This method is fast but lacks of flexibility in boundary delineation and requires domain knowledge to locate the region of interest. Region-growing-based methods have been proposed to optimize boundary delineation by taking texture and shape information into account. However, those methods still need clinicians to locate the seeds for region growing [11].

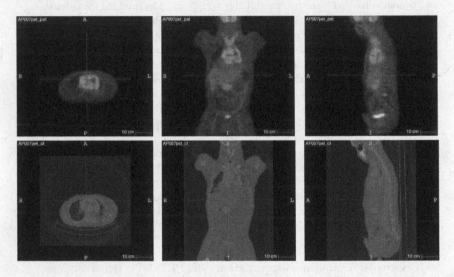

Fig. 1. Examples of patient with lymphomas. The first and second rows show, respectively PET and CT, slices of one patient in axial, sagittal and coronal views. The lymphomas are marked in red. (Color figure online)

CNN-based segmentation methods have recently achieved great success. The UNet architecture [12] has become the most popular medical image segmentation model. Driven by different tasks and datasets, many extended and optimized variants of UNet have been proposed, such as VNet [9], nnUNet [6] and SegResNet [10]. In [7], Li et al. propose a SegResNet-based lymphoma segmentation model with a two-flow architecture (segmentation and reconstruction flows). In [1], Blanc-Durand et al. propose a nnUNet-based lymphoma segmentation network.

Because of low resolution and contrast due to limitations of medical imaging technology, PET/CT image segmentation results are tainted with uncertainty, which greatly limits the segmentation accuracy. Traditional uncertainty measurement methods [8] focus on model rather than information uncertainty to improve the robustness of the model. Belief function (BF) theory [3, 13], also known as Dempster-Shafer theory, is a formal theory for information modeling, evidence combination and decision-making under uncertainty. In this paper, we propose a 3D PET/CT diffuse large b–cell lymphoma segmentation model based on BF theory and deep learning. The proposed deep neural network architecture is composed of a UNet module for feature extraction and a BF module for decision with uncertainty quantification. End-to-end learning is achieved by minimizing a two-part loss function allowing us to increase the Dice score while decrease the uncertainty. The model will first be described in Sect. 2 and experimental results will be reported in Sect. 3.

2   Methods

2.1   Network Architecture

Figure 2 shows the global lymphoma segmentation architecture (ES-UNet). It is composed of (1) an encoder-decoder feature extraction module (UNet), and (2) an evidential segmentation (ES) module comprising a distance activation layer, a basic belief assignment layer and a mass fusion layer. Details about the ES module will be given in Sect. 2.2. Two loss terms are used for optimizing the training process: the *Dice loss*, which quantifies the segmentation accuracy and the *uncertainty loss*, which quantifies the segmentation uncertainty. These loss functions will be described in Sect. 2.3. A "slim UNet" with $(8, 16, 32, 64, 128)$ convolution filters was implemented to reduce computation cost and avoid overfitting.

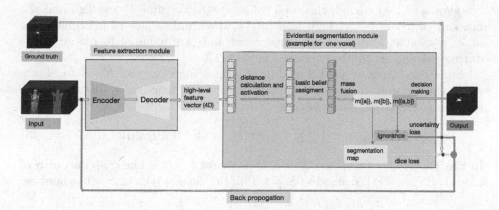

Fig. 2. Global lymphoma segmentation model (ES-UNet).

2.2   Evidential Segmentation Module

A probabilistic network with a softmax output layer may assign voxels a high probability of belonging to one class while the segmentation uncertainty is actually very high because, e.g., the voxel is located close to the fuzzy boundary between the tumor region and other tissues. Based on the evidential neural network model introduced in [2] and using an approach similar to that recently described in [14], we propose a BF theory-based ES module to quantify the uncertainty about the class of each voxel by a Dempster-Shafer mass function. The main idea of the ES module is to assign a mass to each of the K classes and to the whole set of classes Ω, based on the distance between the feature vector of each voxel and I prototype centers. For a given voxel x, each prototype p_i is considered as a piece of evidence, the reliability of which decreases with the Euclidean distance d_i between x and p_i. Each prototype p_i is assumed to have a membership degree u_{ik} to each class ω_k with the constraint $\sum_{k=1}^{K} u_{ik} = 1$. The mass function induced by prototype p_i is

$$m_i(\{\omega_k\}) = \alpha_i u_{ik} \exp(-\gamma_i d_i^2), \quad k = 1, \ldots, K \tag{1a}$$

$$m_i(\Omega) = 1 - \alpha_i \exp(-\gamma_i d_i^2), \tag{1b}$$

The network parameters are the prototypes p_i, the coefficients α_i and γ_i, and the membership degrees u_{ik}. They are learnt by minimizing a loss function [2].

The mass functions induced by the I prototypes are then combined by Dempster's rule [13]

$$m = \bigoplus_{i=1}^{I} m_i. \tag{2}$$

The ES module outputs for each voxel three mass values: two masses corresponding to lymphoma ($\{a\}$) and background ($\{b\}$), and an additional mass corresponding to ignorance (Ω). For the voxels that are easy to classify into lymphoma or background, the mass values $m(\{a\})$ or $m(\{b\})$ are high and the mass $m(\{a, b\})$ is low. A high mass $m(\{a, b\})$ signals a lack of information to make a reliable decision. Thus, some constraints are required to reduce $m(\Omega)$ during training, as will be explained in Sect. 2.3.

Since the output of the ES module is a mass function with $K + 1$ focal sets while there are K classes, we transform the mass function by distributing a fraction ξ of $m(\Omega)$ to each class, as

$$T_{\xi,k} = m(\{\omega_k\}) + \xi\, m(\Omega) \quad \text{with} \quad 0 \leqslant \xi \leqslant 1. \tag{3}$$

In this paper, $\Omega = \{a, b\}$ and $K = 2$, thus we set $\xi = 0.5$. The crisp output S of module ES module is defined as $S = 1$ if $m(\{a\}) < m(\{b\})$ and $S = 0$ otherwise.

2.3 Loss Function Based on Accuracy and Uncertainty for Segmentation

In general, a good segmentation system is expected to make few segmentation errors while providing as informative outputs as possible. Since we quantify uncertainty by the "ignorance class" via the evidential network, we propose to minimize a loss function defined as the sum of two terms: a Dice loss loss_d that measures the discrepancy between the ground truth and segmentation outputs, and an uncertainty loss loss_u that measures the uncertainty of the segmentation outputs. We use the Dice loss instead of the original cross-entropy loss in UNet because the goal of segmentation is to maximize the Dice coefficient. The Dice loss is defined as

$$\text{loss}_d = 1 - \frac{2\sum_{n=1}^N S_n G_n}{\sum_{n=1}^N S_n + \sum_{n=1}^N G_n}, \tag{4}$$

where N is the number of voxels in the image volume, S_n and is the n-th voxel of the segmented output image and G_n is n-th voxel of the ground truth image. The uncertainty loss is defined as

$$\text{loss}_u = \frac{1}{N} \sum_{n=1}^N [m_n(\Omega)]^2, \tag{5}$$

where m_n is the mass function computed for voxel n. With the uncertainty loss, the parameters of the model can be further optimized and more precise segmentation results can be obtained. The total loss function is then

$$\text{loss} = \text{loss}_d + \text{loss}_u + \lambda \|\alpha\|_1, \tag{6}$$

where λ is the regularization coefficient for parameter vector $\alpha = (\alpha_1, \ldots, \alpha_I)$ with the α_i defined in (1). The regularization term allows us to decrease the influence of unimportant prototype centers and avoid overfitting.

3   Experimental Results

3.1   Experimental Settings

The dataset contains 3D images from 173 patients who were diagnosed with large b-cell lymphoma and underwent PET/CT examination. The study was approved as a retrospective study by the Henri Becquerel Center Institutional Review Board. The lymphomas in mask images were delineated manually by experts and considered as ground truth G. The size and spatial resolution of PET and CT images and the corresponding mask images vary due to different imaging machines and operations, from $267 \times 512 \times 512$ to $478 \times 512 \times 512$ and from $276 \times 144 \times 144$ to $407 \times 256 \times 256$, respectively; this makes it difficult to transfer the data into a deep neural model directly. We resized PET, CT and mask images to the same 3D size $256 \times 256 \times 128$, and we applied intensity normalization to both PET and CT images from each patient independently by subtracting the mean and dividing by the standard deviation of the body region only. For data augmentation, we applied a random intensity shift between $[-0.1, 0.1]$ of the standard deviation of each channel, as well as a random scaling intensity of the input between scales $[0.9, 1.1]$.

We randomly selected 80% of the data for training, 10% for validation and 10% for testing. Dice score, sensitivity, specificity, precision and F1 score were used to evaluate the segmentation performance. We first computed the five indices for each test patient and then averaged these indices over the patients. During training, PET and CT images were concatenated as a two-channel input. The number of prototypes was set to 20, taking class number and computation cost into consideration. The prototype vectors and membership degrees were initialized randomly using uniform distributions, while the parameters α and γ were initialized, respectively, at 0.5 and 0.01. The learning rate was set to 10^{-3} during training and the model was trained with 50 epochs using the Adam optimization algorithm. The regularization coefficient λ in (6) was set to 10^{-5}. All methods were implemented in Python with a PyTorch-based, medical image framework MONAI and were trained and tested on a desktop with a 2.20 GHz Intel(R) Xeon(R) CPU E5-2698 v4 and a Tesla V100-SXM2 graphics card with 32 GB GPU memory.

3.2   Results and Discussion

The quantitative results are shown in Table 1. Our model outperforms the baseline model UNet as well as the other state-of-the-art methods. In particular, our model outperforms the best model SegResNet by, respectively, 1.9%, 2.4%, 1.4% in Dice score, Sensitivity and F1 score. It should be noted that the state-of-the-art models were trained with 100 epochs on our dataset because they are slower to converge during training. Figure 3 displays the learning curves of the training loss and validation Dice score for UNet and ES-UNet, showing the

Table 1. Performance comparison with the baseline methods on the test set.

Models	Dice score	Sensitivity	Specificity	Precision	F1 score
ES-UNet (our model)	**0.830**	0.923	0.908	0.912	**0.915**
UNet [12]	0.769	0.798	**0.963**	0.890	0.833
nnUNet [6]	0.702	**0.950**	0.499	0.758	0.807
VNet [9]	0.802	0.882	0.904	0.916	0.909
SegResNet [10]	0.811	0.899	0.942	**0.925**	0.901

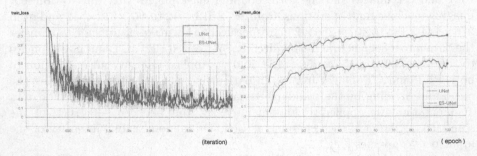

Fig. 3. Training process visualization: training loss (left) and validation Dice score (right).

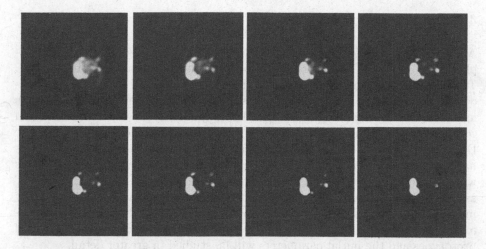

Fig. 4. Uncertainty maps obtained during training, corresponding to different training steps for the same image. For one map, the pixels classified to background, lymphoma, ignorance are marked in purple, yellow and iridescent, respectively. (Color figure online)

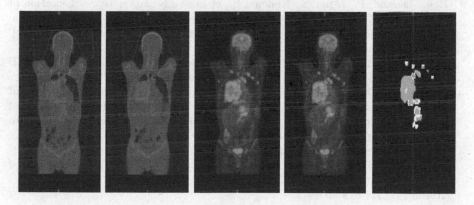

Fig. 5. Segmentation results of ES-UNet. From left to right: ground truth and segmented lymphomas in overlapped CT modality, ground truth and segmented lymphomas overlapped in PET modality, difference map between the ground truth, and segmented lymphomas.

advantage of ES-UNet in terms not only of segmentation accuracy, but also of convergence speed. Figure 4 shows the segmentation and uncertainty maps at different steps during the training of ES-UNet. Our model quantifies the uncertainty of ambiguous pixels instead of classifying them unambiguously into a single class. The uncertainty decreases during the learning process thanks to the minimization of the uncertainty loss term.

Figure 5 shows an example of segmentation results obtained by ES-UNet. Our model can locate and segment most of the lymphomas. The segmentation results were found credible and were confirmed by experts.

4   Conclusion

An evidential segmentation framework (ES-UNet) for segmentation of lymphomas from 3D PET/CT with uncertainty quantification has been introduced. The proposed architecture is based on the concatenation of a UNet and an evidential segmentation layer, making it possible to compute output mass functions for each voxel. The training is performed by minimizing a two-part loss function composed of a Dice loss and an uncertainty loss, with the effect of increasing the Dice score while decreasing the uncertainty. Qualitative and quantitative evaluations show promising results when compared to the baseline model UNet as well as the state-of-the-art methods. While we only concatenated PET and CT as a two-channel input in this work, future research will tackle multi-modality medical image fusion with BF theory by considering PET and CT images separately. Moreover, the sensitivity of the results with respect to the number of prototypes and the initial parameters will be studied in greater detail.

References

1. Blanc-Durand, P., Jégou, S., Kanoun, S., et al.: Fully automatic segmentation of diffuse large B cell lymphoma lesions on 3D FDG-PET/CT for total metabolic tumour volume prediction using a convolutional neural network. Eur. J. Nucl. Med. Mol. Imaging **48**, 1362–1370 (2020)
2. Denœux, T.: A neural network classifier based on Dempster-Shafer theory. IEEE Trans. Syst. Man Cybern. Part A Syst. Hum. **30**(2), 131–150 (2000)
3. Denœux, T., Dubois, D., Prade, H.: Representations of uncertainty in AI: beyond probability and possibility. In: Marquis, P., Papini, O., Prade, H. (eds.) A Guided Tour of Artificial Intelligence Research, pp. 119–150. Springer, Cham (2020). https://doi.org/10.1007/978-3-030-06164-7_4
4. Hu, H., Decazes, P., Vera, P., Li, H., Ruan, S.: Detection and segmentation of lymphomas in 3D PET images via clustering with entropy-based optimization strategy. Int. J. Comput. Assist. Radiol. Surg. **14**(10), 1715–1724 (2019). https://doi.org/10.1007/s11548-019-02049-2
5. Ilyas, H., et al.: Defining the optimal method for measuring baseline metabolic tumour volume in diffuse large B cell lymphoma. Eur. J. Nucl. Med. Mol. Imaging **45**(7), 1142–1154 (2018)
6. Isensee, F., Petersen, J., Klein, A., Zimmerer, D., et al.: nnU-Net: Self-adapting framework for U-Net-based medical image segmentation. arXiv preprint arXiv:1809.10486 (2018)
7. Li, H., Jiang, H., Li, S., et al.: DenseX-Net: an end-to-end model for lymphoma segmentation in whole-body PET/CT images. IEEE Access **8**, 8004–8018 (2019)
8. Mehta, R., Christinck, T., Nair, T., Lemaitre, P., Arnold, D., Arbel, T.: Propagating uncertainty across cascaded medical imaging tasks for improved deep learning inference. In: Greenspan, H., et al. (eds.) CLIP/UNSURE -2019. LNCS, vol. 11840, pp. 23–32. Springer, Cham (2019). https://doi.org/10.1007/978-3-030-32689-0_3
9. Milletari, F., Navab, N., Ahmadi, S.A.: V-Net: fully convolutional neural networks for volumetric medical image segmentation. In: 2016 Fourth International Conference on 3D Vision, pp. 565–571. IEEE (2016)

10. Myronenko, A.: 3D MRI brain tumor segmentation using autoencoder regularization. In: Crimi, A., Bakas, S., Kuijf, H., Keyvan, F., Reyes, M., van Walsum, T. (eds.) BrainLes 2018. LNCS, vol. 11384, pp. 311–320. Springer, Cham (2019). https://doi.org/10.1007/978-3-030-11726-9_28

11. Onoma, D., Ruan, S., Thureau, S., et al.: Segmentation of heterogeneous or small FDG PET positive tissue based on a 3D-locally adaptive random walk algorithm. Comput. Med. Imaging Graph. **38**(8), 753–763 (2014)

12. Ronneberger, O., Fischer, P., Brox, T.: U-Net: convolutional networks for biomedical image segmentation. In: Navab, N., Hornegger, J., Wells, W.M., Frangi, A.F. (eds.) MICCAI 2015. LNCS, vol. 9351, pp. 234–241. Springer, Cham (2015). https://doi.org/10.1007/978-3-319-24574-4_28

13. Shafer, G.: A Mathematical Theory of Evidence, vol. 42. Princeton University Press, Princeton (1976)

14. Tong, Z., Xu, P., Denœux, T.: Evidential fully convolutional network for semantic segmentation. Appl. Intell. **51**(9), 6376–6399 (2021). https://doi.org/10.1007/s10489-021-02327-0

Fusion of Evidential CNN Classifiers for Image Classification

Zheng Tong[1]([✉]) [iD], Philippe Xu[1] [iD], and Thierry Denœux[1,2] [iD]

[1] Université de technologie de Compiègne, CNRS, UMR 7253 Heudiasyc,
Compiègne, France
{zheng.tong,philippe.xu}@hds.utc.fr, thierry.denoeux@utc.fr
[2] Institut universitaire de France, Paris, France

Abstract. We propose an information-fusion approach based on belief functions to combine convolutional neural networks. In this approach, several pre-trained DS-based CNN architectures extract features from input images and convert them into mass functions on different frames of discernment. A fusion module then aggregates these mass functions using Dempster's rule. An end-to-end learning procedure allows us to fine-tune the overall architecture using a learning set with soft labels, which further improves the classification performance. The effectiveness of this approach is demonstrated experimentally using three benchmark databases.

Keywords: Information fusion · Dempster-Shafer theory · Convolutional neural network · Object recognition · Evidence theory

1 Introduction

Deep learning-based models, especially convolutional neural networks (CNNs) [5] and their variants (see, e.g., [9]), have been widely used for image classification and have achieved remarkable success. To train such networks, several image data sets are available, with different sets of classes and different granularities. For instance, a dataset may contain images from dogs and cats, while another one may contain images from several species of dogs. The problem then arises of combining networks trained from such heterogenous datasets. The fusion procedure should be flexible enough to allow the introduction of new datasets with different sets of classes at any stage.

In this paper, we address this classifier fusion problem in the framework of the Dempster-Shafer (DS) theory of belief functions. The DS theory [10], also known as *evidence theory*, is based on representing independent pieces of evidence by mass functions and combining them using a generic operator called Dempster's rule. DS theory is a well-established formalism for reasoning and making decisions with uncertainty [3]. One of its applications is evidential classifier fusion, in which classifier outputs are converted into mass functions and

© Springer Nature Switzerland AG 2021
T. Denœux et al. (Eds.): BELIEF 2021, LNAI 12915, pp. 168–176, 2021.
https://doi.org/10.1007/978-3-030-88601-1_17

fused by Dempster's rule [8,16]. The information-fusion capacity of DS theory makes it possible to combine deep-learning classifiers.

We present a modular fusion strategy based on DS theory for combining different CNNs. Several pre-trained DS-based CNN architectures, also known as *evidential deep-learning classifiers* in [12], extract features from input images and convert them to mass functions defined on different frames of discernment. A fusion module then aggregates these mass functions by Dempster's rule, and the aggregated mass function is used for classification in a refined frame. An end-to-end learning procedure allows us to fine-tune the overall architecture using a learning set with soft labels, which further improves the classification performance. The effectiveness of the approach is demonstrated experimentally using three benchmark databases: CIFAR-10 [4], Caltech-UCSD Birds 200 [15], and Oxford-IIIT Pet [7].

The rest of the paper is organized as follows. DS theory is recalled in Sect. 2. The proposed approach is then introduced in Sect. 3, and the numerical experiment is reported in Sect. 4. Finally, we conclude the paper in Sect. 5.

2 Dempster-Shafer Theory

Let $\Theta = \{\theta_1, \ldots, \theta_M\}$ be a set of classes, called the *frame of discernment*. A (normalized) *mass function* on Θ is a function m from 2^Θ to $[0,1]$ such that $m(\emptyset) = 0$ and $\sum_{A \subseteq \Theta} m(A) = 1$. For any $A \subseteq \Omega$, each mass $m(A)$ is interpreted as a share of a unit mass of belief allocated to the hypothesis that the truth is in A, and which cannot be allocated to any strict subset of A based on the available evidence. Set A is called a *focal set* of m if $m(A) > 0$. A mass function is *Bayesian* if its focal sets are singletons; it is then equivalent to a probability distribution.

A *refining* from a frame Θ to a frame Ω, as defined in [10], is a mapping $\rho : 2^\Theta \to 2^\Omega$ such that the collection of sets $\rho(\{\theta\}) \subset \Omega$ for all $\theta \in \Theta$ form a partition of Ω, and

$$\forall A \subseteq \Theta, \quad \rho(A) = \bigcup_{\theta \in A} \rho(\{\theta\}). \tag{1}$$

The frame Ω is then called a *refinement* of Θ. Given a mass function m^Θ on Θ, its vacuous extension $m^{\Theta \uparrow \Omega}$ in Ω is the mass function defined on frame Ω as

$$m^{\Theta \uparrow \Omega}(B) = \begin{cases} m^\Theta(A) & \text{if } \exists A \subseteq \Theta, \quad B = \rho(A), \\ 0 & \text{otherwise,} \end{cases} \tag{2}$$

for all $B \subseteq \Omega$. Two frames of discernment Θ_1 and Θ_2 are said to be *compatible* if they have a common refinement Ω.

Two mass functions m_1 and m_2 on the same frame Ω representing independent items of evidence can be combined conjunctively by *Dempster's rule* [10] defined as follows:

$$(m_1 \oplus m_2)(A) = \frac{\sum_{B \cap C = A} m_1(B)\, m_2(C)}{\sum_{B \cap C \neq \emptyset} m_1(B)\, m_2(C)} \tag{3}$$

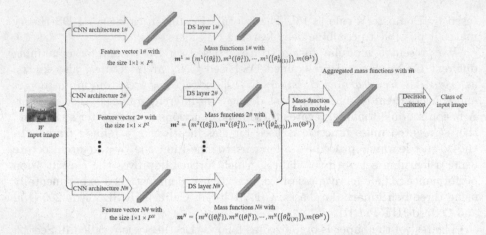

Fig. 1. Architecture of a MF-ECNN classifier.

for all $A \subseteq \Omega$, $A \neq \emptyset$, and $(m_1 \oplus m_2)(\emptyset) = 0$. Mass functions m_1 and m_2 can be combined if and only if the denominator in the right-hand side of (3) is positive. The operator $\oplus$ is commutative and associative. The mass function $m_1 \oplus m_2$ is called the *orthogonal sum* of m_1 and m_2. Given two mass functions m^{Θ_1} and m^{Θ_2} on compatible frames Θ_1 and Θ_2, their orthogonal sum $m^{\Theta_1} \oplus m^{\Theta_2}$ is defined as the orthogonal sum of their vacuous extensions in their common refinement Ω: $m^{\Theta_1} \oplus m^{\Theta_2} = m^{\Theta_1 \uparrow \Omega} \oplus m^{\Theta_2 \uparrow \Omega}$.

3   Proposed Approach

In this section, we describe the proposed framework for fusion of evidential deep learning classifiers. The overall architecture is first described in Sect. 3.1. The end-to-end learning procedure is then introduced in Sect. 3.2.

3.1   Overview

The main idea of the proposed approach is to combine different pre-trained evidential CNN classifiers by plugging a mass-function fusion module at the outputs of these CNN architectures. The architecture of the proposed approach, called *mass-fusion evidential CNN (MFE-CNN) classifier*, is illustrated in Fig. 1 and can be defined by the following three-step procedure.

Step 1: An input image is classified by N pre-trained DS-based CNN architectures [12]. The n-th CNN architecture, $n = 1, \ldots, N$, extracts a feature vector from the input, as done in a probabilistic CNN [5]. The vector is then fed into an evidential distance-based neural-network layer for constructing mass functions, called *the DS layer* [1]. Each unit in this layer computes a mass function on the frame of discernment Θ^n composed of $M(n)$ classes

$\theta_1^n, \ldots, \theta_{M(n)}^n$ and an "anything else" class θ_0^n, based on the distance between the feature vector and a prototype. The mass on Θ^n is larger when the feature vector is further from the prototype. The mass functions computed by each of the hidden units are then combined by Dempster's rule. Given the design of the DS layer, the focal sets of mass function m^n are the singletons $\{\theta_k^n\}$ for $k = 1, \ldots, M(n)$ and Θ^n. The outputs after this first step are the N mass functions $m^1, \ldots, m^N$ defined on N compatible frames $\Theta^1, \ldots, \Theta^N$.

Step 2: A mass-function fusion module aggregates the N mass functions by Dempster's rule. Let Ω be a common refinement of the N frames $\Theta^1, \ldots, \Theta^N$. A combined mass function $\widetilde{m}$ on Ω is computed as the orthogonal sum of the N mass functions $\widetilde{m} = m^1 \oplus \ldots \oplus m^n$. This final output of the mass-function fusion module represents the total evidence about the class of the input image based on the outputs of the N CNN classifiers.

Step 3: The pignistic criterion [2,11] is used for decision-making: the mass function $\widetilde{m}$ is converted into the pignistic probability $BetP_{\widetilde{m}}$ as

$$BetP_{\widetilde{m}}(\{\omega\}) = \sum_{\{A \subseteq \Omega : \omega \in A\}} \frac{\widetilde{m}(A)}{|A|},$$

for all $\omega \in \Omega$, and the final prediction is $\widehat{\omega} = \arg\max_{\omega \in \Omega} BetP_{\widetilde{m}}(\omega)$.

3.2 Learning

An end-to-end learning procedure is proposed to fine-tune all the parameters in a MFE-CNN classifier using a learning set with soft labels, in order to improve the classification performance. In the procedure, the learning sets of different pre-trained CNN architectures are merged into a single one. As some labels become imprecise after merging, they are referred to as *soft labels*. For example, the "bird" label in the CIFAR-10 [4] database becomes imprecise when the database is merged with the Caltech-UCSD Birds 200 database containing 200 bird species [15]. To fine-tune the different classifiers using a learning set with soft labels, we define a label as a non-empty subset $A \in 2^\Omega \setminus \emptyset$ of classes an image may belong to. Label A indicates that the true class is known to be one element of set A, but one cannot determine which one specifically if $|A| > 1$.

In the fine-tuning phase, all parameters in a MFE-CNN classifier are initialized by the parameters in the pre-trained CNNs. Given a learning image with non-empty soft label $A \subseteq \Omega$ and output pignistic probability $BetP_{\widetilde{m}}$, we define the loss as:

$$\mathcal{L}(BetP_{\widetilde{m}}, A) = -\log BetP_{\widetilde{m}}(A) = -\log \sum_{\omega \in A} BetP_{\widetilde{m}}(\omega). \tag{4}$$

This loss function is minimized when the pignistic probability of soft label A equals 1. The gradient of this loss w.r.t all network parameters can be back propagated from the output layer to the input layer.

Table 1. Lists of classes in the CIFAR-10, CUB, Oxford databases. The notations θ_0^2 and θ_0^3 stand for the "anything else" class added to the frames of the CUB and Oxford databases.

Frame	Class
CIFAR-10	airplane, automobile, bird, cat, deer, dog, frog, horse, ship, truck
CUB	cardinal, house wren, ..., (200 species of birds), θ_0^2
Oxford	bengal, boxer, ..., (37 species of cats and dogs), θ_0^3
Common frame	airplane, automobile, deer, frog, horse, ship, truck, cardinal, house wren, ..., (200 species of birds), bengal, boxer, ..., (37 species of cats and dogs)

4 Experiment

In this section, we study the performance of the above fusion method through a numerical experiment. The databases and metrics are first introduced in Sect. 4.1. The results are then discussed in Sect. 4.2.

4.1 Experimental Settings

Three databases are considered in this experiment: CIFAR-10 [4], Caltech-UCSD Birds-200-2010 (CUB) [15], and Oxford-IIIT Pet [7]. The CIFAR-10 database was pre-split into 50k training and 10k testing images. For the CUB (6,033 images) and Oxford (7,349 images) databases, we divided each database into training and testing sets with a ratio of about 1:1. The training and testing sets keep the ratio of about 1:1 in each class. In the fine-tuning procedure, the frames of the three databases are refined into a common one, as shown in Table 1. Thanks to the "anything else" classes, the three frames are compatible. After merging the three databases, there are 56,692 training samples and 16,690 testing samples for, respectively, fine-tuning and performance evaluation.

For a testing set T with soft labels, the *average error rate* is defined as

$$AE(T) = 1 - \frac{1}{|T|} \sum_{i \in T} \mathbb{1}_{A(i)} \left(\widehat{\omega}(i) \right), \tag{5}$$

where $A(i)$ is the soft label of sample i, $\widehat{\omega}(i)$ is the predicted class, and $\mathbb{1}_{A(i)}$ is the indicator function of set $A(i)$.

The three CNNs used for the three datasets have the same FitNet-4 [9] architecture with 128 output units. The numbers of prototypes in the DS layers for the CIFAR-10, CUB, and Oxford databases are, respectively, 70, 350, and 80. We compared the proposed classifier to four classifier fusion systems with the same CNN architectures:

Probability-to-mass fusion (PMF) [16]: we feed the feature vector from each CNN n into a softmax layer to generate a Bayesian mass function on Θ^n.

Table 2. Average test error rates of different classifiers. "E2E" stands for fine-tuned classifiers. E- and P-FitNit-4 are the evidential and probabilistic CNN classifiers before fusion. The lowest error rates are in bold and second low are underlined.

	Classifier	CIFAR-10	CUB	Oxford	Overall
Before fusion	E-FitNit-4	6.4	6.6	10.2	–
	P-FitNit-4	6.5	7.4	10.5	–
After fusion	MFE	<u>5.0</u>	6.6	9.9	<u>6.4</u>
	PMF	5.9	7.3	10.2	7.1
	BF	6.2	8.9	11.1	7.7
	E2E MFE	**4.5**	<u>6.5</u>	<u>9.8</u>	**6.0**
	E2E PMF	5.4	7.3	10.1	6.8
	E2E BF	6.2	8.7	10.9	7.6
	E2E EFC	6.9	7.2	11.3	7.9
	E2E PFC	6.2	**6.4**	**9.7**	7.0

The mass functions from the three CNNs are then combined by Dempster's rule. (It should be noted that the vacuous extension of each Bayesian mass function in the common refinement Ω is no longer Bayesian.)

Bayesian-fusion (BF) [14]: the feature vector from each CNN is converted into a Bayesian mass function on the common frame Ω, by equally distributing the mass of a focal set to its elements; the obtained Bayesian mass functions are combined by Dempster's rule. This procedure is equivalent to Bayesian fusion.

Probabilistic feature-combination (PFC) [6]: the three feature vectors are concatenated to form a new vector of length 384, which is fed into a softmax layer to generate the probability distribution on the common frame.

Evidential feature-combination (EFC): feature vectors are concatenated as in the above PFC approach, but the aggregated vector is fed into a DS layer to generate an output mass function. The dimension of the aggregated vector and the number of prototypes are, respectively, 192 and 400, to obtain optimal performance of the EFC-CNN classifier.

4.2   Results

Table 2 shows the average test error rates of the evidential and probabilistic classifiers trained from each of the three datasets, as well as the performances of the different fusion strategies (with and without fine tuning) on each individual dataset, and on the union of the three datasets.

Looking at the performance of the MFE strategy, we can see that, after fusion, the error rates on the CIFAR-10 and Oxford databases decrease, but the one on the CUB database does not change. As shown in Table 3, the error rates for the "cat", "dog", "bird" classes on the CIFAR-10 database decrease, but the ones of other classes do not change after fusion. These observations

Table 3. Test error rates before and after information fusion on CIFAR-10.

	Classifier	Aero	Mobile	Bird	Cat	Deer	Dog	Frog	Horse	Ship	Truck
Before fusion	E-FitNit-4	2.4	3.9	6.4	13.5	9.0	10.1	5.6	6.8	3.5	2.7
	P-FitNit-4	1.6	2.6	8.7	15.7	9.6	12.5	4.2	5.3	1.9	2.6
After fusion	E2E MFE	2.2	3.9	1.9	6.3	8.5	3.9	5.5	6.5	3.5	2.7
	E2E PMF	1.6	2.5	5.0	12.8	9.0	9.2	4.2	5.3	1.8	2.6
	E2E BF	1.5	2.5	8.1	14.0	9.0	11.0	4.1	5.2	1.8	2.5

Table 4. Examples of mass functions (MF's) before and after fusion by the MFE strategy. Only some masses before and after fusion are shown for lack of space.

Instance/label	Before fusion			MF on Ω after fusion
	MF from CIFAR-10	MF from CUB	MF from Oxford	
/bird	$m(\{\text{airplane}\}) = 0.506$ $m(\{\text{bird}\}) = 0.382$ ... $m(\Theta^1) = 0.065$	$m(\{\text{caspinan}\}) = 0.698$ $m(\{\text{horned grebe}\}) = 0.109$ ... $m(\theta_0^2) = 0.098$	$m(\{\text{samyod}\}) = 0$ $m(\{\text{pyrenees}\}) = 0.001$ ... $m(\theta_0^3) = 0.905$	$m(\{\text{airplane}\}) = 0.101$ $m(\{\text{caspinan}\}) = 0.672$ ... $m(\Omega) = 0.007$
/caspian	$m(\{\text{airplane}\}) = 0.009$ $m(\{\text{bird}\}) = 0.823$ ... $m(\Theta^1) = 0.092$	$m(\{\text{caspinan}\}) = 0.423$ $m(\{\text{horned grebe}\}) = 0.452$ ... $m(\theta_0^2) = 0.084$	$m(\{\text{samyod}\}) = 0$ $m(\{\text{pyrenees}\}) = 0.001$ ... $m(\theta_0^3) = 0.951$	$m(\{\text{caspinan}\}) = 0.415$ $m(\{\text{horned grebe}\}) = 0.450$ ... $m(\Omega) = 0.009$
/byssinian	$m(\{\text{cat}\}) = 0.742$ $m(\{\text{dog}\}) = 0.131$ ... $m(\Theta^1) = 0.032$	$m(\{\text{caspinan}\}) = 0.002$ $m(\{\text{horned grebe}\}) = 0$ ... $m(\theta_0^2) = 0.931$	$m(\{\text{byssinian}\}) = 0.412$ $m(\{\text{bengal}\}) = 0.503$ ... $m(\theta_0^3) = 0.038$	$m(\{\text{byssinian}\}) = 0.414$ $m(\{\text{bengal}\}) = 0.505$ ... $m(\Omega) = 0.005$
/keeshond	$m(\{\text{cat}\}) = 0.158$ $m(\{\text{dog}\}) = 0.705$ ... $m(\Theta^1) = 0.058$	$m(\{\text{albatross}\}) = 0.001$ $m(\{\text{horned grebe}\}) = 0$ ... $m(\theta_0^2) = 0.975$	$m(\{\text{rogdoll}\}) = 0.682$ $m(\{\text{keeshond}\}) = 0.254$ ... $m(\theta_0^3) = 0.001$	$m(\{\text{rogdoll}\}) = 0.369$ $m(\{\text{keeshold}\}) = 0.485$ ... $m(\{\text{cat}\}) = 0.021$

show that the proposed approach makes it possible to combine CNN classifiers trained from heterogenous databases to obtain a more general classifier able to recognize classes from any of the databases, without degrading the performance of the individual classifiers, and sometimes even yielding better results for some classes.

Table 4, which shows examples of mass functions computed by the different classifiers, allows us to explain the good performance of the MFE fusion strategy. The first example from the CIFAR-10 database is misclassified using only the mass function from the classifier trained from this dataset, but the decision is corrected after the evidential fusion because the mass function provided by the classifier trained from the CUB data supports some bird species. In contrast, for images misclassified by the CUB mass function, such as the second instance, the other two individual classifiers do not provide any useful information to correct the decisions. Consequently, the classification performance on the CUB database is not improved. Besides, the mass function from the CIFAR-10 classifier sometimes includes useful information for classifying samples into one species of cat or dog, such as the final instance, even though the number of such examples is small. This phenomenon is responsible for the small change in the performance on the Oxford database.

Comparing the test error rates of the MFE classifiers with and without the end-to-end learning procedure as shown in Table 2, we can see that fine tuning further boosts the overall performance, as well as the performance on the CIFAR-10 and Oxford databases. Thus, the fine-tuning procedure decreases the classification error rate of the proposed architecture, and can be seen as a way to improve the performance of CNN classifiers. This is because the end-to-end learning procedure adapts the individual classifiers to the new classification problem. More specifically, before fusion, the CNN classifiers are pre-trained for the classification tasks with the frames of discernment before refinement. The proposed end-to-end learning procedure fine-tunes the parameters in the CNN and DS layers to make them more suitable to the classification task in the refined frame.

Finally, Table 2 sheds some light on the relative performance of different classifier fusion strategies. The PMF fusion strategy also improves the performance of the probabilistic CNNs trained on each of the three databases, but it is not as good as the proposed method. In contrast, the Bayesian fusion strategy (BF) degrades the performance of the individual classifiers, which shows that the method is not effective when the numbers of classes in the different frames are very unbalanced. The relatively high error rates obtained of the two feature fusion strategies (E2E EFC and E2E PFC) show that this simple fusion method is less effective than the other ones. All in all, the proposed evidential fusion strategy outperforms the other tested methods on the datasets considered in this experiment.

5   Conclusion

In the study, we have proposed a fusion approach based on belief functions to combine different CNNs for image classification. The proposed approach makes it possible to combine CNN classifiers trained from heterogenous databases with different sets of classes. The combined classifier is able to classify images from any of these databases while having at least as good performance as those of the individual classifiers on their respective databases. Besides, the proposed approach makes it possible to combine additional classifiers trained from new datasets with different sets of classes at any stage. An end-to-end learning procedure further improves the performance of the proposed architecture. This approach was shown to outperform other decision-level or feature-level fusion strategies for combining CNN classifiers. Future work will consider combining evidential fully convolutional networks for pixel-wise semantic segmentation [13].

References

1. Denoeux, T.: A neural network classifier based on Dempster-Shafer theory. IEEE Trans. Syst. Man Cybern. Part A Syst. Hum. **30**(2), 131–150 (2000)
2. Denœux, T.: Decision-making with belief functions: a review. Int. J. Approx. Reason. **109**, 87–110 (2019)

3. Denœux, T., Dubois, D., Prade, H.: Representations of uncertainty in AI: beyond probability and possibility. In: Marquis, P., Papini, O., Prade, H. (eds.) A Guided Tour of Artificial Intelligence Research, pp. 119–150. Springer, Cham (2020). https://doi.org/10.1007/978-3-030-06164-7_4
4. Krizhevsky, A., Hinton, G.: Learning multiple layers of features from tiny images. Technical report, University of Toronto (2009)
5. LeCun, Y., Bengio, Y., Hinton, G.: Deep learning. Nature **521**(7553), 436–444 (2015)
6. Nguyen, L.D., Lin, D., Lin, Z., Cao, J.: Deep CNNs for microscopic image classification by exploiting transfer learning and feature concatenation. In: 2018 IEEE International Symposium on Circuits and Systems (ISCAS), pp. 1–5. IEEE (2018)
7. Parkhi, O.M., Vedaldi, A., Zisserman, A., Jawahar, C.V.: Cats and dogs. In: IEEE Conference on Computer Vision and Pattern Recognition, Providence, Rhode Island (2012)
8. Quost, B., Masson, M.H., Denœux, T.: Classifier fusion in the Dempster-Shafer framework using optimized t-norm based combination rules. Int. J. Approx. Reason. **52**(3), 353–374 (2011)
9. Romero, A., Ballas, N., Kahou, S.E., Chassang, A., Gatta, C., Bengio, Y.: FitNets: hints for thin deep nets. In: 3rd International Conference on Learning Representations, San Diego, USA (2015)
10. Shafer, G.: A Mathematical Theory of Evidence. Princeton University Press, Princeton (1976)
11. Smets, P.: Constructing the pignistic probability function in a context of uncertainty. In: Henrion, M., Schachter, R.D., Kanal, L.N., Lemmer, J.F. (eds.) Uncertainty in Artificial Intelligence 5, pp. 29–40. North-Holland, Amsterdam (1990)
12. Tong, Z., Xu, P., Denoeux, T.: An evidential classifier based on Dempster-Shafer theory and deep learning. Neurocomputing **450**, 275–293 (2021)
13. Tong, Z., Xu, P., Denœux, T.: Evidential fully convolutional network for semantic segmentation. Appl. Intell. **51**(9), 6376–6399 (2021). https://doi.org/10.1007/s10489-021-02327-0
14. Wei, Q., Dobigeon, N., Tourneret, J.Y.: Bayesian fusion of multi-band images. IEEE J. Sel. Topics Sig. Process. **9**(6), 1117–1127 (2015)
15. Welinder, P., et al.: Caltech-UCSD Birds 200. Technical report, CNS-TR-2010-001, California Institute of Technology (2010)
16. Xu, P., Davoine, F., Bordes, J.-B., Zhao, H., Denœux, T.: Multimodal information fusion for urban scene understanding. Mach. Vis. Appl. **27**(3), 331–349 (2014). https://doi.org/10.1007/s00138-014-0649-7

Multi-branch Recurrent Attention Convolutional Neural Network with Evidence Theory for Fine-Grained Image Classification

Zhikang Xu[2] , Bofeng Zhang[2(✉)] , Haijie Fu[2] , Xiaodong Yue[1,2] ,
and Ying Lv[2]

[1] Shanghai Institute for Advanced Communication and Data Science,
Shanghai University, Shanghai, China
[2] School of Computer Engineering and Science, Shanghai University,
Shanghai 200444, China
{xuzhikangnba,bfzhang,fhttsfhj,lvying2016,yswantfly}@shu.edu.cn

Abstract. Fine-grained image classification (FGIC) aims to classify subordinate classes belonging to the same meta category. One of the existing FGIC methods is to use attention mechanism to localize and crop a discriminative region from the input image, and this process can be executed recurrently. In this way, the cropped image will progressively focus on a smaller local region containing the object part. However, this may cause the contour information of the object to be incomplete at the finest-scale and thereby the accuracy of the finest-scale is affected. In addition, the fusion strategy of these methods, which generally concatenates the outputs of multiple scales for the final classification, is not sufficient. To tackle the problems, based on the backbone of RA-CNN we first construct a multi-branch attention proposal network (APN) at middle scale of RA-CNN to jointly localize a most discriminative region where multiple APNs can complement each other's incomplete contour information. Moreover, in addition to concatenating the outputs of all scales, we also use the Dempster's combination rule to combine the outputs of all scales. Then, the features of these two parts are further combined for the final classification. Experimental results on the real-world datasets clearly validate the effectiveness of the proposed method.

Keywords: Fine-grained image classification · Attention mechanism · Dempster-Shafer theory · Evidence theory

1 Introduction

In the past few years, deep convolutional neural networks have made remarkable achievements in traditional image classification, such as distinguishing cats from dogs or cars from airplanes. In contrast, fine-grained image classification (FGIC) is a task to distinguish subordinate classes belonging to the same meta

© Springer Nature Switzerland AG 2021
T. Denœux et al. (Eds.): BELIEF 2021, LNAI 12915, pp. 177–184, 2021.
https://doi.org/10.1007/978-3-030-88601-1_18

category [10,16], such as distinguishing the subordinate categories of dogs or the subordinate categories of cars.

Fine-grained image classification still remains challenging for two main reasons. 1) The inter-class variance is small and the intra-class variance is large. 2) Labeled training examples are very limited for each subordinate class. The main idea to alleviate these problems is to find or localize discriminative regions from original images. In recent years, weakly-supervised FGIC becomes a promising method, which can automatically learn to localize discriminative regions [1,9,12,15,17], while strongly-supervised FGIC requires additional object part annotations [11,14].

One of weakly-supervised FGIC methods is to leverage attention mechanism to recurrently localize a discriminative sub-region from the discriminative region obtained in the previous step [4,5,13,17,18]. The results of multiple scales are then combined to obtain the final results. Although promising results have been reported, further improvement suffers from the following limitations. First, the classification accuracy of last scale is lower than that of previous scale [5,13]. The possible reason is that the last scale pays too much attention to the local region of the object part and loses the contour information of the object. Second, the method of fusing multiple scales' results is simple concatenation or sum with different weights.

To tackle these problems, based on the backbone of Recurrent Attention Convolutional Neural Network (RA-CNN), we construct multiple APNs in scale two of RA-CNN to jointly pay attention to a most discriminative region together. Specifically, we propose a novel loss function for optimizing APNs where a regularization term is introduced to make APNs localize similar but complementary region. It is important to note that our method is different from some multiple attentions based FGIC methods, which use multiple attention blocks to localize different discriminative regions. Then, we use Dempster's combination rule [2,3,8] to fuse the results of three scales as an extra information to further improve the classification performance. The contributions of this paper are summarized as follows.

- We construct a multi-branch APN to jointly localize a most discriminative region. By designing a novel loss function, the object contour attended by multiple APNs could be complementary. In addition, compared with the method of using multiple attention blocks to localize different discriminative regions, our proposed framework is intuitive and easy to implement.
- We leverage Dempster's combination rule to fuse the classification results of different scales to improve the overall accuracy.

2   The Proposed Model

In this section, we will introduce our proposed multi-branch RA-CNN, which includes two parts. The first part is shown in Fig. 1, which is similar to RA-CNN. The biggest difference is that we construct a multi-branch APN at scale two

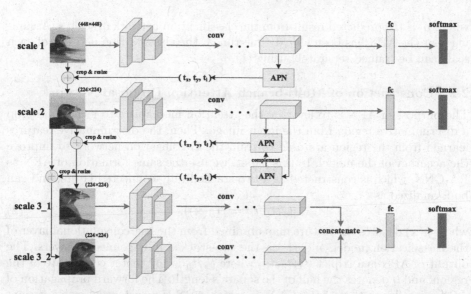

Fig. 1. Overview of the first part of our proposed framework. Each row is a commonly used CNN classification model (connected by blue arrows). (Color figure online)

(here, we set the number of APNs to two for better illustration) to complement the missing information caused by using only one APN at scale two. The second part is a decision fusion of different scales, as shown in Fig. 2. We first use Dempster's combination rule to fuse the results of the three scales as an extra information. Then we concatenate it with the original outputs of the three scales to train a fully-connected layer for the final classification.

2.1   Construction of Classification Network

Given a training set including N labeled images $\{(\boldsymbol{x}_1, \boldsymbol{y}_1), (\boldsymbol{x}_2, \boldsymbol{y}_2), ..., (\boldsymbol{x}_N, \boldsymbol{y}_N)\}$ where $\boldsymbol{y}_i$ is a one-hot vector encoding the ground-truth label of $\boldsymbol{x}_i$, such as $y_{ij} = 1$ and $y_{ik} = 0$ for any $k \neq j$. As shown in Fig. 1, the inputs of three scales use different images, while the targets of all scales are the same. The input image of coarse-scale is cropped and resized by APN to obtain the input image of finer-scale recurrently. Note that scale three has two input images cropped from scale two using two APNs. For scale three, the two images are first fed into two convolutional networks, then the outputs of last convolutional layer of two networks are concatenated and fed into a fully-connected layer.

For each scale s, the classification loss is defined as the sum of the cross-entropy loss of all training images

$$L_{CE}^s = \sum_{i=1}^{N} -\boldsymbol{y}_i \log(\boldsymbol{p}_i), \tag{1}$$

where p_i is the predicted result from the classification network of scale s. Assuming that the APN has been already fine-tuned, the classification network of each scale can be trained separately using (1).

2.2   Construction of Multi-branch Attention Proposal Network

The purpose of APN is to leverage the attention mechanism to locate and crop a discriminative region from the input image. Then, the discriminative features learned from the region are used to train the classification network to improve the accuracy of the model. In this paper, we use the same formation of APN as RA-CNN, which is constructed as a two-stacked fully-connected layers, and can be formalized as

$$[t_x, t_y, t_l] = f(g(\mathbf{x})), \tag{2}$$

where $g(\mathbf{x})$ denotes the feature map obtained from the last convolutional layer of the classification model, and $f(\cdot)$ is the two-stacked fully-connected layers. The output of APN is a triplet $[t_x, t_y, t_l]$, where t_x, t_y is the center coordinate of the region, and t_l denotes the half of the square's length. The forward propagation of APN is similar to that of RA-CNN, in which APN is transformed into a variant of two-dimension boxcar function to ensure it can be optimized in the training stage. In addition, as we mentioned above, the last scale's classification accuracy of RA-CNN is lower than that of previous scale, and the reason is that the last scale pays too much attention to the local region of the object part and thus misses the contour information of the object.

In order to alleviate this problem, we construct a multi-branch APN at scale two to capture the missing information caused by using only one APN. Here, we set the number of APNs to two, and the two APNs are denoted as $\mathrm{APN}_{2,*}$ and $\mathrm{APN}_{2,1}$, respectively. These two APNs need to jointly localize the discriminative region, and the contour of object attended by $\mathrm{APN}_{2,*}$ and $\mathrm{APN}_{2,1}$ should be complementary at the same time. To this end, we first use the original rank loss of RA-CNN to optimize $\mathrm{APN}_{2,1}$

$$L_{rank}^{\mathrm{APN}_{2,1}} = max\{0, p_t^2 - p_t^3 + margin\}, \tag{3}$$

where p_t^2(resp. p_t^3) denotes the prediction probability of scale two (resp. scale three) on the correct category label t. Assume that classification network of scale two and scale three are well-trained, the rank loss will be minimized (which corresponds to $p_t^3 \geq p_t^2 + margin$) when $\mathrm{APN}_{2,1}$ localize a most discriminative region from input image of scale two. Such a design can enable $\mathrm{APN}_{2,1}$ to gradually update the parameters to approach the most discriminative region. Besides, in order to make the region cropped by $\mathrm{APN}_{2,*}$ close to the region cropped by $\mathrm{APN}_{2,1}$, so as to capture the missing information of $\mathrm{APN}_{2,1}$ and maximize p_t^3 with combined information of $\mathrm{APN}_{2,1}$ and $\mathrm{APN}_{2,*}$. We propose a rank loss with regularization term to optimize $\mathrm{APN}_{2,*}$, defined as

$$L_{rank}^{\mathrm{APN}_{2,*}} = max\{0, p_t^2 - p_t^3 + margin\} - \lambda||f_{\mathrm{APN}_{2,*}}(g(\boldsymbol{x}_i)) - f_{\mathrm{APN}_{2,1}}(g(\boldsymbol{x}_i))||_2^2, \tag{4}$$

where λ is a hyper-parameter, $f_{\mathrm{APN}_{2,*}}$ and $f_{\mathrm{APN}_{2,1}}$ denote two triplets given by $\mathrm{APN}_{2,*}$ and $\mathrm{APN}_{2,1}$, respectively.

2.3 Joint Representation of Multi-scale with Dempster's Combination Rule

As we discussed before, concatenating the outputs of different scales to obtain the final classification results does not well represent the multi-scale view. Motivated by [6] and [7], we want to leverage the merit of DS theory to improve the fusion performance. In this section, we first define the mass functions in our model, then introduce how to use Dempster's combination rule in our framework.

According to [3], let Ω be the frame of discernment. A mass function is a mapping m from the power set of Ω to $[0, 1]$ and satisfies following two equations

$$m(\emptyset) = 0 \text{ and } \sum_{A \subseteq \Omega} m(A) = 1, \tag{5}$$

where a focal set of m is defined as the subset A of Ω satisfying $m(A) > 0$.

Here, each class in the dataset is viewed as an element in the frame of discernment Ω, such as $\Omega = \{y_1, y_2, ..., y_k\}$, where k is the number of classes. In addition, we assume that each scale's classification network (corresponding each row in Fig. 1) is treated as an expert's evaluation. Therefore, the output of the softmax layer of each scale's classification network can be used as a mass function m. From the multi-scale classification network, we obtain multiple mass functions. Then, we can use the Dempster's combination rule to combine all mass functions to obtain the fusion result. For all $A_1, A_2, ..., A_s \subseteq \Omega$, the fusion formula can be formalized as

$$(m_1 \oplus m_2 \oplus \cdots \oplus m_s)(A) = \frac{1}{1 - K} \sum_{A_1 \cap A_2 \cap ... \cap A_s = A} m_1(A_1)m_2(A_2) \cdots m_s(A_s), \tag{6}$$

where K is a conflict coefficient defined as

$$K = \sum_{A_1 \cap A_2 \cap ... \cap A_s = \emptyset} m_1(A_1)m_2(A_2) \cdots m_s(A_s). \tag{7}$$

By using Eq. (6), we can obtain the fused prediction result of multi-scale classification network. In order to train and test the final classification layer, the original output of the fully-connected layer at each scale is normalized independently at first. Then we concatenate them with the result derived from Eq. (6) as the input of the final classification layer for training and testing. The overall framework of the final classification layer can be seen in Fig. 2.

2.4 Training Details

The training strategy here is similar to RA-CNN, in addition, we modify some details.

step 1: we replace VGG with pre-trained (on ImageNet) MobileNetV2 to initialize the classification networks of all scales in Fig. 1.

step 2: We use a method similar to RA-CNN to pre-train APN at all scales which constructs a pseudo triplet as the ground-truth, and optimizes the output

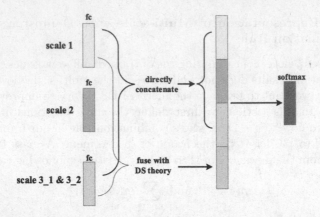

Fig. 2. Decision fusion of multiple scales

of APN to approach it. Different from the method of pseudo triplet construction in RA-CNN, the pseudo triplet is constructed by searching a square in the input image at scale one (or scale two) with the highest response value in the penultimate convolutional layer. In addition, the length of the square is set to half length of the input image. Note that, $APN_{2,*}$ and $APN_{2,1}$ are pre-trained using the same pseudo triplet.

step 3: the parameters of APNs and classification networks are optimized in an alternative way.

3   Experiments

Data Sets: In this section, we conduct experiments on Caltech-UCSD Birds (CUB-200-2011). The dataset contains 5994 training images and 5794 testing images, corresponding to 200 birds species.

Experiments and Results: We compare our proposed method with RA-CNN in terms of top-1 accuracy, and we test two different fusion strategies to combine the output of the three scales, one is Dempster's combination rule (as shown in Fig. 2), and the other is direct concatenation. As shown in Table 1, compared with RA-CNN, our methods achieve better top-1 accuracy with and without Dempster's combination rule, which verifies the effectiveness of our proposed method. In addition as shown in Table 1, compared with the results of direct concatenation of multiple scales' outputs, both our method and RA-CNN can be further improved by using Dempster's combination rule, which validates the superiority of DS theory.

As shown in Table 2, our method outperforms RA-CNN in terms of accuracy at scale three. The superior performance of our method against RA-CNN at scale three clearly verifies the effectiveness of using multi-branch APN.

Table 1. Comparison of different methods on CUB-200-2011

Backbone	Method	DS fusion	Top-1 accuracy (%)
MobileNet-V2	RA-CNN		84.6
	RA-CNN	✓	85.2
	Our method		85.3
	Our method	✓	**85.7**

Table 2. Comparison between RA-CNN and our method at different scales

Scale	*Top-1 accuracy (%)*	
	RA-CNN	Our method
Scale 1	80.4	80.5
Scale 2	82.0	82.0
Scale 3	80.9	**82.3**

4 Conclusion

The existing FGIC methods, which leverage attention mechanism to recurrently localize a discriminative region, are generally constructed as multiple scales. However, the finest-scale of the methods pay too much attention to the local region of object part, and thereby the contour information of the object is lost. In order to alleviate this problem, we constructed a multi-branch APN at middle scale of RA-CNN. By proposing a novel loss function, the multiple APNs can jointly localize the most discriminative region and complement each other's incomplete contour information, so as to improve the classification accuracy of finest-scale. In addition, we utilize Dempster's combination rule to fuse the opinions of different scales to further improve the overall accuracy. Experimental results demonstrate the effectiveness of our proposed method.

References

1. Chang, D., et al.: The devil is in the channels: mutual-channel loss for fine-grained image classification. IEEE Trans. Image Process. **29**, 4683–4695 (2020)
2. Denoeux, T.: Analysis of evidence-theoretic decision rules for pattern classification. Pattern Recogn. **30**(7), 1095–1107 (1997)
3. Denoeux, T.: Decision-making with belief functions: a review. Int. J. Approximate Reasoning **109**, 87–110 (2019)
4. Ding, Y., Zhou, Y., Zhu, Y., Ye, Q., Jiao, J.: Selective sparse sampling for fine-grained image recognition. In: Proceedings of the IEEE/CVF International Conference on Computer Vision, pp. 6599–6608 (2019)
5. Fu, J., Zheng, H., Mei, T.: Look closer to see better: recurrent attention convolutional neural network for fine-grained image recognition. In: Proceedings of the IEEE Conference on Computer Vision and Pattern Recognition, pp. 4438–4446 (2017)

6. Li, F., Qian, Y., Wang, J., Liang, J.: multigranulation information fusion: a Dempster-Shafer evidence theory-based clustering ensemble method. Inf. Sci. **378**, 389–409 (2017)
7. Li, S., Yao, Y., Hu, J., Liu, G., Yao, X., Hu, J.: An ensemble stacked convolutional neural network model for environmental event sound recognition. Appl. Sci. **8**(7), 1152 (2018)
8. Shafer, G.: A Mathematical Theory of Evidence, vol. 42. Princeton University Press, Princeton (1976)
9. Sun, M., Yuan, Y., Zhou, F., Ding, E.: Multi-attention multi-class constraint for fine-grained image recognition. In: Ferrari, V., Hebert, M., Sminchisescu, C., Weiss, Y. (eds.) ECCV 2018. LNCS, vol. 11220, pp. 834–850. Springer, Cham (2018). https://doi.org/10.1007/978-3-030-01270-0_49
10. Wei, X.S., Wu, J., Cui, Q.: Deep learning for fine-grained image analysis: a survey. arXiv preprint arXiv:1907.03069 (2019)
11. Wei, X.S., Xie, C.W., Wu, J., Shen, C.: Mask-CNN: localizing parts and selecting descriptors for fine-grained bird species categorization. Pattern Recogn. **76**, 704–714 (2018)
12. Yang, Z., Luo, T., Wang, D., Hu, Z., Gao, J., Wang, L.: Learning to navigate for fine-grained classification. In: Ferrari, V., Hebert, M., Sminchisescu, C., Weiss, Y. (eds.) Computer Vision – ECCV 2018. LNCS, vol. 11218, pp. 438–454. Springer, Cham (2018). https://doi.org/10.1007/978-3-030-01264-9_26
13. Zhang, L., Huang, S., Liu, W., Tao, D.: Learning a mixture of granularity-specific experts for fine-grained categorization. In: Proceedings of the IEEE/CVF International Conference on Computer Vision, pp. 8331–8340 (2019)
14. Zhang, N., Donahue, J., Girshick, R., Darrell, T.: Part-based R-CNNs for fine-grained category detection. In: Fleet, D., Pajdla, T., Schiele, B., Tuytelaars, T. (eds.) ECCV 2014. LNCS, vol. 8689, pp. 834–849. Springer, Cham (2014). https://doi.org/10.1007/978-3-319-10590-1_54
15. Zhang, Y., et al.: Weakly supervised fine-grained categorization with part-based image representation. IEEE Trans. Image Process. **25**(4), 1713–1725 (2016)
16. Zhao, B., Feng, J., Wu, X., Yan, S.: A survey on deep learning-based fine-grained object classification and semantic segmentation. Int. J. Autom. Comput. **14**(2), 119–135 (2017). https://doi.org/10.1007/s11633-017-1053-3
17. Zheng, H., Fu, J., Mei, T., Luo, J.: Learning multi-attention convolutional neural network for fine-grained image recognition. In: Proceedings of the IEEE International Conference on Computer Vision, pp. 5209–5217 (2017)
18. Zheng, H., Fu, J., Zha, Z.J., Luo, J., Mei, T.: Learning rich part hierarchies with progressive attention networks for fine-grained image recognition. IEEE Trans. Image Process. **29**, 476–488 (2019)

Deep Evidential Fusion Network for Image Classification

Shaoxun Xu[1], Yufei Chen[1(✉)], Chao Ma[2], and Xiaodong Yue[3]

[1] College of Electronics and Information Engineering, Tongji University,
Shanghai 200092, China
yufeichen@tongji.edu.cn
[2] Department of Radiology, Changhai Hospital of Shanghai,
Second Military Medical University, Shanghai 200433, China
[3] School of Computer Engineering and Science, Shanghai University,
Shanghai 200444, China

Abstract. Evidential deep learning (EDL) has been proposed to estimate the uncertainty and the prediction confidence of neural networks. In this paper, we investigate the fusion method based on the EDL model and Dempster's rule of combination. For fusion models, a better uncertainty estimation may be more helpful than high accuracy. To this end, we propose a deep evidential fusion method to best utilize the belief assignment and uncertainty estimation by improving the objective function and introducing the approximation of the base rate distributions. The experimental results show that our proposed method achieves a more reliable fusion result. We also explore the application of belief function and evidence theory in the field of medical image analysis, where multi-modality well fits the framework of belief functions.

Keywords: Evidential fusion · Deep neural network · Belief function · Image classification

1 Introduction

Deep learning models have achieved outstanding performance in a large number of classification tasks in different fields of machine learning. Despite their success in out-performance of traditional methods, the concerns of the poor generalization lead to the research of the ability to estimate the uncertainty and the prediction confidence of the networks. Modeling uncertainty also brings a new perspective for learning on small data without pretraining and other transferred knowledge. Utilizing uncertainty model, with Dempster-Shafer theory (DST) of evidence fusion, we can improve the prediction confidence by combining the evidence from several information sources. This specific kind of task is quite common in medical image analysis, where images are often in several different modalities but difficult to collect. Related research also shows that transfer learning offers limited performance gains in medical tasks [6].

© Springer Nature Switzerland AG 2021
T. Denœux et al. (Eds.): BELIEF 2021, LNAI 12915, pp. 185–193, 2021.
https://doi.org/10.1007/978-3-030-88601-1_19

Dempster-Shafer theory of evidence shows a promising approach to quantify the uncertainty with Basic Belief Assignment (BBA), considering that the uncertainty is caused by a lack of evidence or conflicting evidence. During the past several years, various studies have been conducted to combining belief functions and deep learning [7,9]. The belief theory of Subjective Logic (SL) [4] formalizes the BBA in an equivalent belief/uncertainty representation manner but provides a different interpretation. By formalizing the process of the BBA over the frame of discernment as a (hyper-) Dirichlet model in SL, Evidential Deep Learning (EDL)[3,7,10] may quantify the classification uncertainty in a computationally simple approach, which provides a pipeline flow for information fusion learning, such as multi-view learning [2] and multi-modality learning.

In this paper, we propose a deep evidential fusion method based on the EDL model, and to best utilize the belief assignment and uncertainty estimation, we present that (1)by minimizing an approximation to the expected L^∞ norm of the prediction error, the fusion result will be better and more reasonable despite the small accuracy drop on a single information source, and (2)by introducing the approximation of the base rate distributions, the fusion model would learn the knowledge prior of the task and thereby improve the final fusion result.

The rest of the paper is organized as follows: Sect. 2 reviews the basics concerning the evidence theory and subjective logic. In Sect. 3, the proposed deep evidential fusion learning method is described. In Sect. 4, experiments on public synthetic dataset and real-world medical image dataset are demonstrated. Section 5 concludes this paper.

2   Preliminaries

2.1   Evidential Deep Learning

In the case of the classification task, EDL uses Dirichlet distributions to quantify the belief masses and overall predictive uncertainty. The concentration parameters of a Dirichlet distribution can be viewed as a belief distribution [4]. Specifically, for K mutually exclusive singletons (e.g., class labels), subjective logic assigns a belief mass b_k for every $k = 1, ..., K$ and an uncertainty mass of u. The sum of these $K + 1$ non-negative mass values is one:

$$u + \sum_{k=1}^{K} b_k = 1, \tag{1}$$

where $u \geq 0$ and $b_k \geq 0$. It can be viewed as a particular case of the Dempster-Shafer belief function, where the masses of all conjunction terms except the total set are equal to zero. By limiting the mass of uncertainty terms, we get a more computationally efficient belief function when we don't care about any particular uncertainty term.

A subjective opinion, or also called belief mass assignment in DST, $b = (b_1, ..., b_K)$, can be viewed as a sample of a Dirichlet distribution with concentration parameters $\alpha = (\alpha_1, ..., \alpha_K)$, and the probability density function on vector $\mathbf{p}$ given by

$$Dir(\mathbf{p} \mid \boldsymbol{\alpha}) = \frac{1}{B(\boldsymbol{\alpha})} \prod_{k=1}^{K} p_k^{\alpha_k - 1}, \tag{2}$$

where $B(\cdot)$ is the multivariate Beta function.

To get a non-sparse Dirichlet distribution, we constraint the concentration parameters to be larger than one, $\alpha_k = e_k + 1$, where e_k denotes the non-negative evidence. Accordingly, the belief mass b_k and the uncertainty u are computed as $b_k = e_k / S$ and $u = K / S$, where $S = \sum_{k=1}^{K} (e_k + 1)$. The uncertainty is inversely proportional to the total evidence and the term evidence is a measure of the amount of support learning from data in favor of a sample to be classified into a certain class.

A neural network could capture features from the input and generate subjective opinions if we model the output of the network as a belief distribution. An activation layer, e.g., ReLU, is used to ascertain network output as a nonnegative evidence vector $\boldsymbol{e}$. Consequently, the mean of this distribution is considered as an estimation of the class probabilities $\boldsymbol{p}$. To train an EDL model, the objective can be optimized by minimizing the mean-square-error(MSE) loss. Specifically, for a sample x_i with one-hot labels y_i, the objective reads

$$\mathcal{L}_i(\Theta) = \mathbb{E}_{p_i \sim Dir(\cdot; \alpha_i)}[\|y_i - p_i\|_2^2] = \sum_{j=1}^{K} (y_{ij} - \hat{p}_{ij})^2 + \frac{\hat{p}_{ij}(1 - \hat{p}_{ij})}{(S_i + 1)}. \tag{3}$$

2.2 Dempster's Rule for EDL

The Dempster-Shafer theory (DST) of evidence gives a mathematical framework to combine evidence from different sources. Technically, the basic belief assignment of DST and the belief/uncertainty representation of subjective opinions are equivalent [4]. Through Dempster's rule, we combine the evidence assignments from different EDL models by utilizing all available information sources.

Suppose that we have total T information sources (e.g., T different modalities) on the same K classification task. To combine T basic belief assignments $\{\mathcal{M}^t\}_{t=1}^{T}$, where $\mathcal{M}^t = \{b_1^t, ..., b_K^t, u^t\}$, Dempster's rule of combination shows that $\mathcal{M}^{\oplus} = \oplus_{t=1}^{T} \mathcal{M}^t$.

The fusion of two masses $\mathcal{M}^1 = \{b_1^1, ..., b_K^1, u^1\}$ and $\mathcal{M}^2 = \{b_1^2, ..., b_K^2, u^2\}$ can be formulated as follows:

$$b_k^{1 \oplus 2} = \frac{1}{1 - C} \left(b_k^1 b_k^2 + b_k^1 u^2 + b_k^2 u^1 \right), u^{1 \oplus 2} = \frac{1}{1 - C} u^1 u^2, \tag{4}$$

where $C = \sum_{i \neq j} b_i^1 b_j^2$ is a measure of the degree of relative conflict between the two belief assignments.

After obtaining the fusion result $\mathcal{M}^{\oplus}$ of T information sources, the corresponding concentration of joint Dirichlet distribution is computed by

$$e_k^{\oplus} = \frac{K b_k^{\oplus}}{u^{\oplus}}, \alpha_k^{\oplus} = e_k^{\oplus} + 1. \tag{5}$$

By combining the evidence from all available information sources, we obtain the final joint Dirichlet distribution $Dir(\mathbf{p} \mid \boldsymbol{\alpha}^{\oplus})$ to generate the fusion belief and the overall uncertainty of the classification task.

3   Deep Evidential Fusion

Having reviewed the mathematics of uncertainty estimation and evidence fusion, in this section, we focus on the adaptation of the EDL model to multiple information sources fusion.

3.1   Modeling with Knowledge Prior

A good fusion result with Dempster's rule of combination highly depends on the individual belief assignment and uncertainty estimation. When coming to multiple information sources (e.g., multiple modalities) on the same object, these sources may not share the same background information, which means that different sources have different prior distributions on the confidence and uncertainty estimation.

With the purpose of modeling the different knowledge priors, the concept of *base rate*, also known as prior probabilities, is introduced. For subjective opinions, SL combines the belief mass $\boldsymbol{b}$, uncertain mass u with the base rates $\boldsymbol{a}$ through the projected probability $Pr(y_i = k) = b_k + a_k u$, which represents the probability that sample x_i is assigned to class k. With base rates, SL redefines the concentration parameters $\boldsymbol{\alpha}$ of Dirichlet distribution by introducing the uncertainty evidence through a non-informative prior weight W, formulated as

$$\boldsymbol{\alpha} = \boldsymbol{e} + \boldsymbol{a}W. \tag{6}$$

For belief fusion models, we explicitly model the base rate distribution for each information source t by introducing a series of parameter θ in the learning process, that is, $\boldsymbol{a}^t \sim Dir(\boldsymbol{a}^t \mid \boldsymbol{\theta}^t)$. The final evidence Dirichlet distribution for the observation of information source t will be

$$Dir(\mathbf{p} \mid \boldsymbol{e}^t, \boldsymbol{a}^t) = \frac{1}{B(\boldsymbol{e}^t + \boldsymbol{a}^t W)} \prod_{k=1}^{K} p_k^{e_k^t + a_k^t W - 1}, \tag{7}$$

and the corresponding belief assignment and uncertainty mass are $b_k = e_k/S$ and $u = W/S$, where $S = W + \sum_{k=1}^{K} e_k$.

When applying Dempster's rule for combining evidence from different information sources, Eq. 5 should be modified to

$$e_k^{\oplus} = \frac{W b_k^{\oplus}}{u^{\oplus}}, \alpha_k^{\oplus} = e_k^{\oplus} + a_k^{\oplus} W. \tag{8}$$

The joint base rate distribution $a^{1 \oplus 2}$ of a^1 and a^2 is given by the following belief constraint fusion operation in SL,

$$a_k^{1 \oplus 2} = \begin{cases} \frac{a_k^1 (1 - u^1) + a_k^2 (1 - u^2)}{2 - u^1 - u^2} & \text{for } u^1 + u^2 < 2, \\ \frac{a_k^1 + a_k^2}{2} & \text{for } u^1 = u^2 = 1. \end{cases} \tag{9}$$

where W denotes the non-informative prior weight, which is normally set to $W = 2$. The choice of W would adjust the sensibility of the model to the new observation evidence.

3.2 ˙ Evidential Fusion with Deep Learning

By modeling the knowledge prior, a EDL model could learn the source-related prior distribution to improve the confidence and uncertainty estimation. However, when coming to limited training samples, training a large number of parameters in a EDL model may lead to a high risk of overfitting. The MSE in Eq. 3 will amplify the errors produced by outliers, resulting in over-confident on the observation evidence of training samples and discounting the model's ability of uncertainty estimation.

An intuitive way to improve objective function in Eq. 3 is by minimizing an approximation to the expected L^∞ norm of the prediction error, which is a better representation of the highest prediction error among the classes. In practice, by utilising Jensen's inequality on norm, i.e., $\| \cdot \|_\infty \leq \| \cdot \|_p$, we relax the target L^∞ norm to the L^p space, which provides a tractable upper bound:

$$\mathcal{L}_i(\Theta) = \mathbb{E}_{p_i \sim Dir(\cdot;\alpha_i)}[\|y_i - p_i\|_p]. \tag{10}$$

[11] gives a closed form of Eq. 10 as follows,

$$\mathcal{L}_i(\Theta) = \left(\frac{\Gamma(S_i)}{\Gamma(S_i + p)} \right)^{\frac{1}{p}} \left(\frac{\Gamma\left(\sum_{k \neq c} \alpha_k + p\right)}{\Gamma\left(\sum_{k \neq c} \alpha_k\right)} + \sum_{k \neq c} \frac{\Gamma(\alpha_k + p)}{\Gamma(\alpha_k)} \right)^{\frac{1}{p}}, \tag{11}$$

where $\Gamma(\cdot)$ is the *gamma* function and c is the correct class label of sample x_i. Note that the p in Eq. 11 represents p-norm. The larger p, the tighter the L^p norm approximates the L-infinity of the prediction error, practically p can be adjusted to balance the prediction accuracy and the uncertainty behavior of the incorrect predictions.

The above objective could help the network to discover more patterns to assign evidence to each class label. It guarantees that more evidence will be assigned to the correct label but not ensures less evidence for the incorrect labels. Hence, we put a regularization term to minimize the contribution of the misleading patterns associated with incorrect labels.

Given the auxiliary vector $\tilde{\alpha} = (1 - y) \odot \alpha + y$, where $\odot$ represents Hadamard product, the regularization term is introduced with the following KL divergence:

$$\mathcal{R}_i(\Theta) = KL[Dir(p_i|\tilde{\alpha}_i) \| Dir(p_i|1)]$$

$$= \log \left(\frac{\Gamma\left(\sum_{k=1}^{K} \tilde{\alpha}_{ik}\right)}{\Gamma(K) \prod_{k=1}^{K} \Gamma(\tilde{\alpha}_{ik})} \right) + \sum_{k=1}^{K} (\tilde{\alpha}_{ik} - 1) \left[\psi(\tilde{\alpha}_{ik}) - \psi\left(\sum_{j=1}^{K} \tilde{\alpha}_{ij}\right) \right], \tag{12}$$

where 1 represents vector of K ones, $\psi(\cdot)$ is the *digamma* function.

For N training samples, the total loss to be minimized is:

$$\mathcal{L}(\Theta) = \frac{1}{N} \sum_{i=1}^{N} \mathcal{L}_i(\Theta) + \lambda_t \mathcal{R}_i(\Theta). \tag{13}$$

λ_t increases with an annealing schedule, e.g., $\lambda_t = \lambda_0 min(\frac{t-T_0}{T}, 1)$ for $t > T_0$ for annealing step T and $\lambda_t = 0$ for $t \le T_0$. This annealing factor setting guides the network to learn all available evidence before penalizing the incorrect evidence assignment.

Applying the fusion operation mentioned in Eq. 4, 8 and 9, we extend an EDL model to a Deep Evidential Fusion (DEF) model for combining evidence from different information sources.

4   Experiments

4.1   Experiments on Synthetic Dataset

To illustrate the behavior of the proposed fusion model, we first show a simple example on a synthetic multi-modality MNIST-SVHN dataset proposed by [8], which combines pairs of MNIST and SVHN such that each pair depicts the same digit class. Theoretically, we could treat these two sources as two different *pseudo-modalities* since they share the same intrinsic information on digits but have different data distributions. To simulate the few sample situation, we intentionally limited the number of samples used in training to 50 samples per class, while the validation and test were done on the original set.

A standard ResNet18 was adopted as the backbone for all experiments with same Adam [5] optimizer. Data augmentation, dropout and batch-normalization were used to mitigate overfitting. For proposed DEF model,the results were generated using the norm $p = 4$ and $\lambda_0 = 0.3$. Comparisons were made with following methods: (a) cross entropy (CE) loss as a baseline, (b) cosine loss [1] for few samples, (c) EDL [7], (d) EDL with proposed L^p loss, and (e) DEF is our proposed model. All non-evidential methods were only done for single modality separately and no fusion was made. Table 1 shows the Top-1 test accuracy for these methods.

When given 50 samples per class as training set, as can be seen from Table 1, the original EDL method may have better accuracy on one single modality, but the fusion result suffers a downgrade due to the incorrect predictions on SVHN dataset. By contrast, a more reasonable fusion result was obtained with L^p loss in the case of only a small accuracy drop on a single modality. Our proposed fusion model achieved the best fusion result, and the comparison also shows that the introduction of base rate improves the confidence and uncertainty estimation for fusion.

Table 1. Test accuracy in percent (%) achieved with different methods, trained with 50 samples per class. The best value per row is set in bold.

	(a) CE	(b) Cosine	(c) EDL	(d) EDL(L^p)	(e) DEF (ours)
MNIST	92.8	93.4	**95.2**	91.6	92.6
SVHN	32.2	35.2	**39.6**	38.8	38.6
MNIST-SVHN			93.4	92.8	**95.4**

4.2 Application: Pancreatic Tumor Classification

We applied our proposed method to a real-world medical image analysis task on a dataset of enhanced 3-phase Computed Tomography (CT) images of 5 sub-types of pancreatic tumor patients collected by Changhai Hospital. The dataset consists of 51 patients of pancreatic ductal adenocarcinoma (PDAC), 47 patients of intraductal papillary mucinous neoplasm (IPMN), 28 patients of pancreatic neuroendocrine tumor (NET), 29 patients of serous cystic neoplasm (SCN), 32 patients of adenosquamous carcinoma (ASC), and 187 patients in total. The CT scan was performed in arterial phase (AP), venous phase (VP) and delayed phase (DP). We selected the maximum cross-section of the tumor as the region of interest for network input and did the same comparisons mentioned in Sect. 4.1 with same experiment setups. The fusion was done with any possible combination of the 3 phases. Limited by the total scale of the dataset, all results were obtained by 5-fold cross-validation. Table 2 shows the Top-1 accuracy of different methods.

Table 2. Test accuracy in percent (%) achieved with different methods for pancreatic tumor classification. The best value per row is set in bold.

	(a) CE	(b) Cosine	(c) EDL	(d) EDL(L^p)	(e) DEF (ours)
AP	50.0	45.0	48.0	48.1	**58.0**
VP	53.0	47.5	**53.5**	51.5	53.0
DP	**52.5**	47.5	43.0	43.5	52.3
AP+VP			50.0	52.4	**61.0**
AP+DP			47.2	44.0	**60.4**
VP+DP			44.5	53.1	**55.0**
AP+VP+DP			55.0	55.0	**62.5**

The original EDL method(c) failed to estimate the evidence Dirichlet distribution for tumor subtypes given arterial phase (AP) and delayed phase (DP) CT images, causing a worse prediction result than simple cross entropy loss(a). Our proposed method improved the performance in those phases and obtained the best fusion results on all possible combinations of phases. By evidential fusion of all three phases, the prediction result was 4.5% better than the best prediction in one phase, 9% better than that of original EDL, and 9.5% better than that of non-evidential methods.

5   Conclusion and Future Work

In this work, we propose a deep evidential fusion method for classification tasks on small dataset, based on the belief assignment and evidential fusion. The experimental results show that the evidence theory has a promising application of the fields like medical image analysis, where the explainable interpretation of belief/uncertainty assignment and evidence fusion could be a qualified framework for medical information fusion. The multi-modality of medical imaging makes it a good practical playground for belief functions and evidential fusion theory. In future work, we will focus on better belief assignment and uncertainty estimation with small training samples and, based on the proposed fusion method, promote the application of evidence theory in the field of medical image analysis.

Acknowledgments. This work was supported by National Natural Science Foundation of China (No. 92046008, 62173252), the Shanghai Innovation Action Project of Science and Technology (No. 20Y11912500), and the Fundamental Research Funds for the Central Universities.

References

1. Barz, B., Denzler, J.: Deep learning on small datasets without pre-training using cosine loss. In: Proceedings of the IEEE/CVF Winter Conference on Applications of Computer Vision (WACV), March 2020
2. Han, Z., Zhang, C., Fu, H., Zhou, J.T.: Trusted multi-view classification. In: International Conference on Learning Representations (2021)
3. Hemmer, P., Kühl, N., Schöffer, J.: Deal: deep evidential active learning for image classification. In: 2020 19th IEEE International Conference on Machine Learning and Applications (ICMLA), pp. 865–870 (2020)
4. Jøsang, A.: Subjective Logic. Artificial Intelligence: Foundations, Theory, and Algorithms, Springer, Cham (2016). https://doi.org/10.1007/978-3-319-42337-1
5. Kingma, D.P., Ba, J.: Adam: a method for stochastic optimization. In: Bengio, Y., LeCun, Y. (eds.) 3rd International Conference on Learning Representations, ICLR 2015, San Diego, CA, USA, 7–9 May 2015, Conference Track Proceedings (2015)
6. Raghu, M., Zhang, C., Kleinberg, J., Bengio, S.: Transfusion: understanding transfer learning for medical imaging. In: Wallach, H., Larochelle, H., Beygelzimer, A., d'Alché-Buc, F., Fox, E., Garnett, R. (eds.) Advances in Neural Information Processing Systems, vol. 32. Curran Associates, Inc. (2019)
7. Sensoy, M., Kaplan, L., Kandemir, M.: Evidential deep learning to quantify classification uncertainty. In: Bengio, S., Wallach, H., Larochelle, H., Grauman, K., Cesa-Bianchi, N., Garnett, R. (eds.) Advances in Neural Information Processing Systems, vol. 31. Curran Associates, Inc. (2018)

8. Shi, Y., Siddharth, N., Paige, B., Torr, P.: Variational mixture-of-experts autoencoders for multi-modal deep generative models. In: Wallach, H., Larochelle, H., Beygelzimer, A., d'Alché-Buc, F., Fox, E., Garnett, R. (eds.) Advances in Neural Information Processing Systems, vol. 32. Curran Associates, Inc. (2019)
9. Tong, Z., Xu, P., Denœux, T.: An evidential classifier based on Dempster-Shafer theory and deep learning. Neurocomputing **450**, 275–293 (2021)
10. Tsiligkaridis, T.: Failure prediction by confidence estimation of uncertainty-aware Dirichlet networks. CoRR abs/2010.09865 (2020)
11. Tsiligkaridis, T.: Information aware max-norm Dirichlet networks for predictive uncertainty estimation. Neural Netw. **135**, 105–114 (2021)

Conflict, Inconsistency and Specificity

Conflict Measure of Belief Functions with Blurred Focal Elements on the Real Line

Alexander Lepskiy[(✉)] [iD]

Higher School of Economics, 20 Myasnitskaya Ulitsa, Moscow 101000, Russia
alepskiy@hse.ru
https://www.hse.ru/en/org/persons/10586209

Abstract. The paper studies the variation of the conflict measure with blurring of focal elements and discounting of the masses of the belief functions in the framework of the theory of evidence. Blurring of focal elements is modeled using fuzzy sets. Such properties of the conflict measure as the robustness to transformations of the bodies of evidence, the monotonicity and the direction of change are investigated. A numerical example of calculating the measure of conflict, taking into account the blurring of focal elements and discounting of masses for the selection of bodies of evidence for the aggregation of analysts' forecasts regarding the oil price, is considered.

Keywords: Evidence theory · Conflict measure · Blur focal elements

1 Introduction

Conflict assessment and combining of evidence bodies in the evidence theory is a two-pronged problem. On the one hand, the conflict value must be taken into account when combining evidence. On the other hand, the conflict itself is estimated, as a rule, with the help of aggregation of evidence bodies. The method of conflict estimation and the choice of the combining rule should be consistent with each other in a certain sense [9].

As a rule, the (external) conflict of two bodies of evidence is understood as a quantity proportional to the sum of the products of the masses of non-intersecting (or 'weakly intersecting' with respect to some similarity index) focal elements of this evidence. For example, the conflict in Dempster's rule [2] is the mass of the empty set obtained using the non-normalized conjunctive rule.

Robustness is one of the important requirements for conflict assessment. We will under-stand by robustness the stability of the conflict measure to the 'small' variation of focal elements and their masses. In general, the robustness of calculation the conflict measure can be achieved by applying specialization-generalization procedures [1,5].

The financial support from the Government of the Russian Federation within the framework of the implementation of the 5–100 Programme Roadmap of the National Research University Higher School of Economics is acknowledged.

T. Denœux et al. (Eds.): BELIEF 2021, LNAI 12915, pp. 197–206, 2021.
https://doi.org/10.1007/978-3-030-88601-1_20

We will consider bodies of evidence, the focal elements of which are defined on some set $X \subseteq \mathbb{R}$ [11]. The specialization-generalization procedure for such bodies of evidence can be implemented using fuzzy blurring of focal elements. The concept of fuzzy focal elements was considered, for example, in [12,13].

If the focal elements on the $X \subseteq \mathbb{R}$ are defined by experts, then information on the preferential conservatism or radicalism of expert can also be taken into account using the blur procedure. For example, if one expert predicts the value of shares of a certain company in the interval $A = [40, 45)$, and another in the interval $B = [35, 50)$, then the second evidence can be considered more conservative than the first. If the second expert often gives conservative estimates, then we can assume that the interval B is a support of a fuzzy number-evidence. On the contrary, if the estimates of the first expert are often radical, then we can assume that the interval A is the kernel of a fuzzy number-evidence.

Finally, the blurring procedure of focal elements can be used to account for the reliability of information sources together with a procedure for discounting the masses.

In this paper, we will investigate some properties of the conflict measure of evidence bodies, determined on the $X \subseteq \mathbb{R}$, taking into account blurring of focal elements and discounting of masses. A numerical example of the choice for combining evidence bodies will be considered, taking into account their conflict, reliability and accuracy and using the procedure for blurring of focal elements.

2   Background of the Belief Function Theory

Let X be some set, $\mathcal{A} \subseteq 2^X$ be some finite subset of nonempty sets from X. Some non-negative mass function $m : 2^X \to [0, 1]$, $\sum_{A \in \mathcal{A}} m(A) = 1$ is considered in the theory of evidence [2,10]. Without loss of generality, we can assume that $m(A) > 0$ for all $A \in \mathcal{A}$. In this case, set $\mathcal{A}$ is called the set of focal elements, and a pair $F = (\mathcal{A}, m)$ is called a body of evidence. Let $\mathcal{F}(X)$ be a set of all bodies of evidence on X. There is one-to-one correspondence between the body of evidence $F = (\mathcal{A}, m)$ and the belief function $Bel(A) = \sum_{B \subseteq A} m(B)$ or the plausibility function $Pl(A) = \sum_{B : A \cap B \neq \emptyset} m(B)$.

The body of evidence (evidence) $F_A = (A, 1)$ (i.e., $\mathcal{A} = \{A\}$, $m(A) = 1$) is called categorical. In particular, the body of evidence F_X is called vacuous.

Then any body of evidence $F = (\mathcal{A}, m)$ can be represented as a convex sum of categorical bodies of evidence: $F = \sum_{A \in \mathcal{A}} m(A) F_A$. The body of evidence is called simple, if $F_A^\zeta = (1 - \zeta) F_A + \zeta F_X$, $\zeta \in [0, 1]$.

In this paper, we will consider evidence bodies on $X \subseteq \mathbb{R}$ [11]. Moreover, we assume that all focal elements of evidence are intervals of the form $[a_1, a_2)$. In this case, the intersection of such sets will also have the form $[a_1, a_2)$. Consequently, we get a new set of focal elements of the same kind when combining the bodies of evidence using conjunctive rules.

Suppose there are two bodies of evidence $F_1 = (\mathcal{A}_1, m_1)$ and $F_2 = (\mathcal{A}_2, m_2)$. It is necessary to assess the conflict between these two bodies of evidence. Traditionally this is done most often with the help of the measure

$$Con_0(F_1, F_2) = \sum_{A \cap B = \emptyset} m_1(A)m_2(B).$$

However, this measure does not take into account the 'weakly intersecting' (i.e. pairs of intersecting focal sets of different bodies of evidence for which the external measure (for example, the length of interval on $\mathbb{R}$) the intersection of the sets is small compared to the external measure of each of these sets) focal elements of the bodies of evidence. The value of the conflict should be a decreasing function of the value of the 'strong' intersection of focal elements with large masses in the general case. Therefore, we will use the measure

$$Con^\Gamma(F_1, F_2) = \sum_{A \in \mathcal{A}_1, B \in \mathcal{A}_2} \gamma(A, B)m_1(A)m_2(B), \qquad (1)$$

instead of the measure $Con_0(F_1, F_2)$ to account for 'weakly intersecting' focal elements, where $\Gamma = (\gamma(A, B))_{A, B \in \mathcal{A}}$, $\gamma(A, B) = 1 - s(A, B)$ and $s(A, B)$ is a similarity index satisfying the conditions: 1) $0 \leq s(A, B) \leq 1$; 2) $s(A, B) = 0$, if $A \cap B = \emptyset$; 3) $s(A, A) = 1$ (or weaker condition 3)' $\max_B s(A, B) = s(A, A)$).

An example of such an index is the Jaccard index $s(A, B) = |A \cap B|/|A \cup B|$, which we will mainly consider in this article. Note that if $\gamma(A, B) = 1$ in the case $A \cap B = \emptyset$ and $\gamma(A, B) = 0$ in all other cases, then in (1) we get the measure $Con_0(F_1, F_2)$. Some properties of the bilinear conflict measure of the form (1) were investigated in [6].

The conflict measure (1) will be coordinated with the combination of the bodies of evidence $F_1 = (\mathcal{A}_1, m_1)$, $F_2 = (\mathcal{A}_2, m_2)$, according to the rule $F_{1,2} = (\mathcal{A}, m_{1,2}) = F_1 \otimes F_2$, where

$$m_{1,2}(C) = \frac{1}{K} \sum_{A \cap B = C} s(A, B)m_1(A)m_2(B), \qquad (2)$$

if $K = 1 - Con^\Gamma(F_1, F_2) \neq 0$. This is Zhang's center combination rule [14]. The general structure of the bilinear combination rules was investigated in [7].

The reliability of information sources can be taken into account using Shafer's discounting method [10]: $m^{(\eta)}(A) = \eta m(A)$, if $A \neq X$ and $m^{(\eta)}(X) = 1 - \eta + \eta m(X)$, where $\eta \in [0, 1]$. The change in ignorance after the application of Dempster's rule to the discounted bodies of evidence was studied in [8].

3   Blurring Focal Elements

Let $\widetilde{F} = (\widetilde{\mathcal{A}}, \widetilde{m})$ be a transformation of the body of evidence $F = (\mathcal{A}, m)$, where $\widetilde{\mathcal{A}}$ is a set of fuzzy focal elements, i.e. blurring intervals from the $\mathcal{A}$; $\widetilde{m}$ is a discounted mass function.

We will consider below the important properties of the conflict measure in relation to blur and discounting operations. Let d be some metric on the set of all fuzzy sets (see [4]).

Definition 1. *(robustness).* *The conflict measure Con will be called robust to small transformations* $\tilde{}$ *of the bodies of evidence, if* $\forall F_1, F_2 \in \mathcal{F}(X)$ *and* $\forall \varepsilon > 0 \ \exists \delta_1, \delta_2 > 0$: $\left| Con(\widetilde{F_1}, F_2) - Con(F_1, F_2) \right| < \varepsilon \ \forall \widetilde{F_1} = (\tilde{\mathcal{A}}_1, \widetilde{m_1}) \in \mathcal{F}(X)$:
$d\left(\tilde{\mathcal{A}}_1, \mathcal{A}_1 \right) = \sum_{A \in \mathcal{A}_1} d\left(\tilde{A}, A \right) < \delta_1, \ d\left(\widetilde{m_1}, m_1 \right) = \sum_{A \in \mathcal{A}_1} |\tilde{m}(A) - m(A)| < \delta_2.$

Definition 2. *(monotonicity).* *Let's call the conflict measure Con monotonic (strictly monotonic) with respect to a given transformation* $\tilde{}$ *of the body of evidence, if* $\forall F_1, F_2, F_3 \in \mathcal{F}(X)$: $Con(F_1, F_2) \leq Con(F_1, F_3) \Rightarrow Con(\widetilde{F_1}, F_2) \leq Con(\widetilde{F_1}, F_3)$ *(the corresponding inequalities are strictly).*

If the conflict measure is monotonic, then blurring does not change the order relation with respect to that measure. In particular, in the problem of choosing the least conflicting evidence bodies for combining, the monotonicity of the conflict measure means that these transformations of evidence bodies will not lead to a change in the choice.

It is easy to see that the conflict measure Con_0 is monotonic if the transformation is reduced only to discounting the masses according to Shafer's method.

Definition 3. *(directionality of change).* *A transformation* $\tilde{}$ *is said to be non-increasing (not decreasing) the conflict measure Con, if* $Con(\widetilde{F_1}, F_2) \leq Con(F_1, F_2)$ $(Con(\widetilde{F_1}, F_2) \geq Con(F_1, F_2))$ $\forall F_1, F_2 \in \mathcal{F}(X)$.

It is easy to see that discounting the masses by Shafer's method does not increase the degree of conflict Con_0. This observation is interpreted as follows: if we know that the reliability of the information source is low, then this information itself becomes less conflicting with information from other sources.

The properties of monotony and directionality of transformation may not be satisfied for arbitrary conflict measure and transformation of evidence bodies.

Let the number $\eta \in [0, 1]$ characterize the level of reliability of the information source ($\eta = 1$ corresponds to an absolutely reliable source). If the source of information is not entirely reliable ($\eta < 1$), then we will consider blurring of focal elements together with discounting of masses. If $A = [a_1, a_2]$ is a focal element, then $\tilde{A} = A^{(\eta)}$ is a fuzzy number associated with A. We have that $A^{(1)} = A$.

Let the symmetric (L-R)-type fuzzy number [4] $\tilde{A}$ be the blur of the focal element $A = [a_1, a_2)$. It means that the fuzzy number $\tilde{A}$ has a membership function $\mu_{\tilde{A}}$: $\mu_{\tilde{A}}(x) = 1$ for $x \in [x_1, x_2)$, $\mu_{\tilde{A}}(x) = L(x) = \theta\left(\frac{x_1 - x}{\delta |A|} \right)$ for $x \in [x_1 - \delta |A|, x_1]$, $\mu_{\tilde{A}}(x) = R(x) = \theta\left(\frac{x - x_2}{\delta |A|} \right)$ for $x \in [x_2, x_2 + \delta |A|]$, where $x_1 \leq x_2$, $\delta \in (0, 1)$ and a strictly decreasing integrable function $\theta : [0, 1] \to [0, 1]$ satisfies the conditions $\theta(0) = 1$, $\theta(1) = 0$. The value $\delta = \delta(\eta) > 0$ controls the degree of blur. We assume that $\delta(1) = 0$ and $\delta(\eta)$ are a non-increasing function on $[0, 1]$.

As a result, we get a fuzzy focal element $\tilde{A} = A^{(\eta)}$, which is a blur (more precisely, a δ-blur) of the element A, $\ker \tilde{A} = \left\{ x \in \mathbb{R} : \mu_{\tilde{A}}(x) = 1 \right\} =$

$[x_1, x_2]$ (the core of the fuzzy number $\tilde{A}$), supp $\tilde{A} = \overline{\{x \in \mathbb{R} : \mu_{\tilde{A}}(x) > 0\}} = [x_1 - \delta |A|, x_2 + \delta |A|]$ (the support of the fuzzy number $\tilde{A}$). Let $\left| \tilde{A} \right| = \int_X \mu_{\tilde{A}}(x) dx$ be the cardinality of a fuzzy set $\tilde{A}$, $\mathrm{EI}[\tilde{A}] = [\mathrm{E}[L], \mathrm{E}[R]]$, where $\mathrm{E}[L] = \int_{-\infty}^{x_1} x dL(x)$, $\mathrm{E}[R] = -\int_{x_2}^{+\infty} x dR(x)$ (expected interval of the fuzzy number $\tilde{A}$).

If an expert (decision maker, DM) is a source of information, then different blurring strategies, depending on the information about the degree of caution of the DM, are possible:

1) if the DM estimates are too careful (conservative), then supp $\tilde{A} = A$ (internal blur);
2) if the DM estimates are excessively accurate (radical), then ker $\tilde{A} = A$ (external blur);
3) if the DM estimates are neutral, then $\mathrm{EI}[\tilde{A}] = \bar{A}$, where EI is the expected interval of a fuzzy number (neutral blur).

The meaning of these conditions is as follows. A cautious expert's assessments are often too imprecise. Therefore, they should be made more accurate (supp $\tilde{A} = A$) when blurring. On the other hand, the assessments of an careless expert are often overly precise. Therefore, they must be expanded (ker $\tilde{A} = A$) when blurring.

Lemma 1. *The following properties are valid:*

a) *ker $\tilde{A} = [a_1 + \delta |A|, a_2 - \delta |A|]$, $\delta = \delta(\eta) \in (0, 0.5]$ for internal δ-blur (i.e. supp $\tilde{A} = A$);*
b) *supp $\tilde{A} = [a_1 - \delta |A|, a_2 + \delta |A|]$, $\delta = \delta(\eta) > 0$ for external δ-blur (i.e. ker $\tilde{A} = A$);*
c) *ker $\tilde{A} = [a_1 + \delta |A| \theta_0, a_2 - \delta |A| \theta_0]$, supp $\tilde{A} = [a_1 - \delta |A| (1 - \theta_0), a_2 + \delta |A| (1 - \theta_0)]$ for neutral δ-blur (i.e. $\mathrm{EI}[\tilde{A}] = \bar{A}$), where $\theta_0 = \int_0^1 \theta(s) ds$ and $0 < \delta(\eta) \leq \frac{1}{2\theta_0}$.*

Corollary 1. *If $d\left(\tilde{A}, A\right) = \int_X \left| \mu_A(x) - \mu_{\tilde{A}}(x) \right| dx$, then:*

a) *$d\left(\tilde{A}, A\right) = 2\delta |A| (1 - \theta_0)$ for internal δ-blur of the interval A;*
b) *$d\left(\tilde{A}, A\right) = 2\delta |A| \theta_0$ for external δ-blur of the interval A;*
c) *$d\left(\tilde{A}, A\right) = 4\delta |A| \theta_1$ for neutral δ-blur of the interval A, where $\theta_1 = \int_{\theta_0}^1 \theta(s) ds$.*

Note that $\theta_0 = \frac{1}{2}$, $\theta_1 = \frac{1}{8}$, if $\theta(t) = 1 - t$. The fuzzy number $\tilde{A}$ will be trapezoidal in this case.

Definition 4. *The arrangement of intervals of two sets A_1 and A_2 is called stable if there is such $\delta_0 > 0$ that the nature of the inclusion or intersection of the pairs supp A-supp B; ker A-ker B is preserved for δ-blur $\forall \delta < \delta_0$ and $\forall A \in A_1$, $B \in A_2$. Let's call this δ-blur small.*

4   Conflict Variation When Transforming Evidence Bodies

Let $^\eta F = (\mathcal{A}^{(\eta)}, m^{(\eta)})$, $\eta \in [0, 1]$ be the body of evidence obtained as a result of blurring the focal elements and discounting the masses of evidence $F = (\mathcal{A}, m)$. We have that $^1 F = F$.

The conflict measure Con^Γ will be robust to small transformations in the evidence bodies due to the continuous dependence of the masses on the discounting parameter η and the Jaccard index $s(A, B)$ on the blurring parameter δ.

Let us first consider the variation of the conflict measure Con^Γ when the transformation consists only in discounting the masses. The following statements are true.

Proposition 1. *The condition $Con^\Gamma(^\eta F_1, F_2) \leq Con^\Gamma(F_1, F_2)$, $\eta \in [0, 1]$ is satisfied for the body of evidence F_1 and $\forall F_2$ if and only if the inequality*

$$\sum_{A \in \mathcal{A}_1} s(A, B) m_1(A) \leq s(X, B) \tag{3}$$

is true $\forall B \in \mathcal{A}_2$.

Corollary 2. *The condition $Con^\Gamma(^\eta F_1, F_2) \leq Con^\Gamma(F_1, F_2)$, $\eta \in [0, 1]$ is satisfied $\forall F_1, F_2 \Leftrightarrow s(A_B, B) \leq s(X, B) \; \forall B \in \mathcal{A}_2$, where $A_B = \arg\max_{A \in \mathcal{A}_1} s(A, B)$.*

Corollary 3. *If $s(A, B) \leq s(X, B) \; \forall A, B$ or $s(X, B) = 1 \; \forall B$, then the inequality $Con^\Gamma(^\eta F_1, F_2) \leq Con^\Gamma(F_1, F_2)$, $\forall \eta \in [0, 1]$ is true for arbitrary bodies of evidence F_1 and F_2.*

Example 1. Let $s_0(A, B) = \begin{cases} 1, A \cap B \neq \emptyset, \\ 0, A \cap B = \emptyset. \end{cases}$ Then condition (3) will be true, since $s_0(X, B) = 1 \; \forall B$. In this case, we have $Con^\Gamma(F_1, F_2) = Con_0(F_1, F_2)$ and we will obtain: $Con_0(^\eta F_1, F_2) \leq Con_0(F_1, F_2) \; \forall \eta \in [0, 1]$.

Example 2. Let $s(A, B) = \frac{|A \cap B|}{|X|}$. Then the condition $s(A, B) \leq s(X, B) \; \forall A, B$ is satisfied and also the inequality $Con^\Gamma(^\eta F_1, F_2) \leq Con^\Gamma(F_1, F_2)$, $\forall \eta \in [0, 1]$ is true.

The continuous dependence of the conflict measure $Con^\Gamma(^\eta F_1, F_2)$ on the discount coefficient η implies that the measure Con^Γ will be strictly monotonic for small (close to 1) discounting of the masses.

If focal elements are blurred only, then we have the following proposition.

Proposition 2. *We have in the case of small internal (external) δ-blurring of focal elements: $0 \leq Con^\Gamma(^\eta F_1, F_2) - Con^\Gamma(F_1, F_2) \leq \frac{2\delta(\eta)\theta_0}{1 - 2\delta(\eta)\theta_0}$ ($\left| Con^\Gamma(F_1, F_2) - Con^\Gamma(^\eta F_1, F_2) \right| \leq 2\delta(\eta)\theta_0 \left(1 - Con^\Gamma(F_1, F_2)\right)$).*

5   Conflict Variation When Transforming Categorical Bodies of Evidence

We consider more detailed the conflict variation when transforming categorical bodies of evidence. Let two categorical bodies of evidence F_A and F_B be given. Then $Con^\Gamma(F_A, F_B) = 1 - s(A, B)$. We'll consider blurring and discounting the body of evidence F_A. As a result, we get a simple evidence ${}^\eta F_A = \eta F_{A^{(\eta)}} + (1 - \eta)F_X$, where $A^{(\eta)}$ is some blur of the focal element A. Then

$$Con^\Gamma({}^\eta F_A, F_B) = 1 - s(A^{(\eta)}, B)\eta - s(X, B)(1 - \eta).$$

We will evaluate the change in conflict in the case of discounting and blurring of categorical evidence for substantially 'close' focal elements.

Definition 5. *The focal elements A and B are said to be substantially close to each other with respect to the index s, if $s(A, B) \geq \max\{s(A, X), s(X, B)\}$.*

The substantially closeness of the focal elements A and B suggests that they are not only close to each other (the value $s(A, B)$ is large), but also strongly differ from the entire set X. Below, we will consider relations of substantial closeness only with respect to the Jaccard index.

Suppose now that only the focal element A is blurred in the categorical evidence F_A. In this case, we have the following propositions.

Proposition 3. *We have:*

1) *$Con^\Gamma({}^\eta F_A, F_B) \geq Con^\Gamma(F_A, F_B)$ in the case of small internal blurring of the element A;*
2) *in the case of small external blurring of the element A: $Con^\Gamma({}^\eta F_A, F_B) \geq Con^\Gamma(F_A, F_B)$ if $B \subseteq A$ and $Con^\Gamma({}^\eta F_A, F_B) \leq Con^\Gamma(F_A, F_B)$ in all other cases.*

Proposition 4. *If the focal element A has the same relative position with the elements B and C (with respect to mutual inclusion or intersection), then the conflict measure Con^Γ will be monotonic for small blurring of any nature (internal, external, or neutral) for the triple categorical bodies of evidence F_A, F_B and F_C, i.e. $Con^\Gamma({}^\eta F_A, F_B) \leq Con^\Gamma({}^\eta F_A, F_C)$, if $Con^\Gamma(F_A, F_B) \leq Con^\Gamma(F_A, F_C)$ for any admissible values η.*

Finally, we present some result on the change in the conflict when discounting and blurring categorical bodies of evidence only for the case of a stable location of substantially close focal elements A and B.

Proposition 5. *Let the focal elements A and B be stably located, substantially close and $B \subseteq A$ or $A \subseteq B$. Then the inequality $Con^\Gamma({}^\eta F_A, F_B) \geq Con^\Gamma(F_A, F_B)$ is true for small internal blurs $\Leftrightarrow \delta(\eta) \leq \frac{1-\eta}{2\theta_0} \cdot \frac{s(A,B)-s(X,B)}{s(A,B)-(1-\eta)s(X,B)}$ & $B \subseteq A$ or $\delta(\eta) \leq \frac{1-\eta}{2\theta_0\eta} \cdot \frac{s(A,B)-s(X,B)}{s(A,B)}$ & $A \subseteq B$.*

6 Numerical Example

Let's consider an example of selection of analysts' forecasts on the cost of Brent crude oil in 2021 for aggregation [3]. Forecast prices provided by 7 major investment banks: BNP Paribas, Citigroup, RBC, JPMorgan, Bank of America, Deutsche Bank, Standard Chartered. Each forecast is an interval $A_i = [a_i, b_i)$, where a_i, b_i are the forecasts of the i-th investment bank for Brent crude oil prices in the 4th quarter of 2020 and in the 4th quarter of 2021, respectively (see Table 1). Thus, each prediction is categorical evidence F_{A_i}, $i = 1, ..., 7$.

We will define the reliability η_i of the categorical body of evidence F_{A_i} as inversely proportional to the deviation of the middle of the predicted interval (the 'mean' value of the categorical evidence F_{A_i}) $E(F_{A_i}) = \frac{a_i + b_i}{2}$ from the current value of the Brent oil price c_0. In this case, the formula $\eta_i = \frac{10 + \max d_k - d_i}{15 + \max d_k - \min d_k}$ was used, where $d_i = |E(F_{A_i}) - c_0|$. The value c_0 was taken equal to the price of Brent oil as of the date 1.03.2021: $c_0 = 65$. The values of the lengths $l_i = l(A_i) = b_i - a_i$ of the intervals-focal elements characterize the degree of uncertainty in the forecasts. The interval $X = [20, 80]$ was considered as the base set (upper and lower prices for Brent crude oil for the last three years).

Table 1. Boundaries of focal elements, reliability and uncertainty of evidence bodies.

	Investment banks	a_i	b_i	η_i	l_i
A_1	BNP Paribas	45	59	0.8	14
A_2	Citigroup	44	56	0.71	12
A_3	RBC	41	55	0.63	14
A_4	JPMorgan	39	52	0.53	13
A_5	Bank of America	47	51	0.67	4
A_6	Deutsche Bank	45	50	0.61	5
A_7	Standard Chartered	35	50	0.41	15

Table 2. The values of the conflict measure Con^Γ with discounting and with mixed blur.

	A_1	A_2	A_3	A_4	A_5	A_6	A_7
A_1	0.26	**0.5**	0.64	0.75	0.63	0.69	0.76
A_2	0.5	0.35	**0.55**	0.7	**0.56**	0.61	0.73
A_3	0.64	0.55	0.4	0.59	0.59	**0.51**	0.65
A_4	0.75	0.7	0.59	0.44	0.7	0.58	0.57
A_5	0.63	0.56	0.59	0.7	0.4	0.58	0.76
A_6	0.69	0.61	0.51	0.58	0.58	0.42	0.66
A_7	0.76	0.73	0.65	0.57	0.76	0.66	0.43

It seems advisable to use internal blur for 'large' (in length l_i) focal elements and external blur for 'small' focal elements. It can be seen from Table 1 that the focal elements A_5 and A_6 can be considered small and we will use external blur for them. The rest of the focal elements can be considered large and we will use internal blur for them. The values of the conflict measure with discounting and with the described mixed blurring are presented in Table 2. The blur function $\delta(\eta) = 1 - \eta$ was used. In this case, the following pairs of evidence are prioritized for combining according to the principle of minimum conflict: $(F_1, F_2) \succ (F_3, F_6) \succ (F_2, F_3) \sim (F_2, F_5)$.

Let us now consider the aggregation of the highest priority pairs of bodies of evidence (F_1, F_2) and (F_3, F_6) with mixed blur. We will aggregate simple bodies of evidence $^{\eta_i} F_{A_i^{(\eta_i)}}$ and $^{\eta_j} F_{A_j^{(\eta_j)}}$ using the formula (2): $^{\eta_i} F_{A_i^{(\eta_i)}} \otimes {}^{\eta_j} F_{A_j^{(\eta_j)}} = (\mathcal{A}_{i,j}, m_{i,j}) =: F_{i,j}$.

We will evaluate the quality of the combination by finding changes in the characteristics of the aggregated evidence compared to the same characteristics of the aggregated evidence bodies. We will consider such characteristics of evidence as the degree of imprecision, reliability and conflict.

The degree of imprecision of the body of evidence $F = (\mathcal{A}, m)$ will be estimated using the functional $H(F) = \sum_{A \in \mathcal{A}} m(A)|A|$. Let $H_i = H\left(^{\eta_i} F_{A_i^{(\eta_i)}}\right)$, $H_{i,j} = H(F_{i,j})$.

The reliability of the result of combining the bodies of evidence $^{\eta_i} F_{A_i^{(\eta_i)}}$ and $^{\eta_j} F_{A_j^{(\eta_j)}}$ (we denote it by $\eta_{i,j}$) will be calculated using the above formula, as a normalized estimate of the distance from the 'average' value of the prediction $E(F_{i,j})$ to the current value of c_0. The 'average' value $E(F)$ of the body of evidence $F = (\mathcal{A}, m)$ is calculated by the formula $E(F) = \sum_{A \in \mathcal{A}} m(A)E(F_A)$ (if A is a fuzzy number, then $E(F_A)$ is equal to the center of gravity of this number: $E(F_A) = \int_X x\mu_A(x)dx / \int_X \mu_A(x)dx$).

In addition, we will find the value of the conflict measure Con^Γ between the result of combining $F_{i,j}$ and each of the bodies of evidence $^{\eta_i} F_{A_i^{(\eta_i)}}$ and $^{\eta_j} F_{A_j^{(\eta_j)}}$. The corresponding values of the conflict measure will be denoted by Con_i^Γ and Con_j^Γ. Table 3 shows the changes in the characteristics of imprecision, reliability and conflict after aggregation of simple bodies of evidence $^{\eta_i} F_{A_i^{(\eta_i)}}$ and $^{\eta_j} F_{A_j^{(\eta_j)}}$ for pairs of focal elements A_1, A_2 and A_3, A_6.

Table 3. Changing characteristics after aggregation.

	H_i	H_j	$H_{i,j}$	η_i	η_j	$\eta_{i,j}$	Con^Γ	Con_i^Γ	Con_j^Γ
A_1, A_2	21.11	23.27	13.77	0.8	0.71	0.73	0.5	0.4	0.34
A_3, A_6	27.64	27.51	22.01	0.63	0.61	0.64	0.51	0.65	0.4

It can be seen from this table that the degree of imprecision decreases after combining, reliability increases when combining a pair A_3, A_6 and changes itself in different ways when combining a pair A_1, A_2. The conflict between the result of the combination and the original evidence is reduced. Thus, we get more accurate, equally reliable and less conflicting evidence after combining.

7   Conclusion

The measure of the conflict between the bodies of evidence defined on the real line is considered in this article. The change of this measure in cases of 'blurring' of focal elements and discounting of masses is investigated. These procedures are performed in order to improve the robust properties of the conflict measure, to take into account the caution or optimism of experts as sources of information, and also to take into account the reliability of these sources. Blurring of focal elements is modeled using fuzzy numbers. The properties of the robustness and the monotonicity of the conflict measure, the directionality of change of the conflict measure are considered with regards of small transformation. These properties are being studied for different types of focal element 'blurring', which correspond to varying degrees of expert caution. A numerical example of calculating the values of the conflict measure, taking into account the blurring and discounting, when choosing for the subsequent aggregation of expert forecasts regarding the oil price is considered.

References

1. Bronevich, A., Lepskiy, A., Penikas, H.: The application of conflict measure to estimating incoherence of analyst's forecasts about the cost of shares of Russian companies. Procedia Comput. Sci. **55**, 1113–1122 (2015)
2. Dempster, A.P.: Upper and lower probabilities induced by multivalued mapping. Ann. Math. Stat. **38**, 325–339 (1967)
3. Hodari, D.: Oil prices seen remaining subdued into 2021. The Wall Street Journal, 28 November 2020. https://www.wsj.com/articles/oil-prices-seen-remaining-subdued-into-2021-11606490833. Accessed 1 Mar 2021
4. Klir, G.J., Yuan, B.: Fuzzy Sets and Fuzzy Logic: Theory and Applications. Prentice Hall PTR, Hoboken (1995)
5. Kruse, R., Schwecke, E.: Specialization: a new concept for uncertainty handling with belief functions. Int. J. Gen. Syst. **18**(1), 49–60 (1990)
6. Lepskiy, A.: About relation between the measure of conflict and decreasing of ignorance in theory of evidence. In: Proceedings of the 8th conference of the European Society for Fuzzy Logic and Technology, pp. 355–362. Atlantis Press, Amsterdam (2013)
7. Lepskiy, A.: General schemes of combining rules and the quality characteristics of combining. In: Cuzzolin, F. (ed.) BELIEF 2014. LNCS (LNAI), vol. 8764, pp. 29–38. Springer, Cham (2014). https://doi.org/10.1007/978-3-319-11191-9_4
8. Lepskiy, A.: The qualitative characteristics of combining evidence with discounting. In: Ferraro, M.B., et al. (eds.) Soft Methods for Data Science. AISC, vol. 456, pp. 311–318. Springer, Cham (2017). https://doi.org/10.1007/978-3-319-42972-4_39
9. Lepskiy, A.: On the conflict measures agreed with the combining rules. In: Destercke, S., Denoeux, T., Cuzzolin, F., Martin, A. (eds.) BELIEF 2018. LNCS (LNAI), vol. 11069, pp. 172–180. Springer, Cham (2018). https://doi.org/10.1007/978-3-319-99383-6_22
10. Shafer, G.: A Mathematical Theory of Evidence. Princeton University Press, Princeton (1976)
11. Smets, P.: Belief functions on real numbers. Int. J. Approximate Reasoning **40**(3), 181–223 (2005)
12. Straszecka, E.: An interpretation of focal elements as fuzzy sets. Int. J. Intell. Syst. **18**, 821–835 (2003)
13. Yen, J.: Generalizing the Dempster-Shafer theory to fuzzy sets. In: Yager, R.R., Liu, L. (eds.) Classic Works of the Dempster-Shafer Theory of Belief Functions. Studies in Fuzziness and Soft Computing, vol. 219, pp. 529–554. Springer, Heidelberg (2008). https://doi.org/10.1007/978-3-540-44792-4_21
14. Zhang, L.: Representation, independence and combination of evidence in the Dempster-Shafer theory. In: Yager, R.R., et al. (eds.) Advances in the Dempster-Shafer Theory of Evidence, pp. 51–69. Wiley, New York (1994)

Logical and Evidential Inconsistencies Meet: First Steps

Nadia Ben Abdallah[1], Sébastien Destercke[2(⊠)], Anne-Laure Jousselme[3], and Frédéric Pichon[4]

[1] Clamart, France
[2] UMR CNRS 7253 Heudiasyc, Univ. Technologie de Compiegne, Compiègne, France
sebastien.destercke@hds.utc.fr
[3] NATO STO Centre for Maritime Research and Experimentation, La Spezia, Italy
anne-laure.jousselme@cmre.nato.int
[4] Univ. Artois, UR 3926 LGI2A, 62400 Béthune, France
frederic.pichon@univ-artois.fr

Abstract. Measuring inconsistency has been and is still an active research topic in both logic and evidence theory. However, the two fields have developed distinct notions and measures of inconsistency, following different paths. In this paper, we attempt to build some first bridges between the two trends, suggesting some first means for one to enrich the other, and vice-versa.

Keywords: Inconsistency · Logic · Belief functions

1 Introduction

Evidence theory (a.k.a. Dempster-Shafer theory, Belief function theory) and logic share many common concerns, and one of them is how to deal with inconsistency of information coming from multiple sources. For example, the notion of maximal coherent subsets or its dual, minimal unsatisfiable subset, appear in both settings to deal with inconsistencies [4,18].

One particular problem that has attracted a lot of attention in the two settings is how to measure inconsistency [3,5,6,14,18]. However, as the two fields commonly use different basic models and assumptions (e.g., in the way the set of possible worlds is generated), they have provided different answers to this issue.

Our goal in this paper is not to introduce new ways to measure conflict or inconsistency in belief function theory, as there is already an ample literature on the topic (the reader can check [3,9,14], for instance). Our agenda is rather to explore what logic and belief functions theory can bring to each other when it comes to measure inconsistency. Similarly, while bridges between evidence theory and some logic frameworks such as penalty logic were studied before [16],

© Springer Nature Switzerland AG 2021
T. Denœux et al. (Eds.): BELIEF 2021, LNAI 12915, pp. 207–214, 2021.
https://doi.org/10.1007/978-3-030-88601-1_21

the interconnections of the two settings when it comes to inconsistency is barely mentioned, let alone investigated.

In this paper, we mainly expose why we think tools issued from logic could be interesting for evidential reasoning and inconsistency quantification, and propose a simple way to use them within evidential reasoning. In particular, we think that measures of inconsistency issued from logic can help in identifying the main sources of observed inconsistency, a topic already explored within evidence theory [15], but never by using a logical perspective.

We start by detailing an example motivating the topic considered on this paper in Sect. 2, in which we also provide some notations used in the paper. Section 3 then makes a first simple, yet original proposal[1] to embed inconsistency measures issued from logic within evidential reasoning.

2   A Motivational Example

In this section, we first introduce some notations, before detailing an example motivating the interest of using inconsistency measures within the framework of evidence theory.

2.1   Needed Notations

We consider a finite propositional language $\mathcal{L}$. We denote by Ω the space of all interpretations of $\mathcal{L}$, and by ω an element of Ω. Given a formula ϕ, ω is a model of ϕ if it satisfies it, denoted $\omega \models \phi$. We denote the models of a formula ϕ by E_ϕ, that corresponds to usual subsets of Ω, the set of all interpretations. For convenience and to recall that we are assuming that sets are models of propositional logic formulas, we will also sometimes denote by $\bot$ and $\top$ the empty set $\emptyset$ and Ω, respectively. Since there is no ambiguity in propositional logic, we will also confuse ϕ with its sets of models for convenience. A knowledge base $KB = \{\phi_1, \ldots, \phi_n\}$ is usually formed of a conjunction $\phi_1 \wedge \ldots \wedge \phi_n$ of formulas. We will denote by $\mathbb{K}$ the set of possible knowledge basis.

We consider that uncertain information is modelled by mass functions, i.e., a non-negative and normalised mapping $m : \Omega \to [0, 1]$ with $\sum_{E \subseteq \Omega} m(E)$. Subsets with a strictly positive mass are called *focal elements*. In our case Ω will be the set of interpretations, and focal elements E_ϕ will correspond to sets of models of formulas ϕ. Usually, the inconsistency of a mass function m is measured by the quantity $m(\emptyset)$. While there are good reasons (by which we mean properties and axioms) to consider it as a reasonable inconsistency measure [3], several authors have discussed alternatives [3,9,14]. In this paper, we will not question nor challenge its validity, but will rather increase its expressiveness.

Also, while in propositional logic there is in general no harm in confusing a knowledge base and/or a formula ϕ with its set of interpretations E_ϕ, as two formulas giving rise to the same set are semantically equivalent, in our case it

---

[1] to our knowledge.

will be important, at least in the case of inconsistent knowledge bases and mass functions, to consider not only the sets E_ϕ but also the knowledge bases from which they stem, as we will precisely argue that it can be sensible to distinguish $E_{\phi_1} = \emptyset$ and $E_{\phi_2} = \emptyset$, despite the fact that ϕ_1, ϕ_2 induce the same set of interpretations.

Our discussion is mostly independent of the origin of the mass function, even if such origins may be the reason to use refined inconsistency measures (see Sect. 3.3). Here, we will consider examples where the inconsistency is generated by the merging rule, and we will focus on the standard conjunctive rule to simplify the exposure. Given two mass functions m_1, m_2, the mass m_{12} on a given set C resulting from the conjunctive rule is

$$m_{12}(C) = \sum_{A,B \subseteq \Omega, A \cap B = C} m_1(A)m_2(B). \tag{1}$$

Let us now proceed to an example that will serve as a basis and motivation for our discussion.

2.2   The Example

In maritime security, estimating if a vessel under surveillance is involved in some illicit activity often requires combining different opinions from experts with different expertise, together with artificial intelligence tools. Let us consider the propositional language a, b where a denotes the proposition "fishing vessel" and b the proposition "involved in an illicit activity", so that $\Omega = \{(a,b), (a, \neg b), (\neg a, b), (\neg a, \neg b)\}$. One expert is able to answer about the conjunction of these events with a level of confidence $\alpha \in [0; 1]$, while a classifier trained for recognizing fishing vessels expresses information only about a or $\neg a$ with similar levels of confidence. For instance, the focal elements $\phi_1 = \{a \wedge b\}$, i.e., "fishing vessel involved in an illicit activity" and $\phi_2 = \{\neg a\}$, i.e., "not a fishing vessel" generate $E_{\phi_1} = \{(a,b)\}$ and $E_{\phi_2} = \{(\neg a, b), (\neg a, \neg b)\}$ and the following mass functions:

$$m_1(\phi_1 = \{a \wedge b\}) = \alpha_1 \quad m_1(\top) = 1 - \alpha_1$$

$$m_2(\phi_2 = \{\neg a\}) = \alpha_2 \quad m_2(\top) = 1 - \alpha_2$$

We consider now another observation of that vessel under investigation provided by another expert who thinks that the vessel is indeed a fishing vessel, but that it is not involved in illicit activity, thus partially disagreeing with the first expert, leading to a third mass function m_3:

$$m_3(\phi_3 = \{\neg b \wedge a\}) = \alpha_3 \quad m_3(\top) = 1 - \alpha_3$$

The conjunctive combination of m_3 with m_1 and m_2, if we restrict ourselves to those intersections leading to the empty set, is

$$m_{1...3}(\phi_1 \wedge \phi_2 \wedge \phi_3 = \bot) = \alpha_1 \alpha_2 \alpha_3$$

$$m_{1\ldots3}(\phi_1 \wedge \phi_2 = \bot) = \alpha_1\alpha_2(1 - \alpha_3)$$

$$m_{1\ldots3}(\phi_1 \wedge \phi_3 = \bot) = \alpha_1(1 - \alpha_2)\alpha_3$$

$$m_{1\ldots3}(\phi_2 \wedge \phi_3 = \bot) = (1 - \alpha_1)\alpha_2\alpha_3.$$

The usual measure of inconsistency $m(\emptyset)$ for belief functions is equal to $\alpha_1\alpha_2 + \alpha_1(1 - \alpha_2)\alpha_3 + (1 - \alpha_1)\alpha_2\alpha_3$ as it sums up the masses allocated to the inconsistent formulas. However, it could well be argued (and has been in standard logic setting) that the consistency of knowledge base $\phi_1 \wedge \phi_2 \wedge \phi_3$ is usually perceived as different from the one of $\phi_1 \wedge \phi_2$, $\phi_1 \wedge \phi_3$, or $\phi_2 \wedge \phi_3$. It would therefore be necessary to be able to make such a distinction in the measurement of the inconsistency of m.

In the next section, we make a first proposal as how this could be done by combining inconsistency measures issued from the logic framework to masses of evidence.

3   A First Step Towards a Combination

As we said before, a knowledge base $KB = \{\phi_1, \ldots, \phi_n\}$ is a collection of formulas, which is inconsistent (denoted $KB \vdash \bot$) iff $\{\omega \models \wedge_{i=1}^n \phi_i\} = \emptyset$, i.e., if there is no model/interpretation/world satisfying the set of formulas. An inconsistency measure $\mathcal{I} : \mathbb{K} \rightarrow \mathbb{R}_+^\infty$ for a knowledge base KB associates to KB a positive extended real-value quantifying how much it is inconsistent.

3.1   A Quick Note on Logical Inconsistency Measures

There is a large literature on the properties that $\mathcal{I}$ should satisfy, and we refer to [18] for a rather exhaustive study. In this work, we will only focus on a handful of normalized measures, i.e., $\mathcal{I}(KB) \in [0, 1]$, with the usual requirement that $\mathcal{I}(KB) = 0$ if and only if KB is consistent. While such a normalisation is usually not a pre-requisite in logical settings, it should be satisfied if we want to be able to compare different situations (e.g., different combination rules or subset of sources), and if we want to be consistent with evidence theory. Below we introduce two measures we will consider and discuss in the rest of the paper, as they will allow us to connect logical inconsistency measures with classical evidential ones, as well as illustrate the fact that other classical inconsistency could be useful to better analyse/differentiate various situations.

Definition 1. *Drastic inconsistency measure. The measure $\mathcal{I}_d$ is such that*

$$\mathcal{I}_d(KB) = \begin{cases} 1 & \textit{if } KB \vdash \bot \\ 0 & \textit{else} \end{cases} \tag{2}$$

To introduce the next measure of logical inconsistency, we need to consider probabilities over Ω and the associated probability measure over formulas (or, equivalently, events). If p is a probability mass over the set of interpretations Ω, then the probability associated with a formula ϕ is $P(\phi) = \sum_{\omega \models \phi} p(\omega)$, i.e., the probability of the set of models of this formula. We will denote $\mathcal{P}(\Omega)$ the set of all such probabilities over Ω.

Example 1. Consider the example of Sect. 2.2 with the formula ϕ_2 ("not a fishing vessel"), and the uniform probability over Ω, $p(\omega) = 1/4$. Then

$$P(\phi_2) = p((\neg a, b)) + p((\neg a, \neg b)) = 1/2$$

Definition 2. *Minimal η-inconsistency measure. The measure $\mathcal{I}_\eta$ is such that*

$$\mathcal{I}_\eta(KB) = 1 - \max_{p \in \mathcal{P}(\Omega)} \inf_{\phi \in KB} P(\phi) \qquad (3)$$

$\mathcal{I}_\eta$ corresponds to one minus the probability mass over interpretations that maximises the probability of each formula being true. If KB is consistent, i.e., if all formula have a common model (say ω^*), then $\mathcal{I}_\eta(KB) = 0$ is obtained by $p(\omega^*) = 1$. Otherwise, it is lower than one, and reaches zero if and only if one of the formula ϕ is self-inconsistent (e.g., $\phi = a \wedge \neg a$). It should be noted that the probability in Eq. (3) is a way to measure inconsistency of a single KB, and is not related to the set of probabilities we can associate to the belief function whose inconsistency we are trying to measure. An easy way to see this is that, if we have inconsistency in the belief function, then the corresponding set of probabilities is empty, and therefore replacing $\mathcal{P}(\Omega)$ in Eq. (3) by an empty set would not make much sense.

Example 2. Consider again the example of Sect. 2.2 with the three formulas ϕ_1, ϕ_2, ϕ_3. We do have

$$\mathcal{I}_d(\{\phi_1, \phi_2\}) = \mathcal{I}_d(\{\phi_1, \phi_3\}) = \mathcal{I}_d(\{\phi_2, \phi_3\}) = \mathcal{I}_d(\{\phi_1, \phi_2, \phi_3\}) = 1$$

and

$$\mathcal{I}_\eta(\{\phi_1, \phi_2\}) = \mathcal{I}_\eta(\{\phi_1, \phi_3\}) = \mathcal{I}_\eta(\{\phi_2, \phi_3\}) = 1/2, \quad \mathcal{I}_\eta(\{\phi_1, \phi_2, \phi_3\}) = 2/3$$

That means that according to $\mathcal{I}_d$ there is no difference in inconsistency for instance between "the vessel is not a fishing vessel" (ϕ_2) and "the vessel is a fishing vessel involved in an illicit activity" (ϕ_1) on the one hand and between the same two formulas and "the vessel is a fishing vessel not involved in an illicit activity" (ϕ_3). Using $\mathcal{I}_\eta$ allows discriminating between the two situations.

3.2 A Simple Proposal to Embed Logical Measure in Evidence Framework

Our proposal is quite simple: we consider a mass function m having for "focal elements"[2] a collection of knowledge bases $KB_1, \ldots, KB_n$, among which some

---

[2] Technically speaking, we admit that multiple knowledge bases may lead to the same subset of models, meaning that the same subset may appear multiple times as focal element.

of them are inconsistent. Given a logical inconsistency measure $\mathcal{I}$, and by small abuse of notation, the inconsistency of m according could be computed as

$$\mathcal{I}(m) = \sum_{KB_i} \mathcal{I}(KB_i) \cdot m(KB_i), \tag{4}$$

with $\mathcal{I}(m) = 0$ if and only if $m(\emptyset) = 0$, if $\mathcal{I}$ has the usual requirement of inconsistency measures.

One can also readily see that in the case where $\mathcal{I} = \mathcal{I}_d$, we have

$$\mathcal{I}_d(m) = \sum_{KB_i \vdash \perp} m(KB_i) = m(\emptyset), \tag{5}$$

and we retrieve the usual measure of inconsistency for belief functions. Since $\mathcal{I}_d$ is an upper bound of other normalized inconsistency measures, so will be $\mathcal{I}_d(m)$. Equation (4) is therefore a legitimate and straightforward extension of classical measure of inconsistency for mass functions, that basically consists of a weighted average of masses over the empty set (i.e., KBs having the same empty set of models but are syntactically different). One could however think of different ways to combine the $m(KB_i)$ and $\mathcal{I}(KB_i)$, for example by using a T-norm within the sum, that would not change our observation made for Eq. (5). Replacing the sum seems a bit trickier if we want the classical inconsistency measure $m(\emptyset)$ to still be a particular instance of our framework. Let us now come back to our previous example.

Example 3. If we come back to the example of Sect. 2.2, we can see that using $\mathcal{I}_d(m)$, all focal elements generating some inconsistency will be considered as having the same weight, i.e.,

$$\mathcal{I}_d(m_{1...3}) = \alpha_1 \alpha_2 \alpha_3 + \alpha_1 (1 - \alpha_2) \alpha_3 + \alpha_1 \alpha_2 (1 - \alpha_3) + (1 - \alpha_1) \alpha_2 \alpha_3,$$

in contrast, we will have

$$\mathcal{I}_\eta(m_{1...3}) = \frac{2}{3} \alpha_1 \alpha_2 \alpha_3 + \frac{1}{2} \alpha_1 (1 - \alpha_2) \alpha_3 + \frac{1}{2} \alpha_1 \alpha_2 (1 - \alpha_3) + \frac{1}{2} (1 - \alpha_1) \alpha_2 \alpha_3,$$

indicating that some focal elements leading to inconsistency have more importance than others.

3.3   The Case of Simple Support Mass Functions

In a setting where focal elements are equivalent to logical formulas, a simple support mass function takes the form $m(\phi_i) = \alpha_i$, $m(\top) = 1 - \alpha_i$.

In this case, using logical inconsistency measurements may not only allow one to have a more nuanced analysis of conflict, but also potentially identify which sources are the most responsible for the observed conflict. This would be especially interesting in procedures aiming to restore consistency by, e.g., forgetting some sources.

Indeed, some works exist on identifying the formula bringing the most conflict in a knowledge-base, for instance based on the Shapley value [7] or by considering minimal inconsistent subsets [8]. Since in the case of simple support functions, a formula is equivalent to the information provided by a source (minus its uncertainty), such works could be used to identify which source is to blame for the observed inconsistency. Such values have been used, for instance, to guide the repair of inconsistent knowledge bases [19], i.e., how to remove the inconsistency. One possible use of our proposal would then be to adapt such strategies to the evidential framework, that is to use inconsistency measures to guide discounting strategies or to choose some information item to remove (as a formula ϕ_i in the simple support case is associated with an information source).

4 Conclusion and Discussion

In this paper, we made some first exploration as to how inconsistency handling tools developed in logic and in evidence theory could be combined together, so as to benefits from each others. More precisely, we have pointed out that using inconsistency measurements issued from logic may allow one to have a refined analysis of conflict between mass functions, at least when those are bearing over logical formulas (or their models). To do so, we have made a simple proposal to embed logical inconsistency measures in evidence theory. Note that such mass functions will be typically encountered in applications involving logical settings, such as the semantic web [13] or pattern mining [17] where the extraction of logical association rule is common. Recent works also show that probabilistic answer set programming generates mass functions over possible interpretations [2]

However, and as indicate the title of our paper, these are only the first steps on a possibly long road, as many things remain to be explored, such as studying the meaning of the numerous inconsistency measures for belief functions, or how they can be used in practice (e.g., to repair inconsistent combinations).

In addition to that, we can also point out other ways in which studies about inconsistency in logic and evidence theory could be intertwined:

- A first one is to try to draw connection between logical inconsistency measures and other approaches dealing with conflict in evidence theory. One could think for example of distance-based approaches [14], geometrical approaches [1] or approaches that aim at decomposing the existing conflict into different parts [15];
- A second one would be to develop an axiomatic of inconsistency measures adapted to evidential knowledge bases, for instance in the fashion of [10] where this task was done for prioritized knowledge bases, which are similar to possibilistic knowledge bases. Connections could also be made with the work of Muiño [11,12].

References

1. Burger, T.: Geometric views on conflicting mass functions: from distances to angles. Int. J. Approx. Reason. **70**, 36–50 (2016)

2. Cozman, F.G., Mauá, D.D.: The joy of probabilistic answer set programming: semantics, complexity, expressivity, inference. Int. J. Approx. Reason. **125**, 218–239 (2020)
3. Destercke, S., Burger, T.: Toward an axiomatic definition of conflict between belief functions. Cybern. IEEE Trans. **43**(2), 585–596 (2013)
4. Dubois, D., Prade, H.: A set-theoretic view of belief functions. In: Yager, R.R., Liu, L. (eds.) Classic Works of the Dempster-Shafer Theory of Belief Functions. Studies in Fuzziness and Soft Computing, vol. 219, pp. 375–410. Springer, Berlin, Heidelberg (2008). https://doi.org/10.1007/978-3-540-44792-4_14
5. Grant, J., Hunter, A.: Measuring inconsistency in knowledgebases. J. Intell. Inf. Syst. **27**(2), 159–184 (2006)
6. Hunter, A., Konieczny, S.: Approaches to measuring inconsistent information. In: Bertossi, L., Hunter, A., Schaub, T. (eds.) Inconsistency Tolerance. LNCS, vol. 3300, pp. 191–236. Springer, Heidelberg (2005). https://doi.org/10.1007/978-3-540-30597-2_7
7. Hunter, A., Konieczny, S., et al.: Shapley inconsistency values. KR **6**, 249–259 (2006)
8. Hunter, A., Konieczny, S., et al.: Measuring inconsistency through minimal inconsistent sets. KR **8**(358–366), 42 (2008)
9. Martin, A.: Conflict management in information fusion with belief functions. In: Bossé, É., Rogova, G. (eds.) Information Quality in Information Fusion and Decision Making. Information Fusion and Data Science, pp. 79–97. Springer, Cham (2019). https://doi.org/10.1007/978-3-030-03643-0_4
10. Mu, K., Liu, W., Jin, Z.: Measuring the blame of each formula for inconsistent prioritized knowledge bases. J. Log. Comput. **22**(3), 481–516 (2012)
11. Muiño, D.P.: Measuring and repairing inconsistency in probabilistic knowledge bases. Int. J. Approx. Reason. **52**(6), 828–840 (2011)
12. Muiño, D.P.: Measuring and repairing inconsistency in knowledge bases with graded truth. Fuzzy Sets Syst. **197**, 108–122 (2012)
13. Nagy, M., Motta, E., Vargas-Vera, M.: Multi-agent ontology mapping with uncertainty on the semantic web. In: 2007 IEEE International Conference on Intelligent Computer Communication and Processing, pp. 49–56. IEEE (2007)
14. Pichon, F., Jousselme, A.L., Abdallah, N.B.: Several shades of conflict. Fuzzy Sets Syst. **366**, 63–84 (2019)
15. Roquel, A., Le Hégarat-Mascle, S., Bloch, I., Vincke, B.: Decomposition of conflict as a distribution on hypotheses in the framework on belief functions. Int. J. Approx. Reason. **55**(5), 1129–1146 (2014)
16. Dupin de Saint-Cyr, F.: Penalty logic and its link with dempster-shafer theory. In: Uncertainty in Artificial Intelligence: Proceedings of the Tenth Conference on Uncertainty in Artificial Intelligence, University of Washington, Seattle, 29–31 July 1994, p. 204. Elsevier (2014)
17. Samet, A., Lefèvre, E., Ben Yahia, S.: Evidential data mining: precise support and confidence. J. Intell. Inf. Syst. **47**(1), 135–163 (2016). https://doi.org/10.1007/s10844-016-0396-5
18. Thimm, M.: On the evaluation of inconsistency measures. In: Measuring Inconsistency in Information, vol. 73, chap. 2. College Publications (2018)
19. Yun, B., Vesic, S., Croitoru, M., Bisquert, P.: Inconsistency measures for repair semantics in obda. In: IJCAI-ECAI: International Joint Conference on Artificial Intelligence-European Conference on Artificial Intelligence, pp. 1977–1983 (2018)

A Note About Entropy and Inconsistency in Evidence Theory

Anne-Laure Jousselme[1](✉), Frédéric Pichon[2], Nadia Ben Abdallah[3], and Sébastien Destercke[4]

[1] NATO STO Centre for Maritime Research and Experimentation, La Spezia, Italy
anne-laure.jousselme@cmre.nato.int
[2] Univ. Artois, UR 3926 LGI2A, 62400 Béthune, France
frederic.pichon@univ-artois.fr
[3] Clamart, France
[4] Univ. Technologie de Compiègne, UMR CNRS 7253 Heudiasyc, Compiègne, France
sebastien.destercke@hds.utc.fr

Abstract. Information content is classically measured by entropy measures in probability theory, that can be interpreted as a measure of internal inconsistency of a probability distribution. While extensions of Shannon entropy have been proposed for quantifying information content of a belief function, other trends have been followed which rather focus on the notion of consistency between sets. Relying on previous general entropy measures of probability, we propose in this paper to establish some links between the different measures of internal inconsistency of a belief functions. We propose a general formulation which encompasses inconsistency measures derived from Shannon entropy as well as those derived from the N-consistency family of measures.

Keywords: Information content · Inconsistency · Conflict · Entropy

1 Introduction

In a multi-intelligence context, information generally arises from different systems (or services), each having their own local representation and underlying mathematical formalism. The choice of this formalism is usually driven by the nature of data or information to be handled. For instance, numerical data (when available in large volume) usually summarize in probabilistic models, while human judgments are best handled by logical approaches managing knowledge bases. The underlying mathematical setting constrains not only the internal reasoning of those services, but also their output which includes some meta-information such as the information content or information value. In decision support, conflict or inconsistency measures play an essential role in detecting sources' defect or intentional deception, but also in quantifying information

---

N. Ben Abdallah—Independent researcher.

credibility when no ground truth is available. In probability theory, measuring conflict goes back to Shannon entropy [16] which quantifies an inverse notion of the information contained in a probability distribution, *i.e.* a notion of internal inconsistency (*e.g.*, [10]). Indeed, the state of maximum inconsistency is reached by the uniform distribution while the state of minimal inconsistency is reached whenever an element is assigned a probability of 1. In propositional logic, a belief base (a set of formulas) is inconsistent if it entails the contradiction[1] [6]. In evidence theory, which captures both probabilistic and logical notions, measuring inconsistency of belief functions has thus naturally followed two main trends: on the one hand, some measures extend Shannon entropy (*e.g.*, [5,7–9,21]) and on the other hand inconsistency is measured through the inconsistency between sets (*e.g.*, [3,4,13]).

We propose in this paper to establish some links between the different approaches to inconsistency measurement. In Sect. 2, we introduce basic concepts and notations of belief functions together with two families of measures of entropy for probabilities. In Sect. 3 we survey the different trends followed and propose a general formulation, and highlight the main elementary constructs leading to inconsistency measures. In Sect. 4, after revealing the "hidden" mass of the empty set within Shannon entropy we propose a general formulation which encompasses most of classical existing measures across the different approaches. We conclude in Sect. 5 on perspectives and future work.

2   Background and Notations

2.1   Belief Functions

We consider in this paper the singular interpretation of belief functions as developed by Shafer [15] and Smets and Kennes [17]. Belief functions are thus used to represent and handle subjective uncertainty (or beliefs) of an agent about the actual state of the world. Let us denote by X an uncertain variable defined on frame of discernment $\mathcal{X} = \{x_1, \ldots, x_K\}$ representing the possible values (states) for that variable. A *mass function* is a mapping $m : 2^{\mathcal{X}} \to [0,1]$ satisfying $\sum_{A \subseteq \mathcal{X}} m(A) = 1$. The mass $m(A)$ represents the amount of belief allocated to the fact of knowing only that $x \in A$. We will denote by $\mathcal{M}$ the set of all mass functions on $\mathcal{X}$. Subsets A of $\mathcal{X}$ such that $m(A) > 0$ are called *focal sets* of m, and the set of focal sets of m will be denoted by $\mathcal{F}$. A mass function m is called *categorical* if $m(A) = 1$ for some $A \subseteq \mathcal{X}$, in which case it defines a classical set and will be denoted by m_A in the following. It is called *vacuous* if $m(\mathcal{X}) = 1$ and denoted by $m_{\mathcal{X}}$. It represents total ignorance. The mass function is called *empty* if $m(\emptyset) = 1$ and denoted by $m_{\emptyset}$. It represents total inconsistency in the agent's beliefs about the set of values that are conceivable for x [18]. It is called *Bayesian* if $m(A) \neq 0$ only for $|A| = 1$ and defines a probability distribution. And finally it is called *normalised* if $m(\emptyset) = 0$.

---

[1] Or equivalently, is unsatisfiable or has no model.

We define a *consistency index* between two sets to satisfy minimally:

$$\phi(A, B) = \begin{cases} 0 \text{ if } A \cap B = \emptyset \\ 1 \text{ if } A = B \end{cases} \tag{1}$$

Equivalent representations of a mass function m are the *belief function* Bel and the *plausibility function* Pl which follow the general formulation:

$$f(A) = \sum_{B \subseteq \mathcal{X}} m(B)\phi(A, B) \tag{2}$$

$\text{Pl}(A)$ is obtained with $\phi(A, B) = 1$ if $A \cap B \neq \emptyset$ and 0 otherwise, and is the amount of belief consistent with $x \in A$; $\text{Bel}(A)$ is obtained with $\phi(A, B) = 1$ if $B \subseteq A$ and 0 otherwise, and is the amount of belief implying $x \in A$. The *contour function* $\pi : \mathcal{X} \to [0, 1]$ is such that $\pi(x) = \text{Pl}(\{x\})$, for all $x \in \mathcal{X}$. It is the plausibility function restricted to the singletons of $\mathcal{X}$.

Let m_1 and m_2 be two mass functions representing pieces of evidence about x. Their combination by the *conjunctive rule* [2] is defined by, for all $A \subseteq \mathcal{X}$,

$$m_{1 \cap 2}(A) = \sum_{B \cap C = A} m_1(B)m_2(C). \tag{3}$$

The conflict between m_1 and m_2 can be quantified as $m_{1 \cap 2}(\emptyset)$ [15].

2.2 Generalized Entropy Measures of Probabilities

Rényi Entropy. Given a probability distribution p over $\mathcal{X}$, the Rényi entropy of order α, is defined for a parameter $\alpha \in \mathbb{R}^+ \backslash \{1\}$ as [14]:

$$\delta_R^{(\alpha)}(p) = \frac{1}{1 - \alpha} \log \left(\sum_{x \in \mathcal{X}} p(x)^\alpha \right) \tag{4}$$

For $\alpha = 0$, (4) is Hartley measure $\log(|\mathcal{X}|)$, while Shannon entropy is retrieved whenever $\alpha \to 1$, $\delta_R^{(1)}(p) = -\sum_{x \in \mathcal{X}} p(x) \log p(x)$. For $\alpha = 2$, the collision entropy is defined, $\delta_R^{(2)}(p) = -\sum_{x \in \mathcal{X}} p(x)^2$. Interestingly, $\delta_R^{(\alpha)}(p)$ is a decreasing function of α, and for $\alpha \to +\infty$, we obtain the minimum entropy, $\delta_R^{(\infty)}(p) = -\log \max_{x \in \mathcal{X}} p(x)$.

Power Entropy. Another family of entropy measures which still extends Shannon entropy has been defined by Vajda and Zvárová [20], relying on the decreasing power function $\psi_a :]0; 1] \to \mathbb{R}$, for $a \in \mathbb{R}$:

$$\psi_a(\mu) = \begin{cases} \frac{1}{a-1}\left(1 - \mu^{a-1}\right), \text{ if } a \neq 1 \\ -\log(\mu) \text{ if } a = 1 \end{cases} \tag{5}$$

with $\psi_a(0) = \lim_{\mu \to 0} \psi_a(\mu)$ if $a \neq 1$, $\psi_1(0) = +\infty$ and $0 \cdot \psi_a(0) = 0$. Power entropy measures are thus defined as [20]:

$$\delta_V^{(a)}(p) = \sum_{x \in \mathcal{X}} p(x)\psi_a\left(p(x)\right) \tag{6}$$

Similarly to Rényi entropy, Shannon entropy is obtained for $a = 1$ while $\delta_V^{(0)}(p) = \log\left(|\mathcal{X}|\right) - 1$ is one-to-one related to Hartley entropy. We note that for $a \neq 1$ in Eq. (5), Eq. (6) defines Tsallis entropy [19].

Power functions ψ_a are interesting as they allow to reverse a consistency notion into an inconsistency notion. Indeed, they are decreasing and satisfy $\psi_a(1) = 0$, meaning that if $p(x)$ is a degree of consistency, $\psi_a(p(x))$ is a degree of inconsistency. In the following, we will denote by ϕ indexes or measures of consistency between sets or of a mass function, while the corresponding inconsistency indexes or measures will be denoted by δ. For instance, an inconsistency index corresponding to (1) would satisfy minimally $\delta(A, B) = 1$ if $A \cap B = \emptyset$ and 0 if $A = B$.

3 Inconsistency of Belief Functions

3.1 An Entropy Approach

The first trend followed to quantify the internal inconsistency of a belief function, aims at extending Shannon entropy, focusing on the probabilistic dimension of belief functions. Developed mostly between 1982 and 1992, the measures follow the general formulation:

$$\delta(m) = \sum_{A \subseteq X} m(A)\left(-\log\sum_{B \subseteq X} m(B)\phi(A, B)\right) \tag{7}$$

where $\phi(A, B)$ is a consistency index between the sets A and B satisfying (1). In particular, Yager's dissonance measure is obtained for $\phi(A, B) = 1$ if $A \cap B \neq \emptyset$. Other definitions for ϕ still satisfying (1) lead to the measures of confusion from Höhle [5], from Nguyen [12], of discord from Klir & Ramer [9] and of strife from Klir & Parviz [8]. All these measures degenerate to Shannon entropy when m is a Bayesian mass function and to Hartley measure when m is categorical.

3.2 A Consistency Approach

Another trend followed has given up on the extension of Shannon and Hartley measures, and their additivity property. Yager first defined a measure of consistency [22] while George and Pal defined the measure of total conflict [4] based on Jaccard index. These two measures correspond to the general formulation:

$$\delta(m) = \sum_{A \subseteq X} m(A)\left(1 - \sum_{B \subseteq X} m(B)\phi(A, B)\right) \tag{8}$$

3.3   A N-consistency Approach

Recent work [13] proposes the measure of consistency of m:

$$\phi_N(m) = 1 - m^{(N)}(\emptyset), \tag{9}$$

where $m^{(N)}$ denotes the mass function resulting from the combination of m by itself N times, i.e. $m^{(N)} = m^{(N-1)} \bigcirc m$ with $m^{(0)} := m_\mathcal{X}$. Hence, we have $m^{(1)} = m$, $m^{(2)} = m \bigcirc m$ and more generally $m^{(N)} = \bigcirc_1^N m$. $\phi_N(m)$ measures different "shades" of internal consistency of m as N varies and in particular $\phi_N(m)$ encompasses two forms of consistency already defined in the literature [13]:

$$\phi_1(m) = 1 - m(\emptyset) = \max_{A \subseteq \mathcal{X}} \mathrm{Pl}(A) \tag{10}$$

$$\phi_2(m) = 1 - m^{(2)}(\emptyset) = \sum_{A \subseteq \mathcal{X}} m(A)\mathrm{Pl}(A) \tag{11}$$

where ϕ_1 is the measure of so-called probabilistic consistency defined in [3] and ϕ_2 is the measure of consistency defined in [22]. It has been proved as well that $\phi_{|\mathcal{F}|}$ is an alternative measure of logical consistency to the one proposed in [3] as $\phi_\pi = \max_{x \in \mathcal{X}} \pi(x)$. More details can be found in [13].

All measures introduced in this section are built upon a measure of consistency between sets, ϕ and other elementary constructs such as a reverse function transforming the notion of consistency into inconsistency. Entropy-like measures (Sect. 3.1) as well as consistency-like measures (Sect. 3.2) are all based on pairwise measures of consistency between sets. Instead, the N-consistency derived measures (Sect. 3.3) are based on N-wise measures. In the following Sect. 4, we will thus exploit that extension in order to establish a more general formulation covering the three types of approaches.

4   Extending Inconsistency

Let us introduce the consistency index between N sets as [13]:

$$\phi_N(A_1, \ldots, A_N) = \begin{cases} 1 \text{ if } \bigcap_{i=1,\ldots,N} A_i \neq \emptyset \\ 0 \text{ else} \end{cases} \tag{12}$$

and we explore below the extension from pair-wise index to N-wise index in measuring the inconsistency of m.

4.1   Observation with Probabilities

Let us start by clarifying why Shannon entropy actually quantifies a notion of internal conflict (or inconsistency). Introduced by Shannon as a measure of information [16], the entropy of a probability distribution is the expected information where $I_X(x) = -\log(p(x))$ is the *self-information* associated with the outcome

$x \in \mathcal{X}$. If $p(x) = 0$, then $I_X(x) = -\infty$ and if $p(x) = 1$ then $I_X(x) = 0$. With $0 \cdot \log 0 = 0$, $\delta_{\mathrm{Sh}}(p) = 0$ if and only if the distribution is focused on a single element of $\mathcal{X}$ (*i.e.*, it exists one x such that $p(x) = 1$), and $\delta_{\mathrm{Sh}}(p) = \log(|\mathcal{X}|)$ if and only if p is uniformly distributed over $\mathcal{X}$ (*i.e.*, $p(x) = \frac{1}{|\mathcal{X}|}$ $\forall x \in \mathcal{X}$). Hence, p is the most informative when its entropy is null and it is the least informative when its entropy is maximum. As such, as noticed for instance in [1,11], Shannon entropy is rather a measure of uncertainty, and even a measure of internal *conflict* (or inconsistency) for p. Indeed, when p is uniformally distributed over $\mathcal{X}$ the internal inconsistency of p is maximum since the same confidence is assigned to *inconsistent* hypotheses x_i of $\mathcal{X}$, *i.e.* such that $\delta(x_i, x_j) = 0$ for all $i \neq j$ and $\delta(x_i, x_i) = 1$ for all i, where δ is an inconsistency index satisfying the properties mentioned in Sect. 2. We can thus re-write Shannon entropy making apparent the consistency index:

$$\delta_R^{(1)}(p) = \sum_{x \in \mathcal{X}} p(x) \left(-\log \sum_{y \in \mathcal{X}} p(y)\phi(x,y) \right) \tag{13}$$

Computing the conjunctive self-combination (using (3) for Bayesian mass functions) of p makes it more obvious:

$$(p \varobigcirc p)(\emptyset) = p^{(2)}(\emptyset) = 1 - \sum_{x \in \mathcal{X}} p(x) \sum_{y \in \mathcal{X}} p(y)\phi(x,y) = 1 - \sum_{x \in \mathcal{X}} p(x)^2 \tag{14}$$

which is clearly a measure of the internal inconsistency of p, as $p^{(2)}(\emptyset) = 0$ iff $\exists x \in \mathcal{X}$ such that $p(x) = 1$ and it is maximum for the uniform distribution. Actually, (14) is Gini impurity that can also be written as $\sum_{x \in \mathcal{X}} p(x) \cdot (1 - p(x))$. If we denote by $p^{(N)}$ the conjunctive combination of p with itself N times we obtain before any normalisation:

$$p^{(N)}(\emptyset) = 1 - \sum_{x \in \mathcal{X}} p(x)^N \tag{15}$$

which is also an inconsistency measure such that $p^{(N)}(\emptyset) \geq p^{(M)}(\emptyset)$ if $N > M$. For integer values of α (that we denote by N), Rényi entropy in Eq. (4) can thus be written as:

$$\delta_R^{(N)}(p) = -\log \left(1 - p^{(N)}(\emptyset) \right)^{\frac{1}{N-1}} \tag{16}$$

where $\left(1 - p^{(N)}(\emptyset) \right)^{\frac{1}{N-1}}$ is a measure of consistency for p, and $N \in \mathbb{N}^*$.

While for Bayesian mass functions, the N-wise comparison of focal sets reduces to the pair-wise comparison, it is not true in the general case that we will detail in the next section.

4.2 Extension to Belief Functions

Let us now introduce the function $\phi_m^{(N)}(A)$ which measures the consistency of m relatively to a specific set A of $\mathcal{X}$, so that for $N > 1$:

$$\phi_m^{(N)}(A) = \sum_{B_1 \subseteq \mathcal{X}} m(B_1) \ldots \sum_{B_{N-1} \subseteq \mathcal{X}} m(B_{N-1})\phi_N(A, B_1, \ldots, B_{N-1}) \tag{17}$$

and $\phi_m^{(1)}(A) = \phi_1(A)$ as defined in Eq. (12). Note that for $N = 2$ we get $\phi_m^{(2)}(A) = \mathrm{Pl}(A)$. Then, we define the total consistency of m as:

$$\phi^{(N)}(m) = \sum_{A \subseteq \mathcal{X}} m(A)\phi_m^{(N)}(A) \tag{18}$$

which is the expectation of the local inconsistency of m. or $N = 1$, we get the probabilistic consistency [3] (Eq. (10)), while for $N = 2$ we get Yager's consistency measure [22] (Eq. (11)). We thus propose the following general formulation:

$$\delta_a^{(N)}(m) = \sum_{A \subseteq \mathcal{X}} m(A)\psi_a\left(\phi_m^{(N)}(A)\right) \tag{19}$$

where ψ_a is the power function introduced in (5).

For $a = 1$ (i.e., $-\log(.)$ as inverse function) and $N = 2$ (i.e., pair-wise comparison of focal sets), we retrieve most of the entropy measures introduced earlier, with different consistency indexes ϕ between sets. For $a = 2$ (i.e., $1 - (.)$ as inverse function) and still $N = 2$, we retrieve the consistency-like measures of George and Pal [4], and Yager [22].

Interestingly, this expression allows also capturing the N-consistency approaches focused on the mass of the empty set with N-wise comparison of focal sets. Indeed, if we consider now the case $a = 2$, with a general value of N, (19) becomes simply:

$$\delta_2^{(N)}(m) = 1 - \phi_N(m) = m^{(N)}(\emptyset) \tag{20}$$

As recalled in Sect. 3.3, the function $\phi_{|\mathcal{F}|}(m)$ obtained for $N = |\mathcal{F}|$, the number of focal sets of m, has been proven to satisfy required properties of a logical consistency of m according to the axioms of [3], qualifying itself thus as a valid alternative measure of logical consistency to $\phi_\pi(m) = \max_{x \in \mathcal{X}} \pi(x)$.

Table 1. Entropy and inconsistency measures encompassed by the general expression of Eq. (19), with different values of a, N and ϕ.

		$a = 1$	$a = 2$		
$N = 2$				$N = 1$	N
$\phi(A,B) = \begin{cases} 1 \text{ if } A \cap B \neq \emptyset \\ 0 \text{ else} \end{cases}$		Yager [21]	Yager [22]	Destercke & Burger [3]	Pichon et al. [13]
$\phi(A,B) = \begin{cases} 1 \text{ if } B \subseteq A \\ 0 \text{ else} \end{cases}$		Höhle [5]			
$\phi(A,B) = \begin{cases} 1 \text{ if } A = B \\ 0 \text{ else} \end{cases}$		Nguyen [12]			
$\phi(A,B) = \dfrac{|A \cap B|}{|B|}$		Klir & Ramer [9]			
$\phi(A,B) = \dfrac{|A \cap B|}{|A|}$		Klir & Parviz [8]			
$\phi(A,B) = \dfrac{|A \cap B|}{|A \cup B|}$			George & Pal [4]		

Table 1 summarises the list of measures corresponding to the general expression from Eq. 19 for different values of parameters a and N, and for several consistency indices ϕ satisfying (1) discussed in this paper. For clarity, the consistency indices are provided for $N = 2$ (pair-wise comparison of sets) which corresponds to the most populated case, as displayed in the left part of the table. Other cases for $N = 1$ and general N are displayed in the right part of the table, while the corresponding indices are not defined.

We have thus shown that the expression (19) encompasses not only classical entropy measures in evidence theory, but also some non-additive measures of internal conflict and consistency, and last but not the least, measures of inconsistency derived from the N-consistency family of measures.

5   Conclusions

In this paper, we have firstly shown that most of inconsistency measures defined so far in evidence theory satisfy a general formulation involving a pair-wise consistency index between sets, an inverse function transforming the notion of consistency into inconsistency and some expectation operator. Furthermore, by rendering apparent the underlying inconsistency in Rényi entropy family of measures, we have shown how the inconsistency measure derived from the N-consistency falls also under this general formulation. This preliminary result offers new perspectives on the coherent measurement of inconsistency within and across artificial intelligence systems. In future work, we will study other types of inconsistency indexes as well as possible links with other logical consistency and entropy measures. We will also explore the possible orders induced by such information measures.

References

1. Bronevich, A., Klir, G.J.: Measures of uncertainty for imprecise probabilities: an axiomatic approach. Int. J. Approx. Reason. **51**(4), 365–390 (2010)
2. Dempster, A.P.: Upper and lower probabilities induced by a multivalued mapping. Ann. Math. Stat. **38**, 325–339 (1967)
3. Destercke, S., Burger, T.: Toward an axiomatic definition of conflict between belief functions. IEEE Trans. Syst. Man Cybern. B **43**(2), 585–596 (2013)
4. George, T., Pal, N.R.: Quantification of conflict in Dempster-Shafer framework: a new approach. Int. J. Gen. Syst. **24**(4), 407–423 (1996)
5. Höhle, U.: Entropy with respect to plausibility measures. In: Proceedings of the 12th IEEE International Symposium on Multiple Valued Logic, pp. 167–169. Paris (1982)
6. Hunter, A., Konieczny, S.: Measuring inconsistency through minimal inconsistent sets. In: Proceedings of the Eleventh International Conference on KR, Sydney, Australia, vol. 8, pp. 358–366, 16–19 September 2008
7. Jiroušek, R., Shenoy, P.P.: On properties of a new decomposable entropy of Dempster-Shafer belief functions. Int. J. Approx. Reason. **119**(4), 260–279 (2020)

8. Klir, G.J., Parviz, B.: A note on the measure of discord. In: Dubois, D. (ed.) Proceedings of Eighth Conference on Artificial Intelligence, pp. 138–141. California (1992)
9. Klir, G.J., Ramer, A.: Uncertainty in the Dempster-Shafer theory: a critical re-examination. Int. J. Gen. Syst. **18**(2), 155–166 (1990)
10. Klir, G.J., Smith, R.M.: Recent developments in generalized information theory. Int. J. Fuzzy Syst. **1**(1), 1–13 (1999)
11. Klir, G.J., Yuan, B.: Fuzzy Sets and Fuzzy Logic: Theory and Applications. Prentice Hall International, Upper Saddle River (1995)
12. Nguyen, N.T.: On entropy of random sets and possibility distributions. In: Bezdek, J.C. (ed.) The Analysis of Fuzzy Information, vol. 1. CRC Press, Boca Raton (1986)
13. Pichon, F., Jousselme, A.L., Ben Abdallah, N.: Several shades of conflict. Fuzzy Sets Syst. **366**, 63–84 (2019). https://doi.org/10.1016/j.fss.2019.01.014
14. Rényi, A.: On measures of entropy and information. In: Proceedings of the 4th Berkeley Symposium on Mathematics, Statistics and Probability, vol. 1, pp. 547–561 (1961)
15. Shafer, G.: A Mathematical Theory of Evidence. Princeton University Press, Princeton (1976)
16. Shannon, C.E.: A mathematical theory of communication. Bell Syst. Tech. J. **27**(379–423), 623–656 (1948)
17. Smets, P., Kennes, R.: The transferable belief model. Artif. Intell. **66**, 191–243 (1994)
18. Smets, P.: The nature of the unnormalized beliefs encountered in the transferable belief model. In: Proceedings of the 8th International Conference on Uncertainty in Artificial Intelligence, UAI 1992, pp. 292–297. Morgan Kaufmann Publishers Inc., San Francisco (1992)
19. Tsallis, C.: Possible generalization of Boltzmann-Gibbs statistics. J. Stat. Phys. **52**(1–2), 479–487 (1988)
20. Vajda, I., Zvárová, J.: On generalized entropies, Bayesian decisions and statistical diversity. Kybernetika **43**(5), 675–696 (2007)
21. Yager, R.R.: Entropy and specificity in a mathematical theory of evidence. Int. J. Gen. Syst. **9**, 249–260 (1983)
22. Yager, R.R.: On considerations of credibility of evidence. Int. J. Approx. Reason. **7**(1/2), 45–72 (1992)

An Extension of Specificity-Based Approximations to Other Belief Function Relations

Tekwa Tedjini[✉], Sohaib Afifi, Frédéric Pichon, and Eric Lefèvre

Univ. Artois, UR 3926, Laboratoire de Genie Informatique et d'Automatique
de l'Artois (LGI2A), 62400 Béthune, France
{tekwa.tedjini,sohaib.afifi,frederic.pichon,eric.lefevre}@univ-artois.fr

Abstract. Adopting a general framework to faithfully represent uncertainty, such as belief function theory, usually comes at a cost. In many real-life applications, we are constrained to handle mass functions that have too many focal elements. Fortunately, one can resort to approximation techniques to bypass this issue. In this paper, we extend the classical approximation techniques, which are mainly specificity-based, to other belief function relations such as lattice dominance. This allows to overcome the limits of classical techniques in some applications.

Keywords: Belief function · Approximation · Specificity · Relations

1 Introduction

Belief function theory [15] is a rich and powerful uncertainty reasoning framework as it extends both the set and probability representations of uncertainty. Despite its successful application in many real-life problems, it has been criticized for its high computational complexity. Several techniques have been proposed to simplify the computations pertaining to this theory, either using exact [12] or approximate methods. We are particularly interested in the latter. Approximations can be computed by Monte-Carlo simulations [20], or by replacing the original mass function by a probability measure or a possibilistic one [7,19]. Other approaches can be used where mass functions are combined on a coarsened frame of discernment [3] or where the number of focal sets is reduced [1,2,8,13,14,18]. We draw a particular attention to this last family of methods. Besides simplicity, a good approximation has to be consistent and close enough to the original mass function [8]. Closeness is typically quantified by a distance measure, whereas consistency is unanimously based on comparing the specificity of the informative content of the original mass function and its approximation. Recently, Destercke and al. [4] introduced an approach that extends any set relation to belief functions. This approach generalizes the notion of comparison and allows, along with comparing the informative content of beliefs in terms of specificity, to establish other relations between them such as dominance. In

© Springer Nature Switzerland AG 2021
T. Denœux et al. (Eds.): BELIEF 2021, LNAI 12915, pp. 224–233, 2021.
https://doi.org/10.1007/978-3-030-88601-1_23

this paper, we propose to extend this approach to approximation methods that reduce the number of focal sets of mass functions. We are motivated by the deficiency of classical approximation techniques in some applications. This deficiency arises from the use of approximate beliefs that are more or less specific than the original ones whilst the application requires rather to choose beliefs that are, for instance, dominant. We will develop this idea later in the paper.

The remainder of this paper is organized as follows. Section 2 gives a quick reminder on belief functions and set relations. Section 3 describes the notion of comparison in belief function theory. The proposed generalized approximation and a particular case study are presented in Sect. 4. We conclude the paper in Sect. 5.

2   Basic Definitions

In this section, we provide some basic definitions on belief functions and set relations that are required in our developments.

2.1   Theory of Belief Functions

Let x be an uncertain variable defined on finite set of values $\mathcal{X} = \{x_1, x_2, \ldots, x_n\}$ called the frame of discernment. The available knowledge about x is represented by a mass function $m^{\mathcal{X}} : 2^{\mathcal{X}} \mapsto [0, 1]$ s.t. $\sum_{A \subseteq \mathcal{X}} m^{\mathcal{X}}(A) = 1$ and $m^{\mathcal{X}}(\emptyset) = 0$. $m^{\mathcal{X}}(A)$ quantifies the part of our belief that $x \in A$ without providing any further information about $x \in A' \subset A$. Each subset $A \subseteq \mathcal{X}$ such that $m^{\mathcal{X}}(A) > 0$ is called focal set or focal element of $m^{\mathcal{X}}$. Other knowledge representations can be obtained from $m^{\mathcal{X}}$, such as the *belief* $Bel^{\mathcal{X}}$ and the *plausibility* $Pl^{\mathcal{X}}$ function, defined for all $A \subseteq \mathcal{X}$:

$$Bel^{\mathcal{X}}(A) = \sum_{\emptyset \neq B \subseteq A} m^{\mathcal{X}}(B), \quad Pl^{\mathcal{X}}(A) = \sum_{B \cap A \neq \emptyset} m^{\mathcal{X}}(B). \tag{1}$$

$Bel^{\mathcal{X}}$ is the amount of evidence that supports $x \in A$ and $Pl^{\mathcal{X}}$ is interpreted as the amount of evidence that is consistent with $x \in A$.

2.2   Set Relations

A relation $\mathbf{R}$ between subsets of $\mathcal{X}$ is a subset $\mathbf{R} \subseteq 2^{\mathcal{X}} \times 2^{\mathcal{X}}$ that specifies which pair of subsets are related to each other [4]. Let A and B be two subsets of $\mathcal{X}$. We denote by $A\mathbf{R}B$ whenever $(A, B) \in \mathbf{R}$. A relation may have several properties such as: *reflexivity* ($A\mathbf{R}A, \forall A \subseteq \mathcal{X}$), *transitivity* ($A\mathbf{R}B$ and $B\mathbf{R}C \Rightarrow A\mathbf{R}C$, with $C \subseteq \mathcal{X}$), *antisymmetry* ($A\mathbf{R}B \wedge B\mathbf{R}A \Rightarrow A = B, \forall A, B \subseteq \mathcal{X}$), etc. Note that it is also possible to define more complex relations by combining those properties. For instance, the set-inclusion relation ($A\mathbf{R}B \Leftrightarrow A \subseteq B$) is reflexive, transitive and antisymmetric [4].

3   Comparing Belief Structures

According to the *Least Commitment Principle* [16], if we have to choose among multiple mass functions compatible with a set of constraints, the most appropriate one is the least informative. To use this principle, one has to define tools to compare the content of the available mass functions. This is commonly done via the notion of specialization [6]. Given two mass functions $m_1^{\mathcal{X}}$ and $m_2^{\mathcal{X}}$ defined on $\mathcal{X}$, $m_1^{\mathcal{X}}$ is said to be at least as *informative (specific)* as $m_2^{\mathcal{X}}$, which we denote by $m_1^{\mathcal{X}} \sqsubseteq m_2^{\mathcal{X}}$, if and only if $m_1^{\mathcal{X}}$ can be obtained from $m_2^{\mathcal{X}}$ by sharing each mass $m_2^{\mathcal{X}}(B)$ among subsets $A \subseteq B$. Formally, there exists a non-negative square matrix, known as the specialization matrix $S = [S(A,B)], A, B \in 2^{\mathcal{X}}$, verifying the conditions below:

$$\sum_{A \subseteq \mathcal{X}} S(A,B) = 1, \qquad \forall B \subseteq \mathcal{X}, \tag{2}$$

$$S(A,B) > 0 \Rightarrow A \subseteq B, \quad \forall A, B \subseteq \mathcal{X}, \tag{3}$$

$$m_1^{\mathcal{X}}(A) = \sum_{B \subseteq \mathcal{X}} S(A,B) m_2^{\mathcal{X}}(B), \quad \forall A \subseteq \mathcal{X}. \tag{4}$$

$S(A,B) \in [0,1]$ is the proportion of $m_2^{\mathcal{X}}(B)$ that flows into $A \subseteq B$. Note that if $m_1^{\mathcal{X}} \sqsubseteq m_2^{\mathcal{X}}$ then [6]:

$$[Bel_1^{\mathcal{X}}, Pl_1^{\mathcal{X}}] \subseteq [Bel_2^{\mathcal{X}}, Pl_2^{\mathcal{X}}]. \tag{5}$$

The recent work of Destercke and al. [4] highlighted the relevance of investigating other links, besides specificity, between mass functions, particularly those extending set relations such as equivalence or partial/total order. The authors introduced a more general definition of the comparison of belief function as follows:

Definition 1. *Let $m_1^{\mathcal{X}}$ and $m_2^{\mathcal{X}}$ be two mass functions and let $\mathbf{R}$ be a relation between subsets of $\mathcal{X}$. We say that $m_1^{\mathcal{X}} \widetilde{\mathbf{R}} m_2^{\mathcal{X}}$ if there is a left stochastic matrix S, such that $\forall A, B \subseteq \mathcal{X}$.*

$$m_1^{\mathcal{X}}(A) = \sum_{B \subseteq \mathcal{X}} S(A,B) m_2^{\mathcal{X}}(B), \tag{6}$$

$$\Big(S(A,B) > 0 \Big) \wedge \Big(m_2^{\mathcal{X}}(B) > 0 \Big) \Rightarrow A\mathbf{R}B. \tag{7}$$

$S(A,B)$ is the proportion of $m_2^{\mathcal{X}}(B)$ transferred to A, such that $A\mathbf{R}B$ [4].

Note that when $\widetilde{\mathbf{R}}$ is replaced by $\sqsubseteq$ and $\mathbf{R}$ by $\subseteq$ in (7), we obtain the specialization relation defined earlier. Furthermore, when $\mathcal{X}$ is ordered, it is also possible to recover another relation that was studied in [9], by comparing two subsets $A, B \subseteq \mathcal{X}$ defined as $A = \{\underline{a}, \dots, \overline{a}\}$ ($\underline{a} \leq \overline{a}$) and $B = \{\underline{b}, \dots, \overline{b}\}$ ($\underline{b} \leq \overline{b}$) in terms of lattice dominance [4]. We say then that $m_1^{\mathcal{X}}$ is *at least as small as* $m_2^{\mathcal{X}}$, which we denote by $m_1^{\mathcal{X}} \preceq m_2^{\mathcal{X}}$, with $\widetilde{\mathbf{R}}$ being replaced by $\preceq$ and $\mathbf{R}$ replaced by $\leq_d$ where $A \leq_d B$ if $\underline{a} \leq \underline{b}$ and $\overline{a} \leq \overline{b}$. The following property holds [9]:

$$m_1^{\mathcal{X}} \preceq m_2^{\mathcal{X}} \Rightarrow [Bel_2^{\mathcal{X}}, Pl_2^{\mathcal{X}}] \leq_d [Bel_1^{\mathcal{X}}, Pl_1^{\mathcal{X}}]. \tag{8}$$

4   Generalization of Belief Functions Approximation

Usually, a mass function m is approximated by another mass function m' that is at most as specific as m, i.e., $m \sqsubseteq m'$. Assume that we want to approximate m by reducing the number of its focal sets. m' can be built from m by preserving the most significant focal sets, i.e., those with high mass values, and by aggregating or removing the redundant or the least significant ones as in [13]. It is also possible to reduce the number of focal sets iteratively as in [2,8,14]. These latter methods help to trade-off between the quality and the computational time required to determine m'.

In this section, we extend the previously stated techniques to other possible relations $\tilde{\mathbf{R}}$ between m and m'. Our motivation arises from the fact that specificity-based approximations may be inappropriate in some applications, such as in the combinatorial optimization problem that we studied in [17]. Specifically, we proposed in [17] a belief-constrained programming approach inspired from [9] to model the vehicle routing problem with time windows [11] and evidential service and travel times. In this kind of problems, each vehicle is compelled to start the service at any customer within his time availability interval (window). Arrivals after the closure of time windows are therefore forbidden. To fulfill such particular constraints, given the evidential time parameters, confidence levels are imposed on the belief and the plausibility functions of the arrival times which are combination of service and travel times. For instance, if x is the variable representing the arrival time at a given customer, C is the closure of his time window and $\alpha, \beta \in [0,1]$ ($\alpha \leq \beta$) are two confidence levels, the time constraints for this customer can be expressed as:

$$Bel(x \leq C) \geq \alpha, \qquad Pl(x \leq C) \geq \beta. \tag{9}$$

The use of belief functions adds more complexity to the problem that is already NP-hard. The problem involves indeed costly mass function combinations due to large numbers of focal sets. Consequently, we turned to classical approximation methods to overcome this issue. Nevertheless, we noticed that replacing the original service and travel time mass functions by less specific ones impacts inappropriately the set of feasible solutions, i.e., solutions that satisfy all the problem constraints. Indeed, a solution may be feasible when using approximations while it is rejected when using the original mass functions. Take for instance the variable x defined earlier, and suppose that uncertainty about the value of x is represented by the mass function: $m(\{15,16\}) = 0.9$, $m(\{16,17\}) = 0.05$, $m(\{16.30, 17.30\}) = 0.05$. Suppose that $C = 16$ and that $\alpha = 0.9, \beta = 1$. Using (1), we have $Bel(x \leq 16) = 0.9 = \alpha$ and $Pl(x \leq 16) = 0.95 < \beta$. The confidence level β is not met, thus the customer can not be served. Suppose now that uncertainty about x is represented using an approximation m' such that $m \sqsubseteq m'$. m' is given by $m'(\{15,16\}) = 0.9$, $m'(\{16, 16.30, 17, 17.30\}) = 0.1$. We have $Bel'(x \leq 16) = 0.9 = \alpha$ and $Pl'(x \leq 16) = 1 = \beta$. Note that $Bel' = Bel$ and $Pl' > Pl$, this is due to the relation in (5). In this case, both of the confidence levels are verified and the customer in question can be served. Such a result is quite contradictory with the information we had originally. Hence, it is worthwhile to introduce a more general approach so that one can properly approximate a mass function by another one that is more/less specific or smaller/greater or equivalent, etc., to span a broad range of real-life applications.

4.1 Formalization

Building on the formal definition of approximations given in [2], we can introduce a generalized definition of an approximation as follows:

Definition 2. *Let $\mathcal{P} = \{P_1, P_2, \ldots, P_K\}$ be a partition of the set $\mathbb{N}_n = \{1, \ldots, n\}$, i.e., $P_k \cap P_l = \emptyset$ and $\bigcup_{k=1}^{K} P_k = \mathbb{N}_n$ and let m be a mass function with focal elements $\mathcal{F}(m) = \{A_1, A_2, \ldots, A_n\}$ such that $m(A_i) \geq m(A_{i+1}), \forall i = 1, \ldots, n-1$. Let m' be another mass function with $\mathcal{F}(m') = \{B_1, \ldots, B_K\}$ its focal sets verifying for each $k = 1, \ldots, K$:*

$$A_i \, \mathbf{R} \, B_k, \forall i \in P_k, \tag{10}$$

$$m'(B_k) = \sum_{i \in P_k} m(A_i). \tag{11}$$

m' is called $\widetilde{\mathbf{R}}$-approximation of m.

Definition 2 states that for a given relation $\tilde{\mathbf{R}}$, any mass function m' with fewer focal sets and that is related to m by $\tilde{\mathbf{R}}$, i.e., $m\tilde{\mathbf{R}}m'$, is an approximation of m. Note that m and m' verify the conditions of Definition 1 as it is possible, for any $P_k \in \mathcal{P}$ $(k = 1, \ldots, K)$, to retrieve $m(A_i)$ from $m'(B_k)$ by transferring a proportion $S(A_i, B_k) > 0$ of the mass $m'(B_k) > 0$ from the subset B_k to the subset A_i such that $A_i\mathbf{R}B_k$, with:

$$S(A_i, B_k) = \frac{m(A_i)}{m'(B_k)} = \frac{m(A_i)}{\sum_{j \in P_k} m(A_j)} \tag{12}$$

Particular Cases: Definition 2 covers some well known cases that were already studied in the literature. For instance, if $\mathbf{R}$ is an outer-inclusion relation, i.e., $A_i \subseteq B_k$, with $B_k = \bigcup_{i \in P_k} A_i$, then $\widetilde{\mathbf{R}} = \sqsubseteq$, that is $m \sqsubseteq m'$, which corresponds to the outer approximations of the literature [2, 8, 13, 14].

We can also identify another sub-case when $\mathbf{R}$ is an inner-inclusion relation, i.e., $A_i \supseteq B_k$, where $B_k = \bigcap_{i \in P_k} A_i$. In this case $\widetilde{\mathbf{R}} = \sqsupseteq$ that is $m \sqsupseteq m'$, which is the inner approximation of Denœux [2].

Furthermore, if an order is established on $\mathcal{X}$ and $\widetilde{\mathbf{R}} = \preceq$, it is also possible to approximate m by a mass function m' such that $m \preceq m'$, where $\preceq$ is the generalized lattice dominance relation. This new approximation is detailed in Sect. 4.2.

To use the generalized approximation, one can for instance keep the first $K-1$ most significant focal sets of m and replace the remaining focal sets by a set B such that $A_i\mathbf{R}B, \forall i = K, \ldots, n$. This is the generalization of the summarization

[13]. However, to provide a good quality approximation, we propose to combine the summarization with the hierarchical clustering procedure introduced in [2]. The main idea of our procedure is to preserve the first (most significant) p focal sets ($p < K < n$), then reduce iteratively, starting from $p + 1$, the number of the remaining focal sets, i.e., those with relatively small masses. At each iteration, a similarity measure or a distance is computed between each pair of focal sets A_i and A_j, then the most similar/nearest pair (A_{i^*}, A_{j^*}) is replaced by a set B_{iter}, such that $A_{i^*} \mathbf{R} B_{iter}$ and $A_{j^*} \mathbf{R} B_{iter}$ with B_{iter} being similar to A_{i^*} and A_{j^*}, and where $m(B_{iter}) = m(A_{i^*}) + m(A_{j^*})$. The process is repeated until we reach size K. The pseudo-code of the approach is explained in Algorithm 1 which runs in a time complexity of $\mathcal{O}(n^3)$. The worst case number of iterations in the repeat loop is $(n - 1)$ and the most expensive instruction inside this loop is the update of the similarity matrix $\mathcal{O}(n^2)$, this yields a total complexity of $\mathcal{O}(n^3)$.

Algorithm 1. The generalized approximation procedure $O(n^3)$

Require: a mass function $m = \{A_i, m(A_i), i = \overline{1, n}\}$, two integers $p, K : p < K < n$, a relation $\mathbf{R}$, and a similarity measure S.
Ensure: a mass function m' with K focal sets.
1: Initialization: $m' \leftarrow \emptyset$;
2: Add the most significant p focal sets to m' and remove them from m;
3: Compute the similarity matrix M s.t $M(i, j) \leftarrow S(A_i, A_j) \, \forall i, j = \overline{1, n - p}$;
4: $iter \leftarrow n - p$;
5: **repeat**
6:     Select the most similar pair (A_{i^*}, A_{j^*});
7:     Add B_{iter} to m s.t $A_{i^*} \mathbf{R} B_{iter}$ and $A_{j^*} \mathbf{R} B_{iter}$;
8:     $m(B_{iter}) \leftarrow m(A_{i^*}) + m(A_{j^*})$;
9:     Remove (A_{i^*}, A_{j^*}) from m;
10:    Update the similarity matrix M;
11:    $iter \leftarrow iter - 1$;
12: **until** $(iter = K - p)$;
13: Add all the focal sets of m to m';

4.2   A Lattice Dominance-Based Approximation

This section studies a particular case of the general approximation where $\widetilde{\mathbf{R}} = \preceq$. In other words, we want to approximate m with a mass function m' that is greater than m according to lattice dominance. Note that property (8) holds with this partial order relation.

Consider an ordered set $\mathcal{X} = \{x_1, \ldots, x_n\}$ ($x_1 \leq \ldots \leq x_n$) and a mass function m defined on $\mathcal{X}$ having the following focal sets $\mathcal{F}(m) = \{A_1, \ldots, A_n\}$ s.t $A_i = \{\underline{a_i}, \ldots, \overline{a_i}\}(\underline{a_i} \leq \overline{a_i})$, which we denote by $[\![\underline{a_i}, \overline{a_i}]\!]$, and where $m(A_i) \geq m(A_{i+1}), \forall i = 1, \ldots, n - 1$. Using Definition 2, we can build a lattice dominance-based approximation ($\preceq$-approximation) m' of m such that $m \preceq m'$ and where focal sets of m' are the subsets $B_k = [\![\underline{b_k}, \overline{b_k}]\!] = \{\underline{b_k}, \ldots, \overline{b_k}\}$, with $\underline{b_k} \leq \overline{b_k}$, and verifying for each $i \in P_k$ and $k = 1, \ldots, K$, $A_i \leq_d B_k$, i.e., $\underline{a_i} \leq \underline{b_k}$ and $\overline{a_i} \leq \overline{b_k}$.

To illustrate this approximation, we use Algorithm 1 with the lattice dominance relation $\leq_d$ and Jaccard's similarity measure given by: $S_{\text{Jaccard}}(A_i, A_j) =$

$\dfrac{|A_i \cap A_j|}{|A_i \cup A_j|}$ [10]. The pair of the most similar focal sets (A_{i^*}, A_{j^*}) is replaced by the subset B_{iter} that is the nearest to A_{i^*} and A_{j^*} and which is defined as follows:

$$B_{iter} = [\![\max(\underline{a}_{i^*}, \underline{a}_{j^*}), \max(\overline{a}_{i^*}, \overline{a}_{j^*})]\!] \tag{13}$$

The process is repeated until m' reaches size K. Note that the choice of an adequate measure depends on the relation that is used as well as the application in hand. Jaccard's measure can be replaced by any other similarity measure such as Dice's measure [5] or others. Moreover, if m has disjoint focal sets, one can use a geometric distance [21] instead to capture the nearest focal sets.

Example 1. Let us use Algorithm 1 to build a $\preceq$-approximation for the mass function m defined such that: $\mathcal{F}(m) = \{A_1 = [\![1,3]\!], A_2 = [\![2,7]\!], A_3 = [\![3,9]\!], A_4 = [\![1,6]\!], A_5 = [\![6,8]\!], A_6 = [\![2,4]\!]\}$ with $m(A_1) = 0.4, m(A_2) = 0.3, m(A_3) = 0.1, m(A_4) = 0.1, m(A_5) = 0.05, m(A_6) = 0.05$. Also let $K = 4$ and $p = 2$.

$*$ *Step 1:* Add A_1 and A_2 to m', then remove them from m. In this case, m becomes: $\mathcal{F}(m) = \{A_3 = [\![3,9]\!], A_4 = [\![1,6]\!], A_5 = [\![6,8]\!], A_6 = [\![2,4]\!]\}$ with $m(A_3) = 0.1, m(A_4) = 0.1, m(A_5) = 0.05, m(A_6) = 0.05$.

$*$ *Step 2:* Compute the similarity matrix M for m, *iter* $= n - p = 4$.

F. sets	A_3	A_4	A_5	$A_6 \star$
A_3	-	0.44	0.43	0.25
$A_4 \star$	–	–	0.13	**0.50**
A_5	–	–	–	0
A_6	–	–	–	–

The most similar pair of focal sets (A_4, A_6) is replaced, in m, by the subset $B_{iter} = B_4 = [\![2,6]\!]$ that is computed using Eq. (13), and $m(B_4) = m(A_4) + m(A_6) = 0.15$. Hence m becomes: $\mathcal{F}(m) = \{A_3 = [\![3,9]\!], B_4 = [\![2,6]\!], A_5 = [\![6,8]\!]\}$ with $m(A_3) = 0.1$, $m(B_4) = 0.15$ and $m(A_5) = 0.05$.

$*$ *Step 3:* Update the similarity matrix M, *iter* $=$ *iter* $- 1 = 3$.

F. sets	A_3	$B_4 \star$	A_5
$A_3 \star$	–	**0.50**	0.43
B_4	–	–	0.14
A_5	–	–	–

The pair (A_3, B_4) is replaced, in m, by the subset $B_{iter} = B_3 = [\![3,9]\!]$, given (13), and $m(B_3) = m(A_3) + m(B_4) = 0.25$. Hence m becomes: $\mathcal{F}(m) = \{B_3 = [\![3,9]\!], A_5 = [\![6,8]\!]\}$, with $m(B_3) = 0.25$ and $m(A_5) = 0.05$.

* *Step 4: iter* $= iter - 1 = 2 = K - p$: Stop and add B_3 and A_5 to m'. m' is the $\preceq$-approximation of m where $\mathcal{F}(m') = \{ A_1 = [\![1,3]\!], A_2 = [\![2,7]\!], B_3 = [\![3,9]\!], A_5 = [\![6,8]\!] \}$, with $m'(A_1) = 0.4$, $m'(A_2) = 0.3$, $m'(B_3) = 0.25$ and $m'(A_5) = 0.05$.

4.3   Preliminary Tests

The lattice dominance-based approximation method was incorporated within a meta-heuristic framework to accelerate the solution scheme of the combinatorial optimization problem studied in [17], while preventing the increase of the set of feasible solutions. Tests were conducted on an adaptation of medium to large-sized literature instances. The details about the solution scheme as well as the instances adaptation are explained in [17]. Table 1 presents average cost results for instances *Inst* of 50 customers, after performing 15 executions per instance. Note that the meta-heuristic algorithm stops after 50 iterations without improvement. Columns 2 (resp. 4) and 3 (resp. 5) show costs C (resp. $C_{\preceq}$) without (resp. with) $\preceq$-approximation and the corresponding execution time $CPU(s)$ (resp. $CPU_{\preceq}(s)$) recorded in seconds. The percentage of increase in solution cost induced by $\preceq$-approximation is displayed in column 6. Average costs $C_{\preceq}^*$ of solutions using $\preceq$-approximation for the same amount of time as in column 3 are presented in column 7. The experiments show a significant decrease in $CPU_{\preceq}(s)$ when using the approximation, this is expected since the number of focal sets is reduced. Moreover, the increase in cost values when using approximation is around 7.18% which is quite acceptable given the gain in time. In addition, the highlighted costs in column 7, confirm that incorporating the $\preceq$-approximation in the meta-heuristic scheme helps to enhance the solution quality. Specifically, providing fast solutions helps the meta-heuristic engine to explore, rapidly, further regions of the set of feasible solutions that might contain better quality solutions. Note that we chose to present results on medium-sized instances to highlight the advantage of using the proposed approximation as we were not even able to get results without approximation for large scale instances.

Table 1. Comparing results with and without $\preceq$-approximation.

Inst	C	CPU(s)	$C_{\preceq}$	$CPU_{\preceq}(s)$	Inc(%)	$C_{\preceq}^*$
$C102$	7549.88	260.70	7770.40	26.20	2.92	**7238.41**
$C104$	**6052.45**	668.50	6391.06	48.77	5.59	6421.58
$C204$	3580.75	356.71	3671.60	61.00	2.55	**3474.26**
$R104$	**10479.60**	970.83	11947.63	27.50	9.55	11573.78
$R204$	4026.19	742.13	4509.74	66.50	12.01	**3842.19**
$R207$	5246.60	299.40	5484.06	44.80	4.52	**4294.79**
$R208$	**3399.43**	1512.27	3734.58	70.50	9.85	3411.02
$RC204$	4368.41	494.40	4827.95	60.10	10.51	**3874.12**

5   Conclusions and Perspectives

We proposed a general approach to approximate belief functions. This approach benefits from the generalization of set relations to belief functions and offers to simplify a mass function given any possible relation with its approximation. The presented approach includes some well known sub-cases, such that the inner and outer approximations of the literature. A lattice dominance-based case study was detailed and applied to a combinatorial optimization problem to accelerate the solution search. In future work, we will investigate other possible relations as well as the definition of other similarity measures that are problem-related to get more efficient results. An extension to approximations that are concerned with reducing the size of the frame of discernment is also an interesting perspective.

References

1. Bauer, M.: Approximation algorithms and decision making in the Dempster-Shafer theory of evidence -An empirical study. Int. J. Approx. Reason. **17**(2–3), 217–237 (1997)
2. Denœux, T.: Inner and outer approximation of belief structures using a hierarchical clustering approach. Int. J. Uncertain Fuzz **9**(4), 437–460 (2001)
3. Denœux, T., Yaghlane, A.B.: Approximating the combination of belief functions using the fast Möbius transform in a coarsened frame. Int. J. Approx. Reason. **31**(1), 77–101 (2002)
4. Destercke, S., Pichon, F., Klein, J.: From set relations to belief function relations. Int. J. Approx. Reason. **110**, 46–63 (2019)
5. Dice, L.R.: Measures of the amount of ecologic association between species. Ecology **26**(3), 297–302 (1945)
6. Dubois, D., Prade, H.: A set-theoretic view of belief functions: logical operations and approximations by fuzzy sets. Int. J. Gen. Syst. **12**(3), 193–226 (1986)
7. Dubois, D., Prade, H.: Consonant approximations of belief functions. Int. J. Approx. Reason. **4**(5–6), 419–449 (1990)
8. Harmanec, D.: Faithful approximations of belief functions. In: Proceedings of the Fifteenth Conference on Uncertainty in Artificial Intelligence, pp. 271–278 (1999)
9. Helal, N., Pichon, F., Porumbel, D., Mercier, D., Lefevre, E.: The capacitated vehicle routing problem with evidential demands. Int. J. Approx. Reason. **95**, 124–151 (2018)
10. Jaccard, P.: Étude comparative de la distribution florale dans une portion des alpes et des jura. Bull. Soc. Vaud. Sci. Nat. **37**, 547–579 (1901)
11. Kallehauge, B.: Formulations and exact algorithms for the vehicle routing problem with time windows. Comput. Oper. Res. **35**(7), 2307–2330 (2008)
12. Kennes, R.: Computational aspects of the Möbius transformation of graphs. IEEE Trans. Syst. Man Cybern. Syst **22**(2), 201–223 (1992)
13. Lowrance, J.D., Garvey, T.D., Wilson, T.M.S.N.: A framework for evidential reasoning systems. In: Kehler, T., al. (eds.) Proceedings of the AAAI'86, vol. 2, pp. 896–903. AAAI (August 1994)
14. Petit-Renaud, Simon, Denœux, Thierry: Handling different forms of uncertainty in regression analysis: a fuzzy belief structure approach. In: Hunter, Anthony, Parsons, Simon (eds.) ECSQARU 1999. LNCS (LNAI), vol. 1638, pp. 340–351. Springer, Heidelberg (1999). https://doi.org/10.1007/3-540-48747-6_31

15. Shafer, G.: A Mathematical Theory of Evidence. Princeton University Press, Princeton (1976)
16. Smets, P.: Belief functions: the disjunctive rule of combination and the generalized bayesian theorem. Int. J. Approx. Reason. **9**, 1–35 (1993)
17. Tedjini, T., Afifi, S., Pichon, F., Lefevre, E.: A belief-constrained programming model for the VRPTW with evidential service and travel times. In: Proceedings of the 28es Rencontres Francophones sur la Logique Floue et ses Applications, pp. 217–224. Alès, France (2019)
18. Tessem, B.: Approximations for efficient computation in the theory of evidence. Artif. Intell. **61**(2), 315–329 (1993)
19. Voorbraak, F.: A computationally efficient approximation of Dempster-Shafer theory. Int. J. Man Mach. Stud. **30**(5), 525–536 (1989)
20. Wilson, N.: A Monte-Carlo algorithm for Dempster-Shafer belief. In: D'Ambrosio, B., Smets, P., Bonissone, P. (eds.) Proceedings of the 7th Conference on Uncertainty in AI, UAI 1991, pp. 414–417. Morgan Kaufmann (1991)
21. Zwick, R., Carlstein, E., Budescu, D.V.: Measures of similarity among fuzzy concepts: a comparative analysis. Int. J. Approx. Reason. **1**(2), 221–242 (1987)

Information Fusion

A New Multi-source Information Fusion Method Based on Belief Divergence Measure and the Negation of Basic Probability Assignment

Hongfei Wang[1], Wen Jiang[1,2,3(✉)], Xinyang Deng[1,3], and Jie Geng[1]

[1] School of Electronics and Information, Northwestern Polytechnical University,
Xi'an 710072, China
jiangwen@nwpu.edu.cn
[2] Peng Cheng Laboratory, Shenzhen 518055, China
[3] National Engineering Laboratory for Integrated Aero-Space-Ground-Ocean
Big Data Application Technology, Xi'an 710072, China

Abstract. Dempster-Shafer theory (DST) can effectively distinguish between imprecise information and unknown information, which is widely used in information fusion. However, when the evidence highly contradicts each other, it may lead to counter-intuitive results. In addition, the existing information fusion methods do not take the negation of BPA into consideration, which can be improved. In this paper, we propose a new information fusion method by taking into account not only the information in basic probability assignment (BPA) but also the information contained in the negation of BPA. In the method, the belief divergence measure is not only used to calculate the difference between BPA and its negative BPA to reflect the information volume carried by its initial BPA, but also to calculate the difference between BPA and other BPA to consider the discrepancy between evidence. The efficiency of the method is verified by case studies.

Keywords: Dempster-Shafer theory · Information fusion · Belief divergence measure · Negation

1 Introduction

Multi-source information fusion can intelligently synthesize multiple information of a certain target, resulting in a more accurate and complete evaluation than a single information source. This is a cross-technology, which has received widespread attention from scholars this year [1–3]. However, due to the interference of environmental factors such as noise, data collected from different sources may be imprecise and uncertain.

Supported by the National Science and Technology Major Project (Program No. 2017-V-0011-0062).

T. Denœux et al. (Eds.): BELIEF 2021, LNAI 12915, pp. 237–246, 2021.
https://doi.org/10.1007/978-3-030-88601-1_24

Dempster-Shafer theory (DST) [4,5], also known as evidence theory or belief function theory, can effectively distinguish between imprecise information and unknown information, and provides a powerful tool for the expression of uncertain information. DST has been widely used in information fusion [6,7], fault diagnosis [8], so on [9–11]. However, there are some defects in DST. The biggest controversy of evidence theory is the counter-intuitive result in the combination of highly conflicting evidence. So far, a lot of research has been done to address this problem [6,12], which can be divided into two categories. One is to modify the Dempster's combination rule [13,14]. This idea holds that the Dempster's combination rule will produce paradox because it deducts the part of mass of the empty set after fusion and normalizes the residual parts, so it is necessary to modify the Dempster's combination rules. Yager [15] believed that the conflicting parts could not provide valid information and assigns them to the entire recognition framework. Dubois and Prade [16] redistributed each partial conflict to the union of associated propositions. Smets [17,18] provided the combination rules to assign the conflict to empty set, which is a non-normalized Dempster rule. However, the modification of rules often destroys some good properties of Dempster's combination rule, such as associativity and commutativity. The other one is to modify the original body of evidence before fusing them. In literature [6], xiao used belief Jensen-Shannon (BJS) divergence and Deng entropy [19] to generate revised evidence to deal with evidence conflicts. However, the BJS divergence measure cannot well reflect the influence of different kinds of subsets. Wang et al. [7] proposed a new belief divergence measure (BDM) which has better performance than BJS divergence in calculating the differences between evidence.

All of the above methods only consider the information of basic probability assignment (BPA), but the information contained in the negation of BPA is ignored. Yin et al. [20] proposed a novel method to generate the negation of BPA, and proved that the entropy of negation BPA tends to be maximized with the increase of the number of negation processes. Therefore, the greater the discrepancy between a BPA and its negation BPA, the smaller the information volume carried by its initial BPA. Based on this characteristic, we propose a new information fusion method based on BDM and the negation of BPA. The BDM is used to calculate the differences between BPA and its negation BPA. In addition, this method takes into account the discrepancy between the evidence by calculating the BDM between BPA and other BPAs. The efficiency of the method is verified by case studies.

2   Preliminaries

2.1   Dempster-Shafer Theory

DST is an important method for modeling and processing uncertain information [21–24]. The frame of discernment (FOD) in DST is made up of a limited number of mutually exclusive elements, marked $\Theta = \{\theta_1, \theta_2, \ldots, \theta_n\}$. 2^Θ is a set of all subsets of Θ. If the following formula holds,

$$\sum_{A \subseteq \Theta} m(A) = 1 \quad \text{and} \quad m(\emptyset) = 0. \tag{1}$$

then $m : 2^{\Theta} \to [0, 1]$ is a BPA, or a mass function. If $m(A) > 0$, then A is a focal element.

Dempster's combination rule is used to complete the fusion of evidence from different sources, the calculation formula is as follows,

$$m(A) = \begin{cases} \frac{1}{1-K} \sum\limits_{B \cap C = A} m_1(B)m_2(C) \, , & A \neq \emptyset; \\ 0 \, , & A = \emptyset. \end{cases} \tag{2}$$

with

$$K = \sum_{B \cap C = \emptyset} m_1(B)m_2(C), \tag{3}$$

where K is a conflict.

2.2   Negation of Basic Probability Assignment

Everything in nature has two sides, positive and negative. The negative method provides a way to express the opposite of information, so that more information can be obtained to express knowledge. In uncertainty modeling and knowledge reasoning, how to determine the negation of belief structure is very important. Several methods have been proposed to determine the negation of BPA, such as the work by Dubois & Prade [25], Yin et al. [20], Gao & Deng [26], and recent studies by Deng & Jiang [27]. Among these methods, the negation method proposed by Yin et al. [20] takes into account the number of focal elements, and the negation of a focal element is independent of other focal elements. The calculation formula is as follows,

$$\bar{m}(E_i) = \frac{1 - m(E_i)}{N - 1}, \tag{4}$$

where E_i is the focal element, and N is the number of focal element. Note that when the mass function contains only one focal element, for example, $m(a) = 1$. The negation of the mass function is defined as,

$$\bar{m}(a) = 0, \ \bar{m}(\phi) = 1 \tag{5}$$

where ϕ is used to model the open world. The reason is that due to the lack of complete knowledge, we don't know what focus element can exist (occur), but at least we know that it can't be $m(a)$ [20].

2.3   Belief Divergence Measure

The divergence measure of information theory is used to measure the discrepancy between the two probability distributions. DST is a generalization of Bayesian inference. How to measure the difference between two BPAs in the DST framework remains a problem. Xiao [6] proposed BJS divergence measure, but it cannot well reflect the influence of different kinds of subsets. Wang et al. [7] proposed a new BDM based on the belief and plausibility function of mass function, which is calculated as follows,

$$D(m_1, m_2) = \frac{1}{2}\left[I(PBl_{m_1}, \frac{PBl_{m_1} + PBl_{m_2}}{2}) + I(PBl_{m_2}, \frac{PBl_{m_1} + PBl_{m_2}}{2})\right]$$
(6)

where $PBl_m(\theta_i) = \frac{Bel(\theta_i) + Pl(\theta_i)}{\sum_{\theta_i \in \Theta} Bel(\theta_i) + Pl(\theta_i)}$, which transforms BPA into a probability distribution. $I(PBl_{m_1}, PBl_{m_2}) = \sum_{\theta_i \in \Theta} PBl_{m_1}(\theta_i)\log_2 \frac{PBl_{m_1}(\theta_i)}{PBl_{m_2}(\theta_i)}$, which is the Kullback-Leibler (KL) divergence.

3   A New Multi-source Information Fusion Method

In this paper, we propose a new information fusion method by taking into account not only the information in basic probability assignment (BPA) but also the information contained in the negation of BPA. In the method, the belief divergence measure is not only used to calculate the difference between BPA and its negative BPA to reflect the information volume carried by its initial BPA, but also to calculate the difference between BPA and other BPA to consider the discrepancy between evidence. This method aims to reduce the influence of low reliability information on the decision result in the process of multi-source information fusion. The detailed process of this method is as follows.

Phase 1: *Calculate the credibility weight of evidence*

Suppose there are k evidences, represented by m_i $(i = 1, 2, ..., k)$, which have the same FOD: $\Theta = \{A_1, A_2, ..., A_n\}$.

Step 1: In this method, we use BDM to calculate the difference between BPA and other BPAs to consider the discrepancy between evidence. The BDM between m_i and $m_j (i, j = 1, 2, ..., k)$, denoted as D_{ij}, which can be calculated by Eq. (6). The divergence matrix DMX is represented as

$$DMX = \begin{bmatrix} 0 & \cdots & D_{1i} & \cdots & D_{1k} \\ \vdots & \vdots & \vdots & \vdots & \vdots \\ D_{i1} & \cdots & 0 & \cdots & D_{ik} \\ \vdots & \vdots & \vdots & \vdots & \vdots \\ D_{k1} & \cdots & D_{ki} & \cdots & 0 \end{bmatrix}.$$
(7)

Step 2: For the ith evidence, the average divergence $\tilde{D}(m_i)$ can be calculated by

$$\tilde{D}(m_i) = \frac{\sum_{j=1}^{k} D_{ij}}{k - 1}.$$
(8)

Step 3: Since there is a negative correlation between the support degrees of evidence $Sup(m_i)$ with their divergences. Therefore, $Sup(m_i)$ can be calculated by

$$Sup(m_i) = \frac{1}{\tilde{D}(m_i)}. \tag{9}$$

Step 4: Normalize the support degree to obtain the credibility weight of the evidence $W_c(m_i)$.

$$W_c(m_i) = \frac{Sup(m_i)}{\sum_{i=1}^{k} Sup(m_i)}. \tag{10}$$

Phase 2: *Compute the information volume weights of evidences*

Step 5: This method considers not only the information in BPA, but also the information contained in the BPA negation. The negation BPA $\bar{m}_i$ can be calculated by Eq. (4). The BDM between m_i and $\bar{m}_i$, denoted as $DM(m_i)$, can be calculated by Eq. (6).

$$DM(m_i) = D(m_i, \bar{m}_i). \tag{11}$$

Step 6: Since the negation process is an entropy increasing process [20], the greater the discrepancy between a BPA and its negation BPA, the smaller the information volume carried by its initial BPA. Therefore, the information volume of evidence $IV(m_i)$ is calculated as follows:

$$IV(m_i) = \frac{1}{e^{DM(m_i)}}. \tag{12}$$

Step 7: Normalize the information volume to obtain the information volume weight of the evidence $W_{iv}(m_i)$.

$$W_{iv}(m_i) = \frac{IV(m_i)}{\sum_{i=1}^{k} IV(m_i)}. \tag{13}$$

Phase 3: *Fuse the modified evidence*

Step 8: The comprehensive weight of the evidence $W(m_i)$ is generated by comprehensively considering the credibility weight and information volume weight of the evidence.

$$W(m_i) = \frac{W_c(m_i) \times W_{iv}(m_i)}{\sum_{i=1}^{k} W_c(m_i) \times W_{iv}(m_i)}. \tag{14}$$

Step 9: The modified evidence can be calculated by the weighted average method:

$$\tilde{m}(A) = \sum_{i=1}^{k} W(m_i) \times m_i(A), \ A \subseteq \Theta. \tag{15}$$

Step 10: The final combination result $\tilde{m}_{\mathcal{F}}$ is obtained by fusing the modified evidence $k - 1$ times with Dempster's combination rule.

$$\tilde{m}_{\mathcal{F}} = \underbrace{\tilde{m} \oplus \tilde{m} \oplus \ldots \oplus \tilde{m}}_{k-1 \ times}. \tag{16}$$

4   Experiment

4.1   Application in Fault Diagnosis

A case study from paper [28] is provided to demonstrate the effectiveness of this method. In the case study, there are three kinds of sensors and four states, as shown in Table 1. The objective is to determine what type of failure has occurred in the case.

Table 1. BPAs after modeling from sensors [28].

	F_1	F_2	F_3	F_4	Θ
$S_1 : m_1(\cdot)$	0.06	0.68	0.02	0.04	0.20
$S_2 : m_2(\cdot)$	0.02	0	0.79	0.05	0.14
$S_3 : m_3(\cdot)$	0.02	0.58	0.16	0.04	0.20

First, the divergence matrix DMX can be calculated by Eq. (7) as follows:

$$DMX = \begin{bmatrix} 0 & 0.4149 & 0.0197 \\ 0.4149 & 0 & 0.2838 \\ 0.0197 & 0.2838 & 0 \end{bmatrix}$$

Then the parameter values as mentioned in Sect. 3 are shown in the Table 2.

Table 2. The parameter values in the calculation process

	$\tilde{D}(m_i)$	$Sup(m_i)$	$W_c(m_i)$	$DM(m_i)$	$IV(m_i)$	$W_{iv}(m_i)$	$W(m_i)$
m_1	0.2173	4.6027	0.3274	0.2005	0.8183	0.3348	0.3232
m_2	0.3493	2.8626	0.2037	0.2769	0.7582	0.3102	0.1862
m_3	0.1517	6.5909	0.4689	0.1421	0.8675	0.355	0.4906

Finally, the modified evidence and the final combination result can be calculated by Eq. (15) and Eq. (16), respectively, as follows (Table 3):

Table 3. The modified evidence and the final combination result.

	F_1	F_2	F_3	F_4	Θ
$\tilde{m}(\cdot)$	0.0329	0.5043	0.2321	0.0419	0.1888
$\tilde{m}_{\mathcal{F}}(\cdot)$	0.0102	0.7947	0.1652	0.0135	0.0164

The fusion results of this method are compared with those of other methods, as shown in Table 4. As can be seen from the results of the proposed method in Table 4, the mass value of state F_2 is higher than that of F_1, F_3 and F_4, so it can be determined that the hypothesis F_2 has occurred. In Table 1, because m_2 in the original evidence is a conflicting evidence, the mass value of F_2 in the fusion results of DST is smaller than that of other methods as shown in Table 4, which also indicates that the robustness of DST is poor. In addition, the state F_2 has a mass value of 0.7947 in the proposed method. However, the belief degree of each sensor in state F_2 does not exceed 0.7 before fusion, as shown in Table 1. Compared with other methods, it can be seen that the state F_2 has the highest belief to the correct target in the proposed method, as shown in Table 4. Besides, in order to better show the efficiency of the proposed method, an ablation test for the method is presented. Namely, the combination results just based on credibility weight W_c, the results just based on information volume weight W_{iv}, and the results based on comprehensive weight are comprehensively compared. In Table 4, it can be seen that the mass value of state F_2 within the results based on comprehensive weight is higher than that of results based on credibility weight W_c or information volume weight W_{iv}. Through the above analysis, it is proved that the proposed method is reasonable and effective in information fusion.

Table 4. Comparison of the results of several existing methods.

Method	F_1	F_2	F_3	F_4	Θ	Target
DST [4]	0.0205	**0.5230**	0.3933	0.0309	0.0323	F_2
Jiang et al. [28]	0.0111	**0.7265**	0.2312	0.0144	0.0168	F_2
Song et al. [29]	0.0107	**0.7855**	0.1738	0.0137	0.0163	F_2
Results based on W_c	0.0104	**0.7770**	0.1823	0.0138	0.0165	F_2
Results based on W_{iv}	0.0111	**0.6411**	0.3158	0.0151	0.0169	F_2
Proposed method	0.0102	**0.7947**	0.1652	0.0135	0.0164	F_2

4.2 Application in Target Recognition

To better demonstrate the effectiveness of the proposed method, another case study from paper [30] is provided. This is a target recognition problem based on multi-sensors, which involves sensor reports collected by different types of sensors. The BPAs modeled from sensors are given in Table 5.

Due to space limitations, the calculation results will be listed directly in this case without the calculation process, as shown in Table 6. The calculation results are calculated strictly in accordance with the calculation process in Sect. 3.

The fusion results of this method are compared with those of other methods, as shown in Table 6. As can be seen from the results of the proposed method in Table 6, the mass value of state A is higher than that of B, C, $\{A, B\}$ and

Table 5. BPAs modeled from sensors [30].

	A	B	C	$\{A,B\}$	$\{B,C\}$	Θ
$S_1 : m_1(\cdot)$	0.8	0.1	0	0	0	0.1
$S_2 : m_2(\cdot)$	0.5	0.2	0.1	0.2	0	0
$S_3 : m_3(\cdot)$	0	0.9	0.1	0	0	0
$S_4 : m_3(\cdot)$	0.5	0.1	0.1	0.1	0	0.2
$S_5 : m_3(\cdot)$	0.6	0.1	0	0	0.1	0.2

$\{B,C\}$, so it can be determined that the hypothesis A has occurred. In Table 6, expect for the DST method [4], all other methods can correctly recognize the target A. The main reason is that m_3 in the original evidence is a conflicting evidence, while the DST method cannot handle the conflicting evidence well, which leads to the wrong results. In addition, the state A has a mass value of 0.9624 in the proposed method. However, the belief degree of each sensor in state A does not exceed 0.8 before fusion, as shown in Table 5. Compared with other methods, it can be seen that the state A has the highest belief to the correct target in the proposed method, as shown in Table 6. Besides, in order to better show the efficiency of the proposed method, an ablation test for the method is presented. In Table 6, it can be seen that the mass value of state A within the results based on comprehensive weight is higher than that of results based on credibility weight W_c or information volume weight W_{iv}. Through the above analysis, it is proved that the proposed method is reasonable and effective in information fusion.

Table 6. The results of several existing methods.

Method	A	B	C	$\{A,B\}$	$\{B,C\}$	Θ	Target
DST [4]	0	**0.9922**	0.0078	0	0	0	B
Jiang et al. [28]	**0.9567**	0.0401	0.0015	0.0014	0.0002	0.0001	A
Song et al. [29]	**0.9488**	0.0488	0.0011	0.0011	0.0001	0.0001	A
Results based on W_c	**0.9552**	0.0424	0.0011	0.0011	0.0001	0.0001	A
Results based on W_{iv}	**0.9064**	0.0912	0.0012	0.0009	0.0001	0.0001	A
Proposed method	**0.9624**	0.0350	0.0011	0.0013	0.0002	0.0001	A

5   Conclusion

DST has been widely applied in information fusion. However, when the evidence highly contradicts each other, it may lead to counter-intuitive results. In addition, the existing information fusion methods do not take the negation of BPA into consideration, which can be improved. In this paper, we propose a new information fusion method by taking into account not only the information in BPA but also the information contained in the negation of BPA. The major contributions of this paper are that this study provides a new perspective for information fusion by taking the negation of BPA into consideration. According to the characteristic that the negative process is the entropy increasing process [20], the greater the discrepancy between BPA and its negation BPA, the smaller the information volume carried by its initial BPA. Therefore, in the proposed method, we use BDM to calculate the difference between BPA and its negation BPA to generate the information volume weight. In addition, both information volume weight and credibility weight are used to generate the comprehensive weight, in which the credibility weight is determined by calculating the difference between BPA and other BPAs to consider the discrepancy between evidence. The effectiveness of the method is verified by case studies. The results show that this method is reasonable and effective.

In our future work, we would like to extend the proposed method to the framework of complex mass function [24,31] which is a generalization of DST, so as to make a more complete and accurate information fusion method.

References

1. Rohmer, J.: Uncertainties in conditional probability tables of discrete Bayesian belief networks: a comprehensive review. Eng. Appl. Artif. Intell. **88**, 103384 (2020)
2. Seiti, H., Hafezalkotob, A., Martínez, L.: R-sets, comprehensive fuzzy sets risk modeling for risk-based information fusion and decision-making. IEEE Trans. Fuzzy Syst. **29**, 385–399 (2021)
3. Liu, H., Wang, L., Li, Z., Hu, Y.: Improving risk evaluation in FMEA with cloud model and hierarchical TOPSIS method. IEEE Trans. Fuzzy Syst. **27**(1), 84–95 (2019)
4. Dempster, A.P.: Upper and lower probabilities induced by a multivalued mapping. Ann. Math. Stat. **38**(2), 325–339 (1967)
5. Shafer, G.: A Mathematical Theory of Evidence. Princeton University Press, Princeton (1976)
6. Xiao, F.: Multi-sensor data fusion based on the belief divergence measure of evidences and the belief entropy. Inf. Fusion **46**, 23–32 (2019)
7. Wang, H., Deng, X., Jiang, W., Geng, J.: A new belief divergence measure for Dempster-Shafer theory based on belief and plausibility function and its application in multi-source data fusion. Eng. Appl. Artif. Intell. **97**, 104030 (2021)
8. Zhang, Z., Jiang, W., Geng, J., Deng, X., Li, X.: Fault diagnosis based on non-negative sparse constrained deep neural networks and Dempster-Shafer theory. IEEE Access **8**, 18182–18195 (2020)
9. Jiang, W., Cao, Y., Deng, X.: A novel Z-network model based on Bayesian network and Z-number. IEEE Trans. Fuzzy Syst. **28**, 1585–1599 (2020)
10. Liu, Z., Chen, Z., Linjing, L.: An automatic high confidence sets selection strategy for SAR images change detection. IEEE Geosci. Remote Sens. Lett. 1–5 (2020)
11. Han, D., Dezert, J., Yang, Y.: Belief interval-based distance measures in the theory of belief functions. IEEE Trans. Syst. Man Cybern. Syst. **48**, 833–850 (2018)
12. Kang, B., Zhang, P., Gao, Z., Chhipi-Shrestha, G., Hewage, K., Sadiq, R.: Environmental assessment under uncertainty using Dempster-Shafer theory and Z-numbers. J. Ambient. Intell. Humaniz. Comput. **11**, 2041–2060 (2020)

13. Smets, P.: Analyzing the combination of conflicting belief functions. Inf. Fusion **8**(4), 387–412 (2007)
14. Smarandache, F., Dezert, J.: Advances and Applications of DSmT for Information Fusion, vol. IV: Collected Works. Infinite Study (2015)
15. Yager, R.R.: On the Dempster-Shafer framework and new combination rules. Inf. Sci. **41**(2), 93–137 (1987)
16. Dubois, D., Prade, H.: Representation and combination of uncertainty with belief functions and possibility measures. Comput. Intell. **4**(3), 244–264 (1988)
17. Smets, P.: The combination of evidence in the transferable belief model. IEEE Trans. Pattern Anal. Mach. Intell. **12**(5), 447–458 (1990)
18. Smets, P., Kennes, R.: The transferable belief model. Artif. Intell. **66**(2), 191–234 (1994)
19. Deng, Y.: Deng entropy. Chaos, Solitons Fractals **91**, 549–553 (2016)
20. Yin, L., Deng, X., Deng, Y.: The negation of a basic probability assignment. IEEE Trans. Fuzzy Syst. **27**, 135–143 (2019)
21. Denoeux, T.: Distributed combination of belief functions. Inf. Fusion **65**, 179–191 (2021)
22. Deng, Y.: Information volume of mass function. Int. J. Comput. Commun. Control **15**, 1–13 (2020)
23. Dezert, J., Tchamova, A., Han, D.: Total belief theorem and conditional belief functions. Int. J. Intell. Syst. **33**, 2314–2340 (2018)
24. Xiao, F.: CED: a distance for complex mass functions. IEEE Trans. Neural Netw. Learn. Syst. **32**, 1525–1535 (2021)
25. Dubois, D., View, H.P.S.T.: A set-theoretic view of belief functions logical operations and approximations by fuzzy sets. Int. J. Gen. Syst. **12**, 193–226 (1986)
26. Gao, X., Deng, Y.: The negation of basic probability assignment. IEEE Access **7**, 107006–107014 (2019)
27. Deng, X., Jiang, W.: On the negation of a Dempster-Shafer belief structure based on maximum uncertainty allocation. Inf. Sci. **516**, 346–352 (2020)
28. Jiang, W., Wei, B., Xie, C., Zhou, D.: An evidential sensor fusion method in fault diagnosis. Adv. Mech. Eng. **8**, 1–7 (2016)
29. Song, Y., Deng, Y.: Divergence measure of belief function and its application in data fusion. IEEE Access **7**, 107465–107472 (2019)
30. An, J., Hu, M., Fu, L., Zhan, J.: A novel fuzzy approach for combining uncertain conflict evidences in the Dempster-Shafer theory. IEEE Access **7**, 7481–7501 (2019)
31. Xiao, F.: Generalization of Dempster-Shafer theory: a complex mass function. Appl. Intell. **50**(10), 3266–3275 (2020)

Improving an Evidential Source of Information Using Contextual Corrections Depending on Partial Decisions

Siti Mutmainah[1,2]([⊠]), Samir Hachour[1], Frédéric Pichon[1], and David Mercier[1]

[1] Univ. Artois, UR 3926 LGI2A, 62400 Béthune, France
{siti.mutmainah,samir.hachour,frederic.pichon,
david.mercier}@univ-artois.fr
[2] UIN Sunan Kalijaga, Yogyakarta, Indonesia
siti.mutmainah@uin-suka.ac.id

Abstract. In this paper, an improvement of the quality of an evidential source of information is proposed using contextual corrections depending on partial decisions obtained from an interval dominance relation on the source outputs. Numerical experiments with the EkNN classifier and synthetic and real data allows us to illustrate the performances and the interest of this method.

Keywords: Belief functions · Contextual corrections · Partial decisions · Interval dominance

1 Introduction

In pattern recognition [1,9], the quality of the information provided by a source (*e.g.* a sensor, a classifier, ...) plays an important role in the success of the pattern recognition task as the information may be false, biased or irrelevant.

The belief function framework (or Dempster-Shafer theory [18]) provides a flexible mathematical framework for dealing with imperfect information. In this theory, the quality of a source of information is classically managed by means of the discounting operation introduced by G. Shafer in his seminal book [18, chapter 11, page 251]. This method has since been refined using so-called contextual correction mechanisms [12,16] taking into account more refined knowledge about the quality of a source: its reliability (in the sense of its relevance, meaning the capacity of the source to answer the question of interest) and its truthfulness (meaning its capacity to tell what it knows; this capacity being possibly conscious - as a lie for example - or unconscious - as a bias for example) [15]. More specifically, three mechanisms have been introduced. They are respectively called

Mrs. Mutmainah's research is supported by the overseas 5000 Doctors program of Indonesian Religious Affairs Ministry (MORA French Scholarship).

T. Denœux et al. (Eds.): BELIEF 2021, LNAI 12915, pp. 247–256, 2021.
https://doi.org/10.1007/978-3-030-88601-1_25

Contextual Discounting (CD), Contextual Reinforcement (CR) and *Contextual Negating (CN)*. They all can be mathematically derived [16] from these notions of reliability and truthfulness: CD, which generalizes the discounting operation, can adjust the output of a source in accordance with information about its reliability, while CN, which generalizes the negation of a source [8], can adjust the source according to its truthfulness, and at last, CR is the dual operation of CD, it may reinforce a too cautious source [14,16].

In this paper, we propose to improve the quality of a source of information outputting belief functions regarding a question of interest using contextual corrections CD, CR or CN, and partial decisions computed from the outputs using the relation of interval dominance [19,20], also called strong dominance in [6,11]. More specifically, the source is considered as a black box meaning we have no access to the manner it works to make its evidential outputs. This situation occurs for example when a company buys a sensor from another one to perform a given task, and the decision making process (or the algorithm) used by this sensor is protected [13]. A learning set is available. It is composed of outputs of the source (as mass functions for example), regarding data the ground truth of which is known. As a simple example [10], we may have in the learning set the following information m_S output by the source regarding the true class of an object o, which belongs to a universe $\Omega = \{\omega_1, \omega_2, \omega_3\}$

	ω_1	ω_2	ω_3	$\{\omega_1, \omega_2\}$	$\{\omega_1, \omega_3\}$	$\{\omega_2, \omega_3\}$	Ω	Ground truth of object o
$m_S\{o\}$	0	0	0.5	0	0	0.3	0.2	ω_1

Now, instead of learning a best possible correction among CD, CR and CN from the whole learning set as proposed in [16], we would like to propose a method that takes advantage of all the corrections, and so propose to regroup the outputs leading to the same partial decisions to learn the best possible correction between CD, CR and CN in each of these groups of outputs to reach better global performances (certainly at the cost of learning more models). With this strategy, an output is thus adjusted differently depending on the partial decision it leads to.

This paper is organized as follows. In Sect. 2, the basic concepts and notations on belief functions used in this paper are presented, as well as a reminder on decision making with interval dominance with belief functions. Reminders on contextual corrections are given in Sect. 3. Thereafter, in Sect. 4, the proposed method to learn contextual corrections depending on partial decisions is exposed. It is tested with synthetic and real data in Sect. 5. A discussion is also added in this last Section to conclude the paper.

2   Belief Functions: Necessary Concepts and Notations

In this Section, necessary concepts and notations used in this paper are quickly reminded. Further details can be found for example in [4,17,18].

With $\Omega = \{\omega_1, ..., \omega_K\}$ the universe (or the frame of discernment) representing the finite set of answers to a given question of interest, a piece of evidence regarding the answer to this question of interest induces a *mass function (MF)* m^Ω (or m if no ambiguity) defined from 2^Ω to $[0, 1]$, verifying $\sum_{A \subseteq \Omega} m^\Omega(A) = 1$. The *focal elements* of a MF m are the subsets $A \subseteq \Omega$ s.t. $m(A) > 0$. A MF having only one focal element A is called a *categorical MF* and can be simply denoted by m_A.

A MF m is in one-to-one correspondence with *belief and plausibility functions* Bel and Pl respectively defined for all $A \subseteq \Omega$ by $Bel(A) = \sum_{B \subseteq A} m(B)$, and $Pl(A) = \sum_{B \cap A \neq \emptyset} m(B)$.

The *contour function* pl corresponds to the restriction of the plausibility function to the singletons of Ω, it is defined for all $\omega \in \Omega$ by $pl(\omega) = Pl(\{\omega\})$.

If a source S provides an output m_S, and if it is also known that this source is reliable with a degree of belief $\beta = 1 - \alpha \in [0, 1]$, then this original MF m_S can be *discounted* into a MF m s.t.

$$m = \begin{cases} A \mapsto \beta m_S(A) & \forall A \subset \Omega \\ \Omega \mapsto \beta m_S(\Omega) + \alpha \end{cases} \tag{1}$$

Equation (1) can also be simply rewritten as $m = \beta m_S + \alpha m_\Omega$, with m_Ω the categorical MF defined by $m_\Omega(\Omega) = 1$, and it can also be shown (see for example [16, Prop. 11]) that the contour function associated with the discounted MF is defined for all $\omega \in \Omega$ by $pl(\omega) = 1 - (1 - pl_S(\omega))\beta$, with pl_S the contour function of m_S. Derivations of this operation can be found in [12, 16, 17].

At last, when a decision has to be made [3, 11], if we consider that the set of possible decisions (or acts) is equal to Ω, we can use the following relation of dominance between the singletons of Ω:

$$\omega \succeq \omega' \iff Bel(\{\omega\}) \geq Pl(\{\omega'\}), \tag{2}$$

and make a partial decision composed of the non dominated singletons according to relation (2).

Due to lack of space, details cannot be written, but Eq. (2) comes from for example [6, Equation 43] or [11, Page 6, Strong dominance criterion] with 0–1 utilities and pieces of information represented by belief functions.

3   Contextual Corrections and Learning

In this section, the definitions of CD, CR and CN are recalled as well as the possibilities to learn them from labelled data [16] composed of the outputs of a source regarding objects whose true class is known.

The contour functions resulting from CD, CR and CN are recalled with K parameters, K being the number of elements in Ω. These mechanisms can indeed be used with more parameters [12, 16] but as shown in [16], these configurations with K parameters for each corrections are expressive enough to reach the lowest possible values (with these mechanisms) of the following measure of discrepancy [12, 16] between the source outputs adjusted with these corrections and the ground truth:

$$E_{pl}(\boldsymbol{\beta}) = \sum_{i=1}^{n} \sum_{k=1}^{K} (pl_i(\omega_k) - \delta_{i,k})^2, \tag{3}$$

where n is the number of objects in the learning set, $\boldsymbol{\beta} = (\beta_\omega, \omega \in \Omega)$ is the vector composed of the K parameters of each correction, pl_i is the contour function regarding the class of the object i resulting from a contextual correction (CD, CR or CN) of the MF provided by the source for this object, and $\delta_{i,k}$ is the indicator function of the truth of all the instances $i \in \{1, \ldots, n\}$, meaning $\delta_{i,k} = 1$ if the class of the instance i is ω_k, otherwise $\delta_{i,k} = 0$.

The discrepancy measure E_{pl} yields a linear least-squares optimization problem, which can be then efficiently solved using standard algorithms.

As for the discounting (1), we consider a source of information outputting a MF m_S regarding a question of interest. The corresponding contour functions of each contextual correction of m_S are summed up in Table 1.

Table 1. Contour functions of each contextual correction of a MF m_S given for any $\omega \in \Omega$. Each parameter β_ω may vary in $[0, 1]$.

Corrections	Contour functions
CD	$pl(\omega) = 1 - (1 - pl_S(\omega))\beta_\omega$
CR	$pl(\omega) = pl_S(\omega)\beta_\omega$
CN	$pl(\omega) = 0.5 + (pl_S(\omega) - 0.5)(2\beta_\omega - 1)$

4   Contextual Corrections Depending on Partial Decisions

In this paper, we propose to improve the previous learning method exposed in Sect. 3, meaning we would like to reach better performances. For this, we will consider, in the learning set, groups of distinct partial decisions the outputs of the source lead to.

The idea is to consider that the quality of the source, and then the way we have to adjust it, may depend on the outputs it gives regarding the objects whose true classes are to be found. For example, the source may be quite right when it decides a certain type of class, while having some bias when it decides another class or a group of classes, and another bias for another class or group of classes, etc.

Specifically we investigate the quality of the source depending on groups of partial decisions according to relation (2). CD, CR and CN best parameters $\boldsymbol{\beta}$ according to E_{pl} (3) are then computed in each group. The correction with the lowest value of E_{pl} is kept in each group.

This learning procedure is summarized by Algorithm 1.

5   Numerical Experiments and Discussion

In this Section we present experiments made with the EkNN classifier [2,5] with $k = 5$ as the source of information on synthetic and real data.

Algorithm 1. Learning procedure

Input: A set $\mathcal{I}$ of instances $(m_i\{o_i\}, \omega_i)$, $i \in \{1,\ldots,n\}$, $m_i\{o_i\}$ being the output of the source regarding object o_i whose true class is ω_i.

Outputs: Groups $\mathcal{G}$ of partial decisions and best corrections for each group.

1: **procedure** LEARNINGPROCEDURE
2:     $\mathcal{G}$ initially empty
3:     **for** each instance i in $\mathcal{I}$ **do**
4:         Compute the partial decision coming from m_i using relation (2)
5:         Add i to the group g in $\mathcal{G}$ associated with this partial decision.
6:     **for** each group of partial decisions g in $\mathcal{G}$ **do**
7:         Compute CD, CR and CN best parameters β according to E_{pl} (3) restricted to the instances in this group
8:         Keep for this group the correction reaching the lowest value of E_{pl}.

Synthetic data are composed of 5000 instances, 5 classes and 2 features, which were generated from a multivariate normal distribution with means $\mu_1 = (0,0), \mu_2 = (2,0), \mu_3 = (0,2), \mu_4 = (2,2), \mu_5 = (1,1)$ for respectively class 1, 2, 3, 4 and 5, and the same covariance matrix Σ for each class:

$$\Sigma = \begin{bmatrix} 1 & 0.9 \\ 0.9 & 1 \end{bmatrix} \tag{4}$$

An illustration of a generated synthetic data set with these parameters is given in Fig. 1.

The real data sets used from UCI [7] are described in Table 2.

Table 2. Descriptions of used UCI data sets.

Data sets	# Instances	# Features	# Classes
Breast cancer	569	31	2
Glass	214	10	6
Haberman	306	3	2
Ionosphere	350	34	2
Iris	150	4	3
Liver	345	6	2
Lymphography	146	18	3
Pima	768	8	2
Red wine	1599	12	6
Sonar	208	60	2
Transfusion	748	3	2
Vehicles	846	19	4
Vertebral	310	6	3

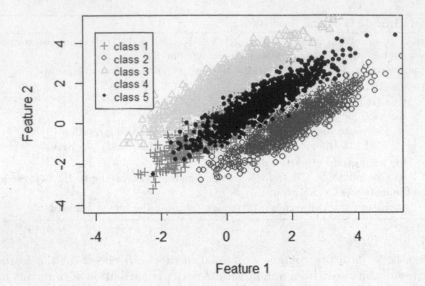

Fig. 1. Example of a generated data set.

For each data set, the following experiment is repeated 10 times: one half of the data $(\mathcal{L}_1)$ is used to learn the EkNN classifier; then a 10-fold cross validation is performed on the second half of the data with 9 folds $(\mathcal{L}_2)$ to learn the best correction in each group of partial decisions using Algorithm 1, and 1 fold for testing (meaning for the test phase).

Some possible partial decisions obtained from the outputs of the source (the EkNN classifier) with the generated data set illustrated in Fig. 1 are given in Fig. 2 for some points in the feature space. Note that these groups of partial decisions are not necessary computed in the training phase $\mathcal{L}_2$. They are just given here as an illustration. As exposed in Algorithm 1, only the points (and their associated partial decisions) in the fold $\mathcal{L}_2$ are considered to learn the corrections. That is why it may happen that during the test phase, a partial decision did not happen in the learning phase. In this situation, we propose to use the best correction for the whole set of $\mathcal{L}_2$. Note that in our experiments presented here, this case very rarely occurred.

To measure the performances, we used the measure E_{pl} (3) with which corrections are learnt, and we also wanted to use another measure with which corrections are not learnt. For this second choice, we opted for the u_{65} utility measure, introduced by Zaffalon et al. [21], which allows one to take into account the advantages of partial decisions concerning the fact of preferring imprecision to being randomly correct. The u_{65} utility measure gives indeed a greater utility

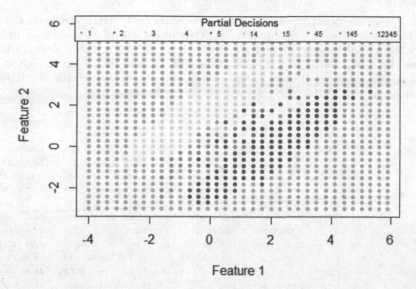

Fig. 2. Points in the feature space belonging to different groups of partial decisions obtained from the generated synthetic data set composed of 5 classes and 2 attributes/features illustrated in Fig. 1. The legend for the partial decisions obtained from the EkNN classifier and relation (2) are given above the figure. We can see that for these points, 10 possible partial decisions have been obtained: 1 (meaning a decision for class 1), 2, 3, 4, 5, 14 (meaning a partial decision in favor of class 1 or class 4), 15, 45, 145 (meaning a partial decision in favor of class 1 or class 4 or class 5), 12345 (meaning a partial decision in favor of all the classes - total uncertainty - each class can be the good one).

to imprecise but correct partial decisions of size n (meaning decisions equal to a set of n singletons one of them being the true class) than precise decisions (in favor of one singleton) only randomly correct with probability $\frac{1}{n}$.

Formally, the U_{65} value of a partial decision d, possibly in favor a set of singletons, is defined by

$$U_{65}(x) = 1.6x - 0.6x^2 \tag{5}$$

with x the so called discounted accuracy of d defined by $\frac{\mathbb{I}(\omega \in d)}{|d|}$, with $\mathbb{I}$ the indicator function, ω the true class of the instance, and $|d|$ the number of elements in d.

Performances according to E_{pl} (3) and the averages of the u_{65} utility measures [21] of the partial decisions obtained from the corrected outputs are regrouped in Table 3.

Table 3. Performances (Average E_{pl} values, the lower the better, and average U_{65} utilities, the greater the better) obtained for EkNN learnt with $\mathcal{L}_1$ (denoted by EkNN), EkNN learnt with groups $\mathcal{L}_1$ and $\mathcal{L}_2$ (denoted by EkNN+), EkNN learnt with $\mathcal{L}_1$ followed by corrections (CD, CR, CN) learnt with $\mathcal{L}_2$, and EkNN learnt with $\mathcal{L}_1$ followed by the new method using groups of partial decisions learnt with $\mathcal{L}_2$. Standard deviations are indicated in parenthesis.

Data	EkNN	EkNN+	CD	CR	CN	Method
Average E_{pl} values (the lower the better)						
Synthetic	188.27 (26.48)	181.27 (16.23)	187.91 (26.56)	134.57 (9.83)	174.06 (21.38)	**106.76 (7.29)**
Breast cancer	3.74 (2.50)	3.59 (2.43)	3.65 (2.35)	3.64 (2.34)	3.56 (2.22)	**3.56 2.21)**
Glass	8.16 (3.61)	8.97 (3.19)	8.13 (3.57)	5.11 (1.40)	6.93 (2.04)	**4.81 (1.49)**
Haberman	8.59 (2.40)	8.79 (2.34)	8.57 (2.32)	5.97 (1.84)	6.61 (1.07)	**5.87 (1.83)**
Ionosphere	2.31 (1.66)	**1.64 (1.18)**	2.31 (1.62)	2.19 (1.49)	2.25 (1.52)	1.81 (1.49)
Iris	0.56 (0.87)	0.59 (1.02)	0.56 (0.85)	**0.55 (0.82)**	0.56 (0.84)	0.56 (0.86)
Lymphography	3.07 (1.44)	**2.32 (1.27)**	3.08 (1.44)	2.74 (1.24)	3.02 (1.18)	2.74 (1.40)
Liver	12.66 (2.08)	12.21 (1.56)	12.66 (2.02)	8.17 (1.08)	8.31 (0.64)	**8.01 (1.12)**
Pima	20.01 (3.32)	18.12 (3.21)	20.01 (3.32)	15.67 (2.20)	16.76 (1.57)	**14.93 (2.40)**
Red wine	162.09 (21.75)	140.60 (13.98)	162.09 (21.75)	48.62 (2.96)	116.83 (4.28)	**45.92 (3.44)**
Sonar	4.04 (1.43)	**3.06 (1.33)**	4.05 (1.43)	3.50 (1.08)	3.80 (1.05)	3.30 (1.37)
Transfusion	20.54 (4.91)	19.77 (4.37)	19.43 (4.19)	15.21 (3.45)	15.44 (2.21)	**13.56 (2.85)**
Vehicles	34.26 (5.66)	28.14 (4.74)	34.26 (5.66)	24.07 (2.35)	32.40 (4.20)	**20.69 (2.96)**
Vertebral	5.08 (2.29)	4.68 (2.05)	5.04 (2.19)	4.64 (1.89)	4.85 (1.89)	**4.31 (1.81)**
Average U_{65} values (the greater the better)						
Synthetic	66.52 (2.68)	66.40 (2.63)	66.49 (2.73)	66.10 (2.69)	65.46 (2.89)	**68.36 (2.59)**
Breast cancer	91.77 (4.65)	**93.74 (3.90)**	92.76 (4.60)	92.75 (4.64)	92.80 (4.62)	92.76 (4.60)
Glass	61.51 (14.47)	63.38 (13.44)	61.56 (14.25)	64.53 (13.87)	45.93 (13.22)	**66.24 (12.35)**
Haberman	74.28 (8.12)	74.45 (8.57)	74.61 (7.97)	74.84 (9.80)	71.69 (8.04)	**75.13 (9.26)**
Ionosphere	93.02 (4.67)	**94.71 (4.16)**	93.02 (4.67)	93.12 (4.62)	92.98 (4.64)	93.86 (4.99)
Iris	96.78 (6.78)	96.51 (6.07)	96.78 (6.78)	96.84 (6.72)	96.78 (6.78)	**97.06 (6.20)**
Liver	66.74 (5.86)	**67.14 (5.53)**	66.73 (5.86)	66.64 (6.82)	64.70 (2.25)	65.45 (7.42)
Lymphography	80.23 (13.13)	**83.84 (12.65)**	80.21 (13.20)	79.42 (13.65)	79.26 (12.26)	79.11 (14.49)
Pima	72.64 (5.06)	73.00 (5.56)	72.64 (5.06)	73.34 (5.50)	71.97 (3.91)	**73.38 (5.02)**
Red wine	45.39 (4.17)	53.29 (4.23)	45.39 (4.17)	57.60 (4.14)	25.00 (0)	**59.12 (4.12)**
Sonar	78.34 (8.53)	**84.50 (7.69)**	78.27 (8.60)	78.96 (8.95)	78.00 (8.03)	79.32 (10.49)
Transfusion	72.76 (5.68)	72.87 (5.41)	74.20 (5.29)	74.39 (6.38)	72.63 (5.42)	**75.91 (6.03)**
Vehicles	61.20 (5.77)	63.46 (5.62)	61.20 (5.77)	60.56 (5.65)	58.11 (5.60)	**63.61 (6.20)**
Vertebral	80.40 (10.14)	**82.09 (8.85)**	80.29 (10.04)	79.95 (10.27)	80.36 (10.19)	81.78 (9.40)

As it can be seen in Table 3, the proposed method using the groups of partial decisions obtains almost in each situation better results than the previous learning using only CD or CR or CN.

Furthermore, the source was supposed to be a black box, but we were also curious to see the performances of this source if we were able to improve its performances by using the data we use to learn the corrections. These performances are given in the column $EkNN+$ in Table 3. We can see that the new learning method succeeds to even improve these results for some data sets.

As summarized in [11] by Ma and Denœux, relation (2) is only one among others. First future works will then consist in testing these other possible relations (s.t. weak dominance, maximality, ...) to compute the groups of partial decisions and see if better performances can be reached.

Acknowledgements. The authors would like to thank the anonymous reviewers for their constructive comments and suggestions that helped them to clarify parts of this paper and will help them for future researches.

References

1. Bishop, C.M.: Pattern Recognition and Machine Learning. Springer, New York (2006)
2. Denœux, T.: A k-nearest neighbor classification rule based on Dempster-Shafer theory. IEEE Trans. Syst. Man Cybern. **25**(5), 804–813 (1995)
3. Denœux, T.: Analysis of evidence-theoretic decision rules for pattern classification. Pattern recognit. **30**(7), 1095–1107 (1997)
4. Denœux, T.: Conjunctive and disjunctive combination of belief functions induced by nondistinct bodies of evidence. Artif. Intell. **172**, 234–264 (2008)
5. Denœux, T.: evclass: Evidential distance-based classification (2017). https://cran.r-project.org/web/packages/evclass/index.html. R package version 1.1.1
6. Denœux, T.: Decision-making with belief functions: a review. Int. J. Approx. Reason. **109**, 87–110 (2019)
7. Dua, D., Graff, C.: UCI Machine Learning Repository. University of California, Irvine (2019). http://archive.ics.uci.edu/ml
8. Dubois, D., Prade, H.: A set-theoretic view of belief functions: logical operations and approximations by fuzzy sets. Int. J. Gen. Syst. **12**(3), 193–226 (1986)
9. Duda, R.O., Hart, P.E., Stork, D.G.: Pattern Classification, 2nd edn. Wiley, New York (2001)
10. Elouedi, Z., Mellouli, K., Smets, P.: Assessing sensor reliability for multisensor data fusion within the transferable belief model. IEEE Trans. Syst. Man Cybern. B **34**(1), 782–787 (2004)
11. Ma, L., Denœux, T.: Partial classification in the belief function framework. Knowl. Based Syst. **214**, 106742 (2021)
12. Mercier, D., Quost, B., Denœux, T.: Refined modeling of sensor reliability in the belief function framework using contextual discounting. Inf. Fusion **9**(2), 246–258 (2008)
13. Mercier, D., Cron, G., Denœux, T., Masson, M.-H.: Decision fusion for postal address recognition using belief functions. Expert Syst. Appl. **36**(3), 5643–5653 (2009)
14. Mercier, D., Lefèvre, E., Delmotte, F.: Belief functions contextual discounting and canonical decompositions. Int. J. Approx. Reason. **53**(2), 146–158 (2012)
15. Pichon, F., Dubois, D., Denoeux, T.: Relevance and truthfulness in information correction and fusion. Int. J. Approx. Reason. **53**(2), 159–175 (2012)
16. Pichon, F., Mercier, D., Lefèvre, E., Delmotte, F.: Proposition and learning of some belief function contextual correction mechanisms. Int. J. Approx. Reason. **72**, 4–42 (2016)
17. Smets, P.: Belief functions: the disjunctive rule of combination and the generalized Bayesian theorem. Int. J. Approx. Reason. **9**(1), 1–35 (1993)
18. Shafer, G.: A Mathematical Theory of Evidence. Princeton University Press, Princeton (1976)

19. Troffaes, M.C.: Decision making under uncertainty using imprecise probabilities. Int. J. Approx. Reason. **45**(1), 17–29 (2007)
20. Yang, G., Destercke, S., Masson, M.H.: Nested dichotomies with probability sets for multi-class classification. In: ECAI, pp. 363–368 (2014)
21. Zafallon, M., Corani, G., Mauá, D.-D.: Evaluating credal classifiers by utility-discounted predictive accuracy. Int. J. Approx. Reason. **53**(8), 1282–1301 (2012)

Elicitation

Validation of Smets' Hypothesis
in the Crowdsourcing Environment

Constance Thierry[✉], Arnaud Martin, Jean-Christophe Dubois,
and Yolande Le Gall

Univ. Rennes, CNRS, IRISA, DRUID, Rennes, France
{constance.thierry,arnaud.martin,jean-christophe.dubois,
yolandele.gall}@irisa.fr
http://www-druid.irisa.fr

Abstract. In the late 1990s, Philippe Smets hypothesizes that the more
imprecise humans are, the more certain they are. The modeling of human
responses by belief functions has been little discussed. In this context, it
is essential to validate the hypothesis of Ph. Smets. This paper focuses
on the experimental validation of this hypothesis in the context of crowdsourcing. Crowdsourcing is the outsourcing of tasks to users of dedicated
platforms. Two crowdsourcing campaigns have been carried out. For the
first one, the user could be imprecise in his answer, for the second one
he had to be precise. For both experiments, the user had to indicate his
certainty in his answer. The results show that by being imprecise, users
are more certain of their answers.

Keywords: Uncertainty · Imprecision · Belief functions ·
Crowdsourcing

1 Introduction

The theory of belief functions is well known for information fusion and for catch-
ing uncertainty and imprecision in machine learning, tracking and data association. A simple mass function allows to represent uncertainty, imprecision and
ignorance at the same time. Of course, when asking someone a question, he or
she does not answer with a belief function.

Few works consider the modeling of uncertain and imprecise responses to
questionnaires. Some works have focused on questionnaires allowing probabilistic [2,7,9,13] or fuzzy [12] answers, but very few allow belief answers. Diaz *et
al.* [5] have proposed a questionnaire to directly build a mass function, but it
remains very unintuitive for any user. Other works [1,3] have considered directly
generated mass functions without worrying about their construction.

In this work, we are interested in modeling the responses of users of a crowdsourcing platform, allowing them to express their uncertainty and imprecision.

This work is supported by the general council of the Côtes d'Armor and the ANR
project HEADWORK.

© Springer Nature Switzerland AG 2021
T. Denœux et al. (Eds.): BELIEF 2021, LNAI 12915, pp. 259–268, 2021.
https://doi.org/10.1007/978-3-030-88601-1_26

These responses will then be modeled by mass functions. However, it is necessary to understand the links between uncertain and imprecise information provided by humans.

Ph. Smets [11] presents in his paper different types of data imperfection and methods to model them. In particular, he presents imprecision as an element relative to an assertion, and uncertainty as the relationship between the information provided by the assertion and the knowledge that the human has of the subject. Ph. Smets then proposes that the more imprecise a person is, the more certain he is, and conversely the more precise, the less certain. This assertion is decomposed here into two hypotheses:

- H1: The more imprecise a person is, the more certain he is.
- H2: The more precise a person is, the less certain he is.

According to Dubois *et al.* [6], there are two types of uncertainty related to human perception. The first one on the realization of an action presenting a risk, the second on the truth of an assertion due to a lack of knowledge. In this article, uncertainty is the consequence of a lack of knowledge that does not allow man to define the truth of an assertion.

H1 and H2 are not the same hypotheses, H2 is the reciprocal of H1. Since the reciprocal of a hypothesis is not always true we work in this paper on the independent validation of H1 and H2. To our knowledge, there is no work to validate these assumptions. In order to perform the experimental validation of the two hypotheses, two crowdsourcing campaigns have been carried out. Crowdsourcing consists in outsourcing tasks on dedicated platforms. The users of the platform perform the tasks for a micro payment. The tasks are very diverse, but generally do not require expertise, so the user profiles are very varied. In this study, the user's task consists of photo annotation through multiple choice questionnaires (MCQs). In the first crowdsourcing campaign the user can be imprecise and choose several answers of the MCQ, in the second one he has to give a precise answer. For both campaigns, the certainty of the user in his answer is required.

The plan of the paper is as follows. Section 2 reviews the theory of belief functions used for data modeling. Section 3 presents the defined crowdsourcing campaigns, the results obtained for the validation of H1 and H2 and the answer modeling. Section 4 concludes the paper.

2   Theory of Belief Functions

The theory of belief functions, introduced by Dempster [4] and formalized by Shafer [10], models the imprecision and uncertainty of imperfect sources. The user u of a platform is a source of information. Considering a question q asked to this user u, the finite set of possible answers to q composes the frame of discernment Ω. A mass function $m_{uq}^{\Omega} : 2^{\Omega} \to [0, 1]$ is defined such that:

$$\sum_{X \in 2^{\Omega}} m_{uq}^{\Omega}(X) = 1,$$

with 2^Ω the set of the disjunctions of Ω.

Let $X \in 2^\Omega$, the mass $m_{uq}^\Omega(X)$ characterizes the belief of the user u in the answer X to the question q. When $m_{uq}^\Omega(X) > 0$, X is called the focal element. A function $m_{uq}^\Omega(X) = 1$, $X \in 2^\Omega$, is a categorical mass function, the user is absolutely certain of this answer which may be imprecise if X is an union of elements of Ω. The set $\Omega \in 2^\Omega$ symbolizes ignorance, if $m_{uq}^\Omega(\Omega) = 1$ then the user is totally unaware of what the right answer is. The element $\emptyset \in 2^\Omega$, under the open world hypothesis, symbolizes a value outside Ω, in the case of normalized mass functions $m_{uq}^\Omega(\emptyset) = 0$.

The response X of u, can be modeled by a simple mass function (X^w):

$$\begin{cases} m_{uq}^\Omega(X) = \omega \text{ with } X \in 2^\Omega \setminus \Omega \\ m_{uq}^\Omega(\Omega) = 1 - \omega \end{cases} \tag{1}$$

This mass function allows to model: the uncertainty ω on the answer, the imprecision of u by the cardinality of X, and the remaining ignorance on the answer. That is the simplest way to model uncertainty and imprecision of the answer of the users. Consonant mass functions that can model different levels of imprecision, have all the focal elements nested. A consonant mass function is a possibility distribution.

In case of doubt about the reliability of a source, a weakening coefficient $\alpha \in [0, 1]$ modeling this reliability can be introduced:

$$\begin{aligned} m_{uq}^{\Omega,\alpha}(X) &= \alpha * m_{uq}^\Omega(X), \forall X \in 2^\Omega \setminus \Omega \\ m_{uq}^{\Omega,\alpha}(\Omega) &= 1 - \alpha * (1 - m_{uq}^\Omega(\Omega)) \end{aligned} \tag{2}$$

If the user u is absolutely unreliable then $\alpha = 0$ and the whole mass is assigned to Ω concluding to total ignorance. Weakening allows to reduce the conflicts occurring during the combination.

The main goal of crowdsourcing platforms is to fuse the answers of the users and decide the best answer to the questions. In the theory of belief function, many combination [8] operators exist. But the most used is the conjunctive rule of combination, given by Eq. (3), which requires that the sources be reliable, distinct and independent.

$$m_{Conj}^\Omega(X) = \sum_{Y_1 \cap \cdots \cap Y_N = X} \prod_{u=1}^N m_u^\Omega(Y_u) \tag{3}$$

This operator reduces the imprecision on the focal elements and increases the belief on the concordant ones. It can generate a non-zero mass on the empty set, so, in order to stay in a closed world, the normalized Dempster operator is preferred:

$$m_D^\Omega(X) = \frac{1}{1-k} m_{Conj}^\Omega(X) \tag{4}$$

with $k = m_{Conj}^\Omega(\emptyset)$ the global conflict from the sum of the partial conflicts. Once the combination of information has been achieved, it is necessary to return to a probabilistic framework to make a decision. To do this, it is possible to calculate the pignistic probability [1,3] on the elements of Ω.

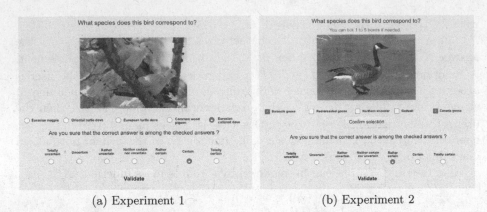

(a) Experiment 1   (b) Experiment 2

Fig. 1. Interfaces used for crowdsourcing campaigns

3   Experimental Validation of Hypotheses H1 and H2

This section presents the crowdsourcing campaigns conducted for data collection. Data is then analyzed to validate hypotheses H1 and H2, and the responses are modeled by the theory of belief functions.

3.1   Crowdsourcing Campaigns

Two crowdsourcing campaigns on the annotation of 50 bird photos have been conducted. For both experiments, and each question, a photo is presented to the user with five bird names as possible answers. The photos to be annotated are the same for both campaigns, as well as the corresponding names in response. All the proposed names are real and the 50 photos are only birds which can be met in natural place in Metropolitan France.

The interfaces of the campaigns are given in Fig. 1. For both campaigns, users specify their degree of certainty in their answer. The proposed degrees of certainty are summarized in Table 1. The user is notified that it is not penalizing to be uncertain in his answers. In the first campaign, corresponding to Experiment 1 in Fig. 1, the user is forced to choose a single answer by checking a radio button. In the second campaign, corresponding to Experiment 2 on Fig. 1, the user can be imprecise if he feels the need, by selecting 1 to 5 bird names by checking a checkbox button. At the beginning of this campaign, the user is informed that there is no penalty for selecting multiple names.

The campaigns were both carried out by 100 different users on the Crowd-panel platform[1], and a user who did the first campaign cannot do the second one. This makes a total of 50 photos × 100 users = 5000 data for each crowdsourcing campaign. The following section analyzes the collected data in order to validate Ph. Smets' hypothesis.

---

[1] https://crowdpanel.io/.

Table 1. Numerical values associated to the certainty scale proposed to the user

Totally uncertain	Uncertain	Rather uncertain	Neither certain nor uncertain	Rather certain	Certain	Totaly certain
0	1	2	3	4	5	6

3.2   Analysis of the Results

In order to validate H1 and H2, we recall that we conducted two experiments: in Experiment 1, the user is required to be precise, whereas in Experiment 2 the user can be imprecise. For both experiments, we associate to each answer of a user u to a question q, a numerical certainty value c_{uq}. The values of c_{uq} are taken in the interval $[0, 6]$, with 0 corresponding to total user uncertainty and 6 to total certainty, as shown in Table 1. Moreover, for Experiment 2, we associate an imprecision degree i_{uq} to each answer, that takes its values in the interval $[1, 5]$ according to the number of selected bird names.

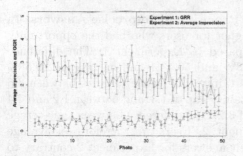
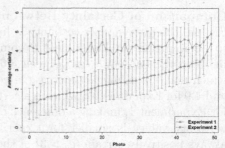

(a) GRR (Experiment 1) and average imprecision (Experiment 2) per photo.

(b) Average certainty per photo for both experiments.

Fig. 2. Comparison of experiments per photo - GRR, imprecision and certainty.

Task Difficulty. Certainty and imprecision of the user's answer may depend on the difficulty of the question according to the proposed bird names. In both crowdsourcing campaigns, the questions are of varying difficulty relative to the user's knowledge of the domain. The more difficult the question is, the more difficult it will be for the contributor to answer it. In order to evaluate the difficulty of each question for Experiment 1, one can calculate a good recognition rate (GRR), given by the average of the photos correctly annotated by the users. To compare the perceived difficulty of the questions by the users of the two campaigns, the GRR of Experiment 1 and the average imprecision of Experiment 2 are calculated for each question with 95% confidence intervals and presented in Fig. 2a. On Fig. 2b, the average user certainty for each question c_q is presented with a 95% confidence intervals. On both figures, bird photos are ordered according to the average certainty of the Experiment 1. The value c_q is increasing for Experiment 1 confirming a variable difficulty between questions. Comparing the two blue curves, we note a link between difficulty and certainty. Experiment 1 users have a GRR between 9% and 91%. The higher the GRR, the simpler the question.

Users who participated in Experiment 2 made good use of the opportunity to be imprecise since the average imprecision varies between 1.4 and 3.7 on Fig. 2a (with a minimum of 1 response to a maximum of 5 and an average of about 2 bird names selected). The more imprecise the user is, the more difficult the question is.

Figure 2a shows that as GRR increases, the average imprecision of users decreases, which means that users in both campaigns had difficulties with the same questions. The users of both experiments therefore have varying levels of knowledge about birds, which is usual in crowdsourcing platforms where the profiles are diverse. Since in both experiments users have varying degrees of knowledge, user confidence in their response will also vary.

Comparison of Certainty Between Precise and Imprecise Answers. In Fig. 2b, the average user certainty is higher for users who had the opportunity to be imprecise in Experiment 2 compared to Experiment 1. The difference between the two curves is between 8.68% and 50.13% and on average 29.68% which makes a significant increase in certainty. The value c_q varies between 3.6 and 4.9 for Experiment 2, which is a certainty gap of 1.3, and between 1.2 and 4.4 for Experiment 1, making a difference of 3.2. The difference between the average certainty values is smaller for Experiment 2. Users are therefore on average always positively certain ($c_q > 3$) of their answers for Experiment 2 contrary to the Experiment 1 where they sometimes uncertain. By having the possibility to be imprecise users are more certain of their answer and this certainty is more constant according to the difficulty. The users to Experiment 1 who were required to give a specific answer are less certain, which confirms hypothesis H2.

Analysis of the Use of Certainty and Imprecision for Experiment 2 Users. In order to understand the relations between certainty and imprecision of the users of Experiment 2, we plot the average user certainty as a function of user imprecision in Fig. 3a, and on the contrary in Fig. 3b, we plot the average user imprecision as a function of user certainty. In these figures a point represents a user positioned according to c_u and i_u.

In Fig. 3a, the values i_u are discretized and the average of c_u is realized for the degrees of imprecision 1 to 5. The red curve is a reference value, it presents the averages, on the answers, of the c_{uq} values for the degrees of imprecision 1 to 5. The average certainty made on the answers is increasing for an imprecision degree ranging from 2 to 5 selected answers. For precise answers, with only

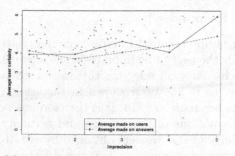
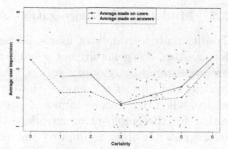

(a) Average certainty as a function of the degree of imprecision.

(b) Average imprecision as a function of the certainty.

Fig. 3. Relation between certainty and imprecision for Experiment 2.

one name chosen, the certainty is on average higher than for an imprecision degree of 2, but remains lower or equal to the imprecision degrees 3 to 5. This slightly higher certainty value for accurate responses is explained by the fact that some users in the crowd have more knowledge about birds. These qualified users manage to give accurate answers, while being certain. For the user curve, the certainty increases with the imprecision, except for a degree of imprecision of 4 where the certainty is lower than the reference value. This is due to three users who have a very low certainty c_u while they are on average very imprecise.

As in Fig. 3a, the c_u values are discretized and the averages of i_c is performed for each certainty value to obtain the green curve Fig. 3b. The red curve is a reference calculated by averaging the imprecision i_{uq} of the answers for the certainty values in Table 1. For positive certainty values, the average imprecision of the data is increasing. Symmetrically, the average imprecision is also increasing for negative certainty values, whereas one would expect the imprecision to decrease. Nevertheless, the certainty values 0 to 2 are represented by only 16% of the answers, so the variations in this part of the graph are less relevant than for the certainty values 3 to 6. Likewise, for the majority of users $c_u \geq 3$, for these values, the average imprecision of users increases with the certainty.

The obtained results show that allowing the user to be imprecise makes him more certain of his answers and requiring him to be precise less certain. It is notably observed that the more imprecise the user is, the more certain he is, which validates hypothesis H1. When the user chooses a single answer, if he has domain knowledge, he is certain, but if he does not have domain knowledge his answer will be uncertain. Therefore, in the absence of a qualification for the task, the more precise the user is the less certain he is, which validates H2. On the basis of the validated hypotheses H1 and H2, one can model the imprecision and uncertainty of the answers by belief functions.

3.3   Modeling and Aggregation of Responses

Traditionally, certainty of user's answer is not required in crowdsourcing platforms. Also, it is generally not possible for the user to be imprecise. The answers
are aggregated by majority voting, which consist of selecting the answer given
by the majority of the crowd. However, as it has been shown in the section
above, when they can be imprecise, users are more certain. It is therefore interesting for the employer to introduce the notions of imprecision and certainty
in the tasks in order to improve the quality of the answers obtained by crowdsourcing campaigns. Indeed, in crowdsourcing campaigns, sometimes the user
is in a situation of indecision but has to give a precise answer while hesitating
between different choices. The user is required to select an answer among all his
hesitations. This answer is then not very certain or even random which is not
desirable for the employer. On the contrary, by offering the user the possibility
to select the set of choices that he considers correct, the certainty of the latter is
higher. The collected responses can be modeled and aggregated using the theory
of belief functions that takes into account the imprecision and the certainty of
the answer.

For the conducted experiments, the user associates to his answer X to the
question q a certainty value c_{uq}. In the case where the answer is imprecise (such
as in Experiment 2), the imprecision correspond to the selected bird names. We
therefore propose to model the answers thanks to belief functions by a simple
mass function $X^{\omega_{uq}}$ with $\omega_{uq} = \frac{c_{uq}}{c_{MAX}}$, in this paper $c_{MAX} = 6$ and with X the
subset of selected bird names. Hence, $|X|$ is the number of selected bird names
and is equal to i_{cq}. With this mass function, the more certain the user is of his
answer, the higher the value of ω_{uq}. The mass function is then weakened by a
coefficient $\alpha = 0.8$ and aggregated by questions by the Dempster conjunctive
combination operator. Then, the decision on the answer is made by the pignistic
probability on the answer.

After the decision phase, a correct answer rate of 84% is obtained for Experiment 1, while for Experiment 2, this value is 90%. This 6% increase in the
correct answer rate between the two experiments is interesting for the employer
because it shows an improvement in the quality of the data collected. Moreover,
when the data is aggregated by majority voting, the correct answer rate is 70%
for Experiment 1 and 83% for Experiment 2. Even with majority voting it is
interesting to allow the contributor to be imprecise in case of indecision. The
modeling and aggregation of the answers is even more interesting with belief
functions, because it offers better results than the majority vote. By allowing
users to be imprecise, the employer can be more confident that the data collected
is of good quality.

4   Conclusion

Ph. Smets hypothesizes that the more imprecise one is, the more certain one
is and reciprocally, the more certain one is, the less precise one is. We carried
out two crowdsourcing campaigns, one where the user is required to be precise

(Experiment 1) and the other where he could be imprecise (Experiment 2). For both campaigns, the certainty of the contributor in his answer is required.

The users in Experiment 2 made good use of the opportunity to be imprecise. The analysis of the collected data shows that the average certainty per question is quite stable and higher for users who were able to be imprecise compared to those who had to be precise. In addition, when we plot the average of users' certainty according to the average of users' imprecision, we find that certainty increases with imprecision. The experimental analysis thus validates the hypothesis of Ph. Smets.

In order to model the uncertainty and imprecision of responses, the theory of belief functions is used. Currently we use simple support mass functions to represent the user's response. A mass equivalent to the user's certainty is associated to his answer. This gives more weight to answers that users are certain of when aggregating the data. Modeling and aggregating responses using belief function theory offers better results than the majority voting commonly used in crowdsourcing platforms.

In future work we would like to offer the contributor the possibility to select several sets of answers with different degrees of certainty in order to consider more than two focal elements. Consistent mass functions can then be used to model the responses.

References

1. Abassi, L., Boukhris, I.: A worker clustering-based approach of label aggregation under the belief function theory. Appl. Intell. **49**(1), 53–62 (2018). https://doi.org/10.1007/s10489-018-1209-z
2. Ambroise, C., Denœux, T., Govaert, G., Smets, P.: Learning from an imprecise teacher: probabilistic and evidential approaches. Appl. Stochastic Models Data Anal. **1**, 100–105 (2001)
3. Ben Rjab, A., Kharoune, M., Miklos, Z., Martin, A.: Characterization of experts in crowdsourcing platforms. In: Vejnarová, J., Kratochvíl, V. (eds.) BELIEF 2016. LNCS (LNAI), vol. 9861, pp. 97–104. Springer, Cham (2016). https://doi.org/10.1007/978-3-319-45559-4_10
4. Dempster, A.P.: Upper and lower probabilities induced by a multivalued mapping. Ann. Math. Stat. **38**, 325–339 (1967)
5. Diaz, J., Rifqi, M., Bouchon-Meunier, B., Jhean-Larose, S., Denhiére, G.: Imperfect answers in multiple choice questionnaires. In: Dillenbourg, P., Specht, M. (eds.) EC-TEL 2008. LNCS, vol. 5192, pp. 144–154. Springer, Heidelberg (2008). https://doi.org/10.1007/978-3-540-87605-2_17
6. Dubois, D., Prade, H., Smets, P.: Representing partial ignorance. IEEE Trans. Syst. Man Cybern. Part A: Syst. Humans **26**(3), 361–377 (1996)
7. Ipeirotis, P.G., Provost, F., Wang, J.: Quality management on Amazon mechanical Turk. In: KDD-HCOMP'10 (2010)
8. Martin, A.: Conflict management in information fusion with belief functions. In: Bossé, E., Rogova, G.L. (eds.) Information Quality in Information Fusion and Decision Making, pp. 79–97. Information Fusion and Data Science (2019)
9. Raykar, V.C., Yu, S.: Annotation models for crowdsourced ordinal data. J. Mach. Learn. Res. (2012)

10. Shafer, G.: A Mathematical Theory of Evidence. Princeton University Press, Princeton (1976)
11. Smets, P.: Imperfect information: imprecision and uncertainty. In: Motro, A., Smets, P. (eds.) Uncertainty Management in Information Systems: From Needs to Solutions, pp. 225–254. Springer, Boston (1997). https://doi.org/10.1007/978-1-4615-6245-0_8
12. Wagner, C., Anderson, D.T.: Extracting meta-measures from data for fuzzy aggregation of crowd sourced information. In: 2012 IEEE International Conference on Fuzzy Systems, pp. 1–8. IEEE (2012)
13. Wang, J., Ipeirotis, P.G., Provost, F.: Managing crowdsourcing workers. In: The 2011 Winter Conference on Business Intelligence, pp. 10–12 (2011)

Quantifying Confidence of Safety Cases with Belief Functions

Yassir Idmessaoud[1]([✉]), Didier Dubois[2], and Jérémie Guiochet[1]

[1] LAAS-CNRS, University of Toulouse, Toulouse, France
{yassir.id-messaoud,jeremie.guiochet}@laas.fr
[2] IRIT, University of Toulouse, Toulouse, France
dubois@irit.fr

Abstract. Structured safety argument based on graphical representations such as GSN (Goal Structuring Notation) are used to justify the certification of critical systems. However, such approaches do not deal with uncertainties that might affect the merits of arguments. In the recent past, some authors proposed to model the confidence in such arguments using Dempster-Shafer theory. It enables us to determine the confidence degree in conclusions for some basic GSN patterns. In this paper, we refine this approach and improve the elicitation method for expert opinions used in previous papers.

Keywords: Safety cases · Goal Structuring Notation (GSN) · Dempster–Shafer theory (DST) · Belief elicitation · Confidence assessment

1 Introduction

GSN (Goal Structuring Notation) is a graphical formalism used to represent argument structures (assurance cases, dependability cases, etc.). Originally, GSN structures do not include a representation of uncertainty in the arguments. Several independent works, e.g., [2,13] proposed to augment this approach to argumentation with confidence assessment methods. They design numerical confidence propagation models for some GSN patterns. However, in [2,15], the data collection method, that enable these mathematical models to be fed with initial confidence values, allowing the computation of the overall confidence in the system, needs improvement. The previous elicitation methods also present some technical defects. In this paper, after reviewing previous work in Sect. 2, we introduce an extensive confidence propagation method in Sect. 3, starting with a brief description of argument types. In Sect. 4, we present an improved expert opinion elicitation method. Finally, in Sect. 5, we illustrate our approach on a small example.

2 Related Work

The issue of confidence assessment in argument structures has been addressed in multiple works. Safety cases, which represent one type of such structures, aim

© Springer Nature Switzerland AG 2021
T. Denœux et al. (Eds.): BELIEF 2021, LNAI 12915, pp. 269–278, 2021.
https://doi.org/10.1007/978-3-030-88601-1_27

at proving the safety of a system by producing several pieces of documented evidence. In this paper, we focus on safety cases modeled by the so-called Goal Structuring Notation (GSN) defined in [9]. Safety argumentation is a core activity in safety critical systems development. Such an argumentation can be carried out using structured notations. It decomposes the safety requirements of the system into elementary pieces as presented in Fig. 1, called *goals*. *Strategies* are the components that justify this decomposition. Each goal is supported by one or multiple pieces of evidence called *Solutions*. The example of GSN in Fig. 1, is a classical pattern of an argument that treats the hazards existing in a system, and listed in the context box. However, many other patterns exist. This type of representation does not consider the uncertainty that may pervade each premise or the support relation between solutions and goals. Moreover, it is important to note that GSN models are non formal, and no explicit formal logical relations is expressed between elements.

In [6], we compared some works that deal with confidence assessment in GSN and give some recommendation to improve these methods. For instance, we showed why it is more adequate to use implication instead of equivalence (used in [13–15]) to represent argument types. We also discussed why Dempster rule of combination is more suitable for combining evidence, in our case, than other methods used in [2] for instance. Building such a confidence model relies on input values, usually provided by experts in qualitative form, and transformed into quantitative values. Such an activity could be called "Expert opinion elicitation". This method is more often used with probabilistic models. For instance, in [3], authors used an expert elicitation procedure in a risk assessment approach in fault trees. However, it can also be used in evidence theory. Ben Yaghlane et al. [16], generate belief functions from a preference relation between events provided by experts. In relation to our framework, few authors augmented their confidence assessment method by such a data elicitation procedure in order to

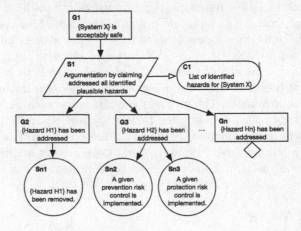

Fig. 1. GSN example adapted from Hazard Avoidance Pattern [10]

provide quantitative values for their models. Only some authors such as [2,15] used an elicitation method that transforms expert opinions given in the form of qualitative values, into quantitative ones.

Uncertainty propagation can be addressed by standard existing belief function software based on results in [11] (e.g., the belief function machine implemented in MatLab), but the GSNs we study have a particular tree-like structure that enable an explicit symbolic calculation of the belief function on the conclusion space. The explicit formulas, that can be obtained from approaches such as the one we propose, make the calculation more efficient and we can predict the effect of changing selected inputs, thus better explaining the obtained results, and validating the approach.

3   Argument Types

In this section, we define the argument types used in our confidence assessment method. Remember that an argument type represents the relationship between premises and a conclusion. For instance, in Fig. 1, (G_2) to (G_n) are the premises of (G_1) and they are all necessary to support it. This logical relation could be assimilated to the strategy component in a GSN (e.g., S1). When adopting a logical viewpoint [6], we then speak of a rule. Unlike the types of arguments proposed in [14], which use the equivalence connective to model such rules, we decided to break down this equivalence into two implications. Each implication brings a single piece of information about the conclusion, given premises. For instance, in the case of a single premise (P) supporting one conclusion (C), $P \Rightarrow C$ (that we call *direct rule*) can only infer the acceptance of the conclusion if the premise is true. On the other hand, the rule $\neg P \Rightarrow \neg C$ (that we call *reverse rule*) can only infer the rejection of the conclusion when the premise is false. We believe that, when assessing uncertainty of the relationship between P and C, this separate handling makes the resulting models more accurate and easy to assess. We also decided to associate a simple support mass function to each rule to avoid dependence in the confidence between premises. Below are uncertainty propagation formulas for various argument types; all calculation details are available in [7].

D-Arg (Disjunctive): In this situation, each premise can support alone the whole conclusion. We formally define this argument by : $\wedge_{i=1}^{n}(p_i \Rightarrow C)$. Rules that infer the rejection of the conclusion $(\neg C)$ can be deduced from this argument type by reversing this rule to obtain : $(\wedge_{i=1}^{n}\neg p_i) \Rightarrow \neg C$. To get formulas (1) and (2), we first assign to each rule a simple support mass function (resp. m_{dir}^i for $p_i \Rightarrow C$ and m_{rev} for the reverse rule). We also assign one mass function m_p^i to each premise p_i. This function uses three masses on p_i, $\neg p_i$ and the tautology $(\top)$ summing to one. Then, we combine, using DS rule of combination, all masses on premises together $(m_p = m_p^1 \oplus m_p^2 \oplus ... \oplus m_p^n)$ and masses on rules together $(m_r = [m_{dir}^1 \oplus m_{dir}^2 \oplus ... \oplus m_{dir}^n] \oplus m_{rev})$. Finally, we combine the resulting masses on the rules and premises ($m = m_p \oplus m_r$). We obtain degrees of belief and disbelief in C:

$$Bel_C(C) = 1 - \prod_{i=1}^{n}[1 - Bel_p^i(p_i)Bel_\Rightarrow^i(p_i \Rightarrow C)] \tag{1}$$

$$Disb_C(C) = Bel_\Leftarrow(\wedge_{i=1}^{n}[\neg p_i] \Rightarrow \neg C) \prod_{i=1}^{n} Disb_p^i(p_i). \tag{2}$$

We can notice that (1) expresses a *"Multivalued Disjunction"*. To have maximal belief in the conclusion, it is enough that the degree of belief in one single premise equals 1 (assuming that the mass on the direct rule also equals 1). Formula (2), in contrast, expresses a "Multivalued Conjunction". To have a maximal disbelief in the conclusion, all the disbelief degrees on premises should be equal to 1, supposing that the mass on the reverse rule equals 1 too. We can also notice that, when $Bel_C(C)$ is maximal, $Disb_C(C)$ is minimal.

C-Arg (Conjunctive): This argument type describes the situation when two premises or more are jointly needed to support a conclusion. Following the same reasoning as in the previous type, we define it formally by two rules: $(\wedge_{i=1}^{n}p_i) \Rightarrow C$ and its reverse $\wedge_{i=1}^{n}(\neg p_i \Rightarrow \neg C)$. Following the same calculation in the disjunctive type, we get the formulas below:

$$Bel_C(C) = Bel_\Rightarrow([\wedge_{i=1}^{n}p_i] \Rightarrow C) \prod_{i=1}^{n} Bel_p(p_i) \tag{3}$$

$$Disb_C(C) = 1 - \prod_{i=1}^{n}[1 - Disb_p^i(p_i)Bel_\Leftarrow^i(\neg p_i \Rightarrow \neg C)] \tag{4}$$

We can notice that, in contrast with formulas obtained for D-Arg, (3) and (4) respectively express a *"Multivalued Conjunction"* and *"Multivalued Disjunction"*.

H-Arg (Hybrid): This argument describes the case when it is difficult to choose between the conjunctive or disjunctive types. Each premise supports the conclusion to some extent, and the conjunction of the premises does it to a larger extent. We obtain degrees of belief and disbelief in C:

$$Bel_C(C) = Bel_\Rightarrow([\wedge_{i=1}^{n}p_i] \Rightarrow C) \times \prod_{i=1}^{n} Bel_p^i(p_i)[1 - Bel_\Rightarrow^i(p_i \Rightarrow C)]$$

$$+ \{1 - \prod_{i=1}^{n}[1 - Bel_p^i(p_i)Bel_\Rightarrow^i(p_i \Rightarrow C)]\}. \tag{5}$$

$$Disb_C(C) = Bel_{\Leftarrow}([\wedge_{i=1}^n \neg p_i] \Rightarrow \neg C) \times \prod_{i=1}^n Disb_p^i(p_i)[1 - Bel_{\Leftarrow}^i(\neg p_i \Rightarrow \neg C)]$$

$$+ \{1 - \prod_{i=1}^n [1 - Disb_p^i(p_i)Bel_{\Leftarrow}^i(\neg p_i \Rightarrow \neg C)]\}. \tag{6}$$

We can notice from (5) and (6) that these formulas subsume those in conjunctive and disjunctive types. On the one hand, if masses on $p_i \Rightarrow C$ are zero, it becomes the formula of the conjunctive type. On the other hand, if the mass on $[\wedge_{i=1}^n p_i] \Rightarrow C$ is zero, we get the disjunctive type formula.

This argument provides a general framework that allow as to calculate belief and disbelief values in different situations. D-Arg and C-Arg represent extreme cases where the value of some rules is null.

Note that moving away from these extreme cases may lead to encounter situations of conflict. A contradiction may appear when we have opposite opinion about two premises along corresponding direct and reverse rules. Formally, it always takes the form of a combination of four items of the form: $\{p_i, p_i \Rightarrow C, \neg p_j, \neg p_j \Rightarrow \neg C\}$. The sum $(Bel_C(C) + Disb_C(C))$ is then greater than 1. This may indicate something wrong in the GSN or in the way the experts replied questions, or yet on the reported experiments. Equation (7) represents the conflict calculation formula:

$$m(\perp) = \sum_{i=1, j \neq i}^n [Bel_p^i(p_i)Bel_{\Rightarrow}^i(p_i \Rightarrow C) \times Disb_p^j(p_j)Bel_{\Leftarrow}^j(\neg p_j \Rightarrow \neg C)] \tag{7}$$

To address this issue, we choose to subtract the mass of the conflict $m(\perp)$ from $Bel_C(C)$ and $Disb_C(C)$ in (5) and (6), and get contradiction-free degrees $bel(C) = Bel_C(C) - m(\perp)$ and $disb_C(C) = Disb_C(C) - m(\perp)$. We choose not to normalize the results (dividing by $1 - m(\perp)$) as proposed in the usual DS rule of combination because this operation will eliminate the conflict and proportionally increase the contradiction-free degrees of beliefs and disbelief $bel_C(C)$ and $disb_C(C)$ in a misleading way in the case of strong conflict. On the other hand, keeping $m(\perp)$ and subtracting it from $Bel_C(C)$ and $Disb_C(C)$ will legitimately increase uncertainty (i.e., $bel_C(C) + disb_C(C)$ is small) and show that the system is not that safe because of the presence of a conflict.

4 Expert Opinion Elicitation

In the previous section, we defined three argument types and proposed analytical formulas to calculate the belief and disbelief degrees in conclusions. However, using these models requires the presence of two types of information: belief degrees in premises (e.g., Bel_p^i) and the belief degrees for rules (e.g., $Bel_{\Rightarrow}^i$). In this section, we are first going to see how we can transform an expert opinion about a premise into belief, uncertainty and disbelief degrees. Then, in the second part, we provide some hints on how we can identify masses on rules.

4.1   Elicitation of Belief and Disbelief on Premises

In order to directly obtain belief and disbelief degrees in a premise p from an expert, authors in [2] consider asking two pieces of information: one, called decision index $Dec(p)$, describes which side the expert leans towards, acceptance or rejection of p; the other, called confidence $Conf(p)$, reflects the amount of information an expert possesses that can justify his opinion. Namely $Dec(p) = 1$ (resp., 0) indicates the certainty that p is true (resp false), while $Conf(p) = 1$ (resp., 0) indicates the expert has full (resp. no) information supporting the choice of $Dec(p)$. This is represented on Fig. 2.

The problem is then to define the belief and disbelief degrees in a proposition p in terms of $Dec(p)$ and $Conf(p)$. In [2], it is proposed to let $Bel(p) = Dec(p) \cdot Conf(p)$ and $Disb(p) = (1 - Dec(p)) \cdot Conf(p)$, which implies a natural result:

$$Conf(p) = Bel(p) + Disb(p). \tag{8}$$

However, it also implies that $Dec(p) = \frac{Bel(p)}{Bel(p)+Disb(p)}$. Note that this formula, presents a discontinuity in case of no information $(Bel(p) + Disb(p) = 0)$. The expression is then completed by assuming $Dec(p) = 1$ [2] or 0 [15], in this case, which sounds arbitrary.

It is more convincing to use the *Pignistic* transform [12] that turns a mass function m on a set Ω (the frame of discernment) into a probability, changing the focal sets into uniform distributions. When $\Omega = \{p, \neg p\}$ has two possible states, $Dec(p)$ is the midpoint between belief and plausibility of p, which reads:

$$Dec(p) = \frac{1 + Bel(p) - Disb(p)}{2} \tag{9}$$

Note that when $Bel(p) = Disb(p) = 0$, we get $Dec(p) = 1/2$.

Some authors suggest to define $Dec(p)$ by renormalising the pair $(Pl(p), Pl(\neg p)$, where $Pl(p) = 1 - Bel(\neg p)$, dividing them by $Pl(p) + Pl(\neg p)$ (plausibility transformation method [1]). This method is in agreement with Dempster rule of combination. However, we do not get the midpoint between belief and plausibility, which is intuitively surprizing, and in case of more than 3 elements in the frame, such a transformation may give probability values outside the range $[Bel, Pl]$ [5].

Using Eqs. (8) and (9) and the knowledge of $Dec(p)$ and $Conf(p)$, we can calculate belief and disbelief values: $Bel(p) = \frac{Conf(p)-1}{2} + Dec(p)$, $Disb(p) = \frac{Conf(p)+1}{2} - Dec(p)$. Viewing $Bel(p)$ as a lower probability, the pignistic transformation computes the center of gravity of the convex set of probabilities $\{P : P \geq Bel\}$.

However, the pignistic transform also presents one issue for the elicitation procedure. Some values of the pair $(Dec, Conf)$ provided by the expert may lead to negative values of belief $Bel(p)$ or disbelief degrees $Disb(p)$, which makes no sense. This is because there are constraints relating $Conf(p)$ and $Dec(p)$: (8) and (9) imply $1 - Conf(p) \leq \min(2Dec(p), 2(1 - Dec(p))$, which is known as Josang triangle [8]. To fix this problem, we can express the range of $Dec(p)$ for a given confidence level as:

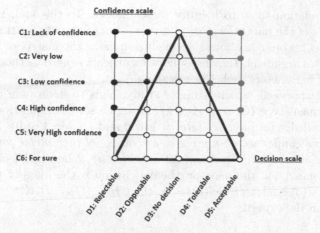

Fig. 2. Expert opinion extraction matrix

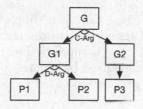

Fig. 3. GSN toy example

$$\frac{1 - Conf(p)}{2} \leq Dec(p) \leq \frac{1 + Conf(p)}{2} \tag{10}$$

For instance, a strong decision (full acceptance or rejection) should only be made when we have a very high level of confidence, since when $Conf(p) = 1$, $Dec(p)$ is not constrained and ranges on $[0, 1]$. In contrast, under ignorance $Conf(p) = 0$ enforces $Dec(p) = 1/2$. So, when the pair $(Dec(p), Conf(p))$ is situated outside the triangle, and $Dec(p) < \frac{1-Conf(p)}{2}$ (rejection: black dots on Fig. 2), we set $Dec(p) = \frac{1-Conf(p)}{2}$. On the other hand, when $Dec(p) > \frac{1+Conf(p)}{2}$ (acceptance: grey dots on Fig. 2), we set $Dec(p) = \frac{1+Conf(p)}{2}$. Choosing scales for $(Conf, Dec)$ and translating such pairs into numerical degrees is not trivial, we thus make the equidistance assumption for simplicity and to be comparable to previous works.

4.2   Determination of Belief Weights for Rules

Now, consider the mass functions for rules. Unlike belief and disbelief degrees in premises, belief degrees in rules are more difficult to obtain directly from an expert. Remember that a rule is representing a support relation between a conclusion and its premises. As a first approach, Wang et al. [14] proposed to exploit this relation so as to identify these masses. To this end, they propose certain values of the pair $(Dec, Conf)$ on premises, as inputs for the model, and ask the expert his opinion about the conclusion using the matrix of Fig. 2. Then, for each type of argument, they use a non-linear least square method to estimate the values of parameters (belief in rules).

A second approach, which is under study, aims to determine these masses and the argument type (C, D or H-Arg) through a series of questions, assuming clear-cut knowledge for premises ($Bel = 1$ or 0 and $Disb = 1$ or 0). For example, *If decision and confidence on a premise are maximal (acceptable for sure), what is your decision and confidence on the conclusion?* When belief or disbelief in premises are maximal, the mass on the conclusion is the mass of the rule. For instance, if in (3) we let $\forall i, Bel(p_i) = 1$, then $Bel_C(C) = Bel([\wedge_{i=1}^{n} p_i] \Rightarrow C)$ is obtained from the expert.

5   Toy Example

We apply our approach in this section to a simple GSN example presented in Fig. 3, including two types of arguments. It presents a top goal (G) supported by two sub-goals (G_1) and (G_2). (G_1) is supported by two premises (P_1) and (P_2). On the other hand, (G_2) is supported by a single premise (P_3). For simplicity, we chose a *C-Arg* for the argument type used to calculate (G) and a *D-Arg* in the calculation of (G_1).

In order to see how the variation of belief and disbelief degrees in the premises affect the conclusion, we show six different configurations in Table 1, where we set for each premise $(P_1, P_2$ and $P_3)$ a qualitative pair (decision, confidence) (and the corresponding pair (Bel, Disb)), and calculate the conclusion (G) by means of formulas (1), (2), (3) and (4). We also set the values of the masses on the rules to 1. As a result, the values of belief and disbelief in the conclusion will depend only on the masses on the premises.

Table 1. Qualitative (decision, confidence) and quantitative (belief, disbelief) pairs for the example (see Fig. 2 for the meaning of symbols)

	1st	2nd	3rd	4th	5th	6th
P_1	(R;C_6)	(A;C_5)	(A;C_5)	(T;C_5)	(A;C_5)	(T;C_6)
	(0 ; 1)	(0.8 ; 0)	(0.8 ; 0)	(0.65 ; 0.15)	(0.8 ; 0)	(0.75 ; 0.25)
P_2	(R;C_5)	(A;C_6)	(R;C_5)	(T;C_6)	(A;C_6)	(T;C_1)
	(0 ; 0.8)	(1 ; 0)	(0 ; 0.8)	(0.75 ; 0.25)	(1 ; 0)	(0 ; 0)
P_3	(R;C_6)	(A;C_6)	(A;C_6)	(O;C_6)	(A;C_2)	(T;C_6)
	(0 ; 1)	(1 ; 0)	(1 ; 0)	(0.25 ; 0.75)	(0.2 ; 0)	(0.75 ; 0.25)
G	(R;C_6)	(A;C_6)	(A;C_5)	(O;C_6)	(A;C_2)	(T;C_4)
	(0 ; 1)	(1 ; 0)	(0.8 ; 0)	(0.23 ; 0.76)	(0.2 ; 0)	(0.56 ; 0)

We can notice, on Table 1, that when we have either three rejectable or three acceptable premises with high levels of confidence (1st and 2nd columns), the models maintain the same decision with the same high level of confidence. On the other hand, when we have divergent opinions on the premises, either by opposite decisions (3rd and 4th columns) or opposite confidence levels (5th and 6th columns), the results will depend on the nature of the argument involved. In the 3rd and 6th column, decision levels (resp. acceptable and tolerable) were maintained because the divergence is located in a D-Arg. Due to its disjunctive nature, this argument favors the propagation of the premises that maximally support the conclusion. However, confidence levels were slightly decreased because of a C-Arg, which cumulates the uncertainty present in each premise and propagates it to the conclusion. In the 4th and 5th columns the divergence is located in a C-Arg. Unlike D-Arg, this argument favors the propagation of the premises that support the conclusion with the least strength. Thus, we end up with a mildly negative ("opposable") decision level in the 4th column and a very low level of confidence in the 5th column.

6 Conclusion

In this article, we propose a method for confidence assessment in GSN. It covers both the definition of argument types (belief propagation formulas) and data transformation (from elicited qualitative data to belief and disbelief pairs). We also illustrate this approach on a toy example. First results show that it was possible to improve previous work on uncertainty propagation and elicitation issues. We still need to conduct a full experiment for assessing beliefs in rules. We will investigate the expert questionnaire. We also want to propose an approach for automatic rule type identification. In the long range, we also plan to do away with the qualitative to quantitative transformation that contains some arbitrariness, by developing the purely qualitative approach to information fusion outlined in [4], and compare it to the quantitative one.

References

1. Cobb, B.R., Shenoy, P.P.: On the plausibility transformation method for translating belief function models to probability models. Int. J. Approx. Reason. **41**(3), 314–330 (2006)
2. Cyra, L., Górski, J.: Support for argument structures review and assessment. Reliab. Eng. Syst. Saf. **96**(1), 26–37 (2011)
3. De Persis, C., Bosque, J.L., Huertas, I., Wilson, S.P.: Quantitative system risk assessment from incomplete data with belief networks and pairwise comparison elicitation. arXiv preprint arXiv:1904.03012 (2019)
4. Dubois, D., Faux, F., Prade, H., Rico, A.: A possibilistic counterpart to shafer evidence theory. In: IEEE International Conference on Fuzzy Systems (FUZZ-IEEE), New Orleans, LA, USA, 23–26 June, pp. 1–6. IEEE (2019)
5. Dubois, D., Prade, H.: Practical methods for constructing possibility distributions. Int. J. Intell. Syst. **31**(3), 215–239 (2016)

6. Idmessaoud, Y., Dubois, D., Guiochet, J.: Belief functions for safety arguments confidence estimation: a comparative study. In: Davis, J., Tabia, K. (eds.) SUM 2020. LNCS (LNAI), vol. 12322, pp. 141–155. Springer, Cham (2020). https://doi.org/10.1007/978-3-030-58449-8_10

7. Idmessaoud, Y., Guiochet, J., Dubois, D.: Calculation of aggregation formulas for GSN argument types using belief functions. Technical report, LAAS-CNRS (2021). https://hal.laas.fr/hal-03210201

8. Jøsang, A.: Subjective Logic. Springer, Heidelberg (2016). https://doi.org/10.1007/978-3-319-42337-1

9. Kelly, T.: Arguing safety - a systematic approach to safety case management. Ph.D. thesis, Department of Computer Science, University of York, UK (1998)

10. Kelly, T.P., McDermid, J.A.: Safety case construction and reuse using patterns. In: Daniel, P. (ed.) Safe Comp 97, pp. 55–69. Springer, Heidelberg (1997). https://doi.org/10.1007/978-1-4471-0997-6_5

11. Shenoy, P.P., Shafer, G.: Axioms for probability and belief-function propagation. In: Shachter, R.D., Levitt, T.S., Kanal, L.N., Lemmer, J.F. (eds.) Uncertainty in Artificial Intelligence, Machine Intelligence and Pattern Recognition, vol. 9, pp. 169–198. North-Holland (1990)

12. Smets, P.: Decision making in the TBM: the necessity of the Pignistic transformation. Int. J. Approx. Reason. **38**, 133–147 (2005)

13. Wang, R., Guiochet, J., Motet, G., Schön, W.: D-S theory for argument confidence assessment. In: Vejnarová, J., Kratochvíl, V. (eds.) BELIEF 2016. LNCS (LNAI), vol. 9861, pp. 190–200. Springer, Cham (2016). https://doi.org/10.1007/978-3-319-45559-4_20

14. Wang, R., Guiochet, J., Motet, G., Schön, W.: Modelling confidence in railway safety case. Saf. Sci. **110**, 286–299 (2018)

15. Wang, R., Guiochet, J., Motet, G., Schön, W.: Safety case confidence propagation based on Dempster-Shafer theory. Int. J. Approx. Reason. **107**, 46–64 (2019)

16. Yaghlane, A.B., Denœux, T., Mellouli, K.: Elicitation of expert opinions for constructing belief functions. In: Uncertainty and Intelligent Information Systems, pp. 75–89. World Scientific (2008)

Algorithms and Computation

Discussions on the Connectedness of a Random Closed Set

Juan Jesús Salamanca[✉] [ID]

Universidad de Oviedo, Gijón 33203, Spain
salamancajuan@uniovi.es

Abstract. This work studies the connectedness of a random closed set in a Euclidean space. The well-known Choquet-Kendall-Matheron theorem states that a random closed set is characterized by its capacity functional. Consequently, any topological property must be also determined by such functional. In this work we consider connectedness. Under mild conditions, this property can be determined by taking into account only the capacity functional valued on predetermined finite families of compact and convex sets. The technique is based on the construction of an abstract simplicial complex associated with a cover of the support of the random closed set. We consider a new application in Probability, where we are able to approximate some probability computations.

Keywords: Random closed set · Connectedness · Capacity functional · Abstract simplicial complex

1 Introduction

Quite often data is vague, contains error or simply is imprecise. The random closed sets are nice models to deal with this imprecision. Instead of an exact output of the random experiment as in classical probability, now the output is a set. In general lines, a random closed set is a random element with values on the family of closed sets of the underlying space, [6,7].

The study of random closed sets has a long history, [6]. One of the main results related to this topic is the Choquet-Kendall-Matheron theorem, that states that a random closed set is characterized by its capacity functional -recall that the capacity functional T of a random closed set $\widehat{X}$ is a map which associates each compact set K with $\mathbb{P}\left(\widehat{X} \cap K \neq \emptyset\right)$. As a consequence, any topological property of the random closed set must be characterized by its capacity functional. In this work we focus on connectedness.

Observe that the T is supported on a very large family, the class of compact sets. We will not use the entire capacity functional T, but only some finite evaluations of it. More precisely, the problem of determining connectedness can be solved requiring that T is valued only on a predetermined finite family of

By project PGC2018-098623-B-I00.

T. Denœux et al. (Eds.): BELIEF 2021, LNAI 12915, pp. 281–290, 2021.
https://doi.org/10.1007/978-3-030-88601-1_28

compact sets. Then, the characterization of the connectedness needs to be based on a (simple) equation that such values must satisfy. In this framework, a suitable characterization of the connectedness of a random closed set can be found in [8] in the case of $\mathbb{R}^1$ and in [9] for higher dimension. There, the main idea is reducing the underlying topological space to a finite algebraic structure, an abstract simplicial complex, [5]. Then, a set induces a sub-abstract simplicial complex in a natural way, [5,9]. As a consequence, under certain assumptions on the random closed set, the associated random sub-abstract simplicial complex can describe the topology of $\widehat{X}$, partially at least.

This work tries to clarify several mathematical aspects of [8] and [9]. Moreover, we provide a nice application of its theoretical results. More concretely, we show how to compute $\mathbb{P}(d(X, Y) \leq \delta)$, where X and Y are random points of a metric space $(\mathbb{R}^n, d)$.

2   Preliminaries

Let us start with some basic definitions.

Definition 1 ([6]). *A random closed set $\widehat{X}$ on $\mathbb{R}^n$ is a map from a complete probability space $(\Omega, \sigma, \mathbb{P})$ to the class of closed sets of $\mathbb{R}^n$ such that for any compact set K of $\mathbb{R}^n$ it holds:*

$$\left\{ w \in \Omega : \widehat{X}(w) \cap K \neq \emptyset \right\} \in \sigma.$$

The measurability assumption in the previous definition implies that the map T that associates a compact set K with $T(K) := \mathbb{P}\left(w : \widehat{X} \cap K \neq \emptyset\right)$ is well-defined. This map is the *capacity functional* of $\widehat{X}$. Note that this map can be canonically extended to non-necessarily compact sets, [6]. According to the Choquet-Kendall-Matheron theorem, the capacity functional characterizes the random closed set, [6,7].

The main aim of this manuscript is to characterize the connectedness of a random closed set. This notion is defined as follows:

Definition 2. *Let $\widehat{X}$ be a random closed set of $\mathbb{R}^n$. This random closed set is connected if $\mathbb{P}\left(\widehat{X} \text{ is connected}\right) = 1$.*

We need to require some topological considerations on the random closed set $\widehat{X}$. Let us say that $\widehat{X}$ is *continuous* if any open set O of the Fell topology with $\widehat{X}^{-1}(O) \neq \emptyset$ satisfies $\mathbb{P}(\widehat{X}^{-1}(O)) > 0$, [6,9]. In particular, if the probability space Ω is discrete, then the random closed set is continuous. The same occurs when the probability space is a subset of a Euclidean space endowed with an absolutely continuous probability measure $\mathbb{P}$.

Assume further that the random closed set $\widehat{X}$ is contained in a convex and compact set K (that is, $\mathbb{P}(\widehat{X} \subseteq K) = 1$). Actually, here we are assuming that the random closed set is bounded (has compact support), [6,7]. The necessity to enlarge this support to a convex set comes from topological considerations. Following the literature, let us say that K is a *good support* of $\widehat{X}$.

2.1 Abstract Simplicial Complexes Associated with Convex Coverings

Now, consider a *convex covering* on the previous convex set K, [5,9]. That is, a finite covering $\mathcal{U} = \{U_i\}_{i=1}^r$ of K such that each U_i is a convex set. For example, a square grid. That is, $\mathcal{U} = \{U_{j_1,\dots,j_n} = \prod_{i=1}^n [a_i + h\, j_i, a_i + h\, (j_i+1)]\}_{(j_1,\dots,j_n) \in J \subset \mathbb{Z}^n}$, where each $U_{j_1,\dots,j_n}$ is a (topological) n-cube.

Recall that a covering of a topological space induces an abstract simplicial complex, which is an algebraic structure modelling the topology of the space, [5]. The abstract simplicial complex has as vertices (or 0-simplex set) the sets of the covering, $\mathcal{U} = \{U_i\}_{i\in I}$; now, $k+1$ vertices span a $k+1$-simplex if their corresponding U_j's have non-empty intersection (in such a case, the $k+1$-simplex is identified with the corresponding intersection), [3,5]. For instance, the above square grid in $\mathbb{R}^2$ yields to an abstract simplicial complex. All the details for a toy example are described below.

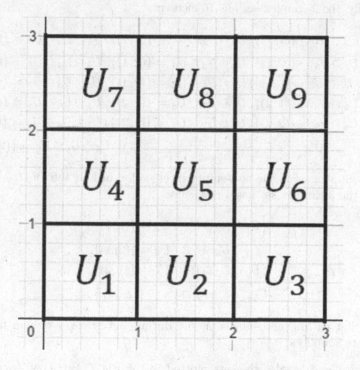

Fig. 1. A convex cover of $[0,3] \times [0,3]$.

Example 1. In Fig. 1 there are plotted 9 squares in $\mathbb{R}^2$: $\mathcal{U} := \{U_1 = [0,1] \times [0,1], U_2 = [1,2] \times [0,1], U_3 = [2,3] \times [0,1], U_4 = [0,1] \times [1,2], U_5 = [1,2] \times [1,2], U_6 = [2,3] \times [1,2], U_7 = [0,1] \times [2,3], U_8 = [1,2] \times [2,3], U_9 = [2,3] \times [2,3]\}$. These 9 sets form a convex covering of $[0,3] \times [0,3]$. These sets lead to the abstract simplicial complex characterized as follows:

- The 0-simplex set is $\mathcal{U}$. There are 9 0-simplices.
- The 1-simplex set is the non-empty intersection of two sets of $\mathcal{U}$. More precisely, the 1-simplex set has 20 elements:

$$\Big\{ U_1 \cap U_2 = \{1\} \times [0,1],\, U_1 \cap U_4 = [0,1] \times \{1\},\, U_1 \cap U_5 = \{(1,1)\},$$
$$U_2 \cap U_3 = \{2\} \times [0,1],\, U_2 \cap U_4 = \{(1,1)\},\, U_2 \cap U_5 = [1,2] \times \{1\},$$
$$U_2 \cap U_6 = \{(2,1)\},\, U_3 \cap U_5 = \{(2,1)\},\, U_3 \cap U_6 = [2,3] \times \{1\},$$
$$U_4 \cap U_5 = \{1\} \times [1,2],\, U_4 \cap U_7 = [0,1] \times \{2\},\, U_4 \cap U_8 = \{(1,2)\},$$
$$U_5 \cap U_6 = \{2\} \times [1,2],\, U_5 \cap U_7 = \{(1,2)\},\, U_5 \cap U_8 = [1,2] \times \{2\},$$
$$U_5 \cap U_9 = \{(2,2)\},\, U_6 \cap U_8 = \{(2,2)\},\, U_6 \cap U_9 = [2,3] \times \{2\},$$
$$U_7 \cap U_8 = \{1\} \times [2,3],\, U_8 \cap U_9 = \{2\} \times [2,3] \Big\}.$$

- The 2-simplex set is the non-empty intersection of three sets of $\mathcal{U}$. More precisely, the 2-simplex set has 16 elements:

$$\Big\{ U_1 \cap U_2 \cap U_4 = \{(1,1)\},\, U_1 \cap U_2 \cap U_5 = \{(1,1)\},\, U_1 \cap U_4 \cap U_5 = \{(1,1)\},$$
$$U_2 \cap U_3 \cap U_5 = \{(2,1)\},\, U_2 \cap U_3 \cap U_6 = \{(2,1)\},\, U_2 \cap U_4 \cap U_5 = \{(1,1)\},$$
$$U_2 \cap U_5 \cap U_6 = \{(2,1)\}\, U_3 \cap U_5 \cap U_6 = \{(2,1)\},\, U_4 \cap U_5 \cap U_7 = \{(1,2)\},$$
$$U_4 \cap U_5 \cap U_8 = \{(1,2)\},\, U_4 \cap U_7 \cap U_8 = \{(1,2)\},\, U_5 \cap U_6 \cap U_8 = \{(2,2)\},$$
$$U_5 \cap U_6 \cap U_9 = \{(2,2)\},\, U_5 \cap U_7 \cap U_8 = \{(1,2)\},\, U_5 \cap U_8 \cap U_9 = \{(2,2)\},$$
$$U_6 \cap U_8 \cap U_9 = \{(2,2)\} \Big\}.$$

- The 3-simplex set is the non-empty intersection of four sets of $\mathcal{U}$. More precisely, the 3-simplex set has 4 elements:

$$\Big\{ U_1 \cap U_2 \cap U_4 \cap U_5 = \{(1,1)\},$$
$$U_2 \cap U_3 \cap U_5 \cap U_6 = \{(2,1)\},$$
$$U_4 \cap U_5 \cap U_7 \cap U_8 = \{(1,2)\},$$
$$U_5 \cap U_6 \cap U_8 \cap U_9 = \{(2,2)\} \Big\}.$$

Note that with some changes on the initial convex cover we can find more appropriate simplices:

Example 2. Consider also the sets plotted in the Fig. 1. But now, consider the following convex cover of $[0,3] \times [0,3]$:

$$\mathcal{V} = \Big\{ V_1 = U_1 \cup U_2 \cup U_4 \cup U_5 = [0,2] \times [0,2],\, V_2 = U_2 \cup U_3 \cup U_5 \cup U_6 = [1,3] \times [0,2]$$
$$V_3 = U_4 \cup U_5 \cup U_7 \cup U_8 = [0,2] \times [1,3],\, V_4 = U_5 \cup U_6 \cup U_8 \cup U_9 = [1,3] \times [1,3] \Big\}.$$

This new convex cover leads to the abstract simplicial complex characterized as follows:

- The 0-simplex set is $\mathcal{U}$. There are 4 0-simplices.
- The 1-simplex set has 6 elements:

$$\Big\{ V_1 \cap V_2 = [1,2] \times [0,2],\ V_1 \cap V_3 = [0,2] \times [1,2],$$
$$V_1 \cap V_4 = [1,2] \times [1,2],\ V_2 \cap V_3 = [1,2] \times [1,2],$$
$$V_2 \cap V_4 = [1,3] \times [1,2],\ V_3 \cap V_4 = [1,2] \times [1,3] \Big\}.$$

- The 2-simplex set has 4 elements:

$$\Big\{ V_1 \cap V_2 \cap V_3 = [1,2] \times [1,2],$$
$$V_1 \cap V_2 \cap V_4 = [1,2] \times [1,2],$$
$$V_1 \cap V_3 \cap V_4 = [1,2] \times [1,2],$$
$$V_2 \cap V_3 \cap V_4 = [1,2] \times [1,2] \Big\}.$$

- The 3-simplex set has 1 element: $\Big\{ V_1 \cap V_2 \cap V_3 \cap V_4 = [1,2] \times [1,2] \Big\}$.

Note that all the simplices of this example are formed by geometric rectangles. Clearly, this example can be extended trivially to an arbitrary number of simplices that are also geometric rectangles.

Under mild conditions, the topology of the abstract simplicial complex coincides with the topology of the topological space, thanks to the nerve theorem [2–5]. Since any compact, convex set in $\mathbb{R}^n$ is contractible to a point [5], the abstract simplicial complex associated to a convex covering must also be topologically equivalent to the trivial abstract simplicial complex (that is, the latter contains a single vertex).

Now, we recall the Euler-Poincaré characteristic, [3,5]. This is a well-known topological invariant which describes the holes of a topological space. It can be proved that the Euler-Poincaré characteristic χ must be 1 for both, K and the abstract simplicial complex, [3,5]. For an abstract simplicial complex, it is the number of vertices (0-simplex) minus the number of edges (1-simplex) plus the number of faces (2-simplex) minus the number of 3-simplices, ..., [3,5].

A closed subset A induces a sub-abstract simplicial complex by restriction on the original structure: a k-simplex V of the abstract simplicial complex belongs to the sub-abstract simplicial complex if $A \cap V \neq \emptyset$.

The main topological problem is that the original abstract simplicial complex cannot describe the topology of subsets of the covered space, in general (see [5,9] for further details and examples). This is the main drawback with this approach. We simplify $\mathbb{R}^n$ to a finite algebraic structure but we lose the topology of the subsets of the original space. However, there exists a distinguished case when the previous fact occurs. In [9], it is proved that, for a *convex* set A (with $A \subseteq K$), the topology of the sub-abstract simplicial complex associated with a convex cover (of K) coincides with the topology of A. Moreover, if A is the union of r disjoint convex components, then the Euler-Poincaré characteristic of the sub-abstract simplicial complex must be between 1 and r.

3   Main Results

Now, let $\widehat{X}$ be a random convex closed set of $\mathbb{R}^n$. The crucial step is to show how $\widehat{X}$ induces a random sub-abstract simplicial complex. Let us clarify this idea. Let $\mathcal{C}$ denote an abstract simplicial complex. Denote also the set of sub-abstract simplicial complexes of $\mathcal{C}$ by $\mathcal{S}(\mathcal{C})$. That is, $\mathcal{S}(\mathcal{C})$ is constituted by those abstract simplicial complexes that are formed from $\mathcal{C}$ by restriction. Then, a random sub-abstract simplicial complex is a map from a complete probability space to $\mathcal{S}(\mathcal{C})$.

It can be proved the following key fact: if $\widehat{X}$ is connected, then the expected Euler-Poincaré characteristic of the induced random sub-abstract simplicial complex must be 1.

The collection of arguments in [8] and in [9] shows that the expectation of the number of k-simplices of the induced random sub-abstract simplicial complex coincides with $\sum_{v \in S^k} T(v)$, where S^k denotes the finite collection of k-simplices of the abstract simplicial complex (associated to the compact set K) and T is the capacity functional of the random closed set.

The main result is the following:

Theorem 1 ([9]). *Let $\widehat{X}$ be a continuous random closed set, bounded and convex by components, with K a good support of $\widehat{X}$. The random set $\widehat{X}$ is connected if and only if the following equation holds for any convex cover of K:*

$$\sum_i \sum_{v \in \mathcal{G}^{(i)}} (-1)^i T(v) = 1. \tag{1}$$

Note that $T(\emptyset) = 0$. This means that if some sets of the convex cover have empty intersection, then the capacity functional valued on that intersection must vanish identically. Therefore Eq. (1) can be written in a more tractable way: For a convex cover $\mathcal{U} = \left\{ U_i \right\}_{i=1}^m$ of the good support K of $\widehat{X}$, the following equation must hold if $\widehat{X}$ is connected:

$$\sum_{i=1}^m \sum_{1 \le j_1 < j_2 < ... < j_i \le m} (-1)^i T\left(U_{j_1} \cap U_{j_2} \cap ... \cap U_{j_i}\right) = 1. \tag{2}$$

Example (Continuation of example 2). Let $\widehat{X}$ be a random closed set of $[0,3] \times [0,3]$. The Eq. (1) associated to the abstract simplicial complex adopts the following form:

$$T([0,2] \times [0,2]) + T([1,3] \times [0,2]) + T([0,2] \times [1,3]) + T([1,3] \times [1,3])$$
$$+2\,T([1,2] \times [1,2]) - T([0,2] \times [1,2]) - T([1,2] \times [1,3]) = 1. \quad (3)$$

If $\widehat{X}$ is connected, then the Eq. (3) must be satisfied. Observe that Eq. (3) coincides with the Eq. (2) applied to the same convex cover.

4   An Application to Probability

Let us apply the previous theoretical tools to a concrete problem. Let X and Y be two independent random points of $\mathbb{R}^2$. Assume that there exists a convex set K such that $\mathbb{P}(X \in K) = \mathbb{P}(Y \in K) = 1$. Endow $\mathbb{R}^2$ with a distance map d. We need to compute $\mathbb{P}(d(X,Y) \leq \delta)$. This problem has different motivations and applications. For instance, the well-known case of the travelling salesman problem. Other examples of application can be found in astronomy or in optimization theory. For a nice reference discussing about these topics we may recommend [1] and references therein.

If X or Y has a complex structure, sampling could not be a good option. For instance, when there are known the marginal distributions and the copula has a complex analytic structure. Let see how the previous theoretical reasoning induces an algorithm to estimate this probability. Begin noting that $\mathbb{P}(d(X,Y) \leq \delta)$ coincides with the probability that the random closed set $\widehat{Z} := \mathbb{B}_X(\delta) \cup \{Y\}$ is connected, where $\mathbb{B}_p(r)$ denotes the d-ball with centre p and radius r. It is easy to see that the random closed set $\widehat{Z}$ is convex by components (under simple assumptions on d). Moreover, it has only two connected components. Consequently, $1 + \mathbb{P}(\widehat{Z}$ is not connected) can be estimated by the expectation of the Euler-Poincaré characteristic associated to a convex cover of K. Consequently, from the Eq. (1), $\mathbb{P}(d(X,Y) \leq \delta)$ can be estimated from a convex cover of K by:

$$2 - \sum_i \sum_{v \in \mathcal{G}^{(i)}} (-1)^{i+1} \, T(v), \quad (4)$$

where T is the capacity functional associated to $\mathbb{B}_X(\delta) \cup Y$. Analytically, T satisfies: $T(K) = \mathbb{P}(X \in K^\delta \vee Y \in K)$, where $K^\delta := \{p \in \mathbb{R}^n : \exists q \in K, d(p,q) \leq \delta\}$.

Let see briefly a more concrete example. Consider a Cartesian coordinate system of $\mathbb{R}^2$ and take the following distance map: $d((x_1, x_1), (y_1, y_2)) = \max\{|x_1 - y_1|, |x_2 - y_2|\}$. It is clear that $([a,b] \times [c,d])^\delta = [a-\delta, b+\delta] \times [c-\delta, d+\delta]$ for any real numbers a, b, c, d with $a \leq b$ and $c \leq d$.

Let X and Y be two independent random points of $\mathbb{R}^2$, such that $\mathbb{P}(X \in [A, B] \times [C, D]) = \mathbb{P}(Y \in [A, B] \times [C, D]) = 1$ (or ≈ 1). Here, $[A, B] \times [C, D]$ plays the role of the support K as above. Fix two integers m_x and m_y, the number of vertices in the x-axis and the number of vertices in the y-axis. Define the following covering $\mathcal{U}$ of $[A, B] \times [C, D]$: $\mathcal{U} = \{[A + (B - A) \cdot (i - 1)/m_x, A + (B - A) \cdot i/m_x] \times [C + (D - C) \cdot (j - 1)/m_y, C + (D - C) \cdot j/m_y]\}_{i \in \{1, \ldots, m_x\}, j \in \{1, \ldots, m_y\}}$. This covering induces an abstract simplicial complex. Due to limitations in the page format, we left the full details to the reader. For instance, the 4-simplex set is $\{(A + (B - A) \cdot i/m_x, C + (D - C) \cdot j/m_y)\}_{i \in \{1, \ldots, m_x - 1\}, j \in \{1, \ldots, m_y - 1\}}$; the elements of the 3-simplex set are the elements of the 4-simplex set with multiplicity 4; the elements of the 2-simplex set are segments joint with the elements of the 4-simplex set with multiplicity 2.

With these elements, the right side of the Eq. (4) reads as follows (we left the details about $J_1, \ldots, J_4$ below to the reader):

$$\sum_{(i,j) \in J_2 \subset \mathbb{Z}^2} T(\{(A + (B - A) \cdot i/m_x, C + (D - C) \cdot j/m_y)\})$$

$$+ \sum_{(i,j) \in J_1 \subset \mathbb{Z}^2} T(\{A + (B - A) \cdot i/m_x\} \times [C + (D - C) \cdot (j - 1)/m_y, C + (D - C) \cdot j/m_y])$$

$$- \sum_{(i,j) \in J_3 \subset \mathbb{Z}^2} T(\{A + (B - A) \cdot i/m_x\} \times [C + (D - C) \cdot (j - 1)/m_y, C + (D - C) \cdot j/m_y])$$

$$- \sum_{(i,j) \in J_4 \subset \mathbb{Z}^2} T(\{(A + (B - A) \cdot (i - 1)/m_x, C + (D - C) \cdot j/m_y)\}) ,$$

where the related capacity functional T satisfies

$$T([a, b] \times [c, d]) = \mathbb{P}(X \in [a - \delta, b + \delta] \times [c - \delta, d + \delta]) \cdot \mathbb{P}(Y \in [a, b] \times [c, d]) .$$

Note.- It can be proved that the previous algorithm overestimates $\mathbb{P}(d(X, Y) \leq \delta)$. In the underground machinery, the convex cover may not detect whether $\mathbb{B}_X(\delta) \cup \{Y\}$ is connected if $d(X, Y)$ is a little bigger than δ. For that cases, the computation of the Euler-Poincaré characteristic leads to 1, instead of 2 as corresponds. This fact tells how the different parameters should be configured: m_x and m_y such that $(B - A)/m_x \ll \delta$ and $(D - C)/m_y \ll \delta$. It can also be proved that when m_x and m_y tends to infinity, the previous algorithm tends to $\mathbb{P}(d(X, Y) \leq \delta)$. The details may be published elsewhere.

Let us close this application section providing explicitly a program in R to compute $\mathbb{P}(d(X, Y) \leq \delta)$, when X and Y are random points of $\mathbb{R}^2$ endowed with the maximum distance (that is, $d((x_1, x_2), (y_1, y_2)) = \max\{|x_1 - y_1|, |x_2 - y_2|\}$ for any $(x_1, x_2), (y_1, y_2) \in \mathbb{R}^2$). This program has been configured for easy probability laws of the random points, because of clarity in the lecture.

```
##Probability law first random point X = (X1, X2)
 # The marginal distribution of X1
FX1 <-function(x){pnorm(x,mean=50,sd=4) }
# The other marginal distribution of X: X2
FX2 <-function(x){pnorm(x,mean=50,sd=4) }
# The copula for the marginal distributions
tnorm1 <-function(p,q){p*q}
# The probability law of X = (X1, X2)
FX1X2 <- function(a,b,c,d){
  tnorm1(FX1(b),FX2(d))+tnorm1(FX1(a),FX2(c))
                 -tnorm1(FX1(b),FX2(c))-tnorm1(FX1(a),FX2(d))}


##Probability law second random point Y = (Y1, Y2)
# The marginal distribution of Y1
FY1 <-function(x){pnorm(x,mean=50,sd=4) }
# The other marginal distribution of Y: Y2
FY2 <-function(x){pnorm(x,mean=50,sd=4) }
 # The copula for the marginal distributions
tnorm2 <-function(p,q){p*q}
#Probability law of (Y1,Y2)
FY1Y2 <- function(a,b,c,d){
  tnorm2(FY1(b),FY2(d))+tnorm2(FY1(a),FY2(c))
   -tnorm2(FY1(b),FY2(c))-tnorm2(FY1(a),FY2(d))}


#Capacity law of the random closed set
# |B_(X1,X2) (delta) U { (Y1,Y2) }; X and Y independent
capacity <- function(a,b,c,d,delta){
  prob1 = FX1X2(a-delta,b+delta,c-delta,d+delta)
  prob2 = FY1Y2(a,b,c,d)
  prob3 = prob1+prob2-prob1*prob2
  return(prob3)
}

#Distribution law:: calculates P( max{ |X1-Y1|, |X2-Y2|} < delta )
# when support [A,B] x [C, D] and step h
distributionlaw <- function(delta,A,B,C,D,h){
euler =0
mx=ceiling((B-A)/h)
my = ceiling((D-C)/h)
for(i in 1:mx){for(j in 1:my){
  euler=euler+ capacity(A+h*i ,A+h*(i+1) , C+h*j ,C+h*(j+1),delta)+
  capacity(A+h*i ,A+h*i , C+h*j ,C+h*j,delta)
    -capacity(A+h*i ,A+h*i , C+h*j ,C+h*(j+1),delta)
    -capacity(A+h*i ,A+h*(i+1) , C+h*j ,C+h*j,delta)
}}
probconnected = 2- euler
return(probconnected)
}
```

5   Conclusions

We have characterized the connectedness of a random closed set in $\mathbb{R}^n$. We have
that if the random closed set is connected, then it must satisfy an equation asso-

ciated to any convex cover of its support. We have computed that equation for a concrete convex cover, illustrating the general procedure. Furthermore, we have provided a nice application in Probability. More precisely, we have approximated the probability that two random points distance less than a given number.

Note that other applications may arise in other contexts.

As a future work, it is worth to study the connectedness of a random closed set in a more general space than $\mathbb{R}^n$. For instance, in the n-dimensional sphere $\mathbb{S}^n$, the torus $\mathbb{S}^n \times \mathbb{S}^m$ or the cylinder $\mathbb{R}^n \times \mathbb{S}^m$. In those spaces, an extra difficulty appears, since the topology of the space must have distinguished from the topology of the random closed set.

References

1. Bhattacharyya, P., Chakrabarti, B.K.: The mean distance to the nth neighbour in a uniform distribution of random points: an application of probability theory. Eur. J. Phys. **29**, 639–645 (2008)
2. Cavanna, N.J., Sheehy, D.R.: The generalized persistent nerve theorem. arXiv preprint arXiv:1807.07920 (2008)
3. Fritsch, R., Piccinini, R.: Cellular Structures in Topology. Cambridge University Press, Cambridge (1990)
4. Govc, D., Skraba, P.: An approximate nerve theorem. Found. Comput. Math. **18**(5), 1245–1297 (2018)
5. Hatcher, A.: Algebraic Topology. Cambridge University Press, Cambridge (2005)
6. Molchanov, I.: Theory of Random Sets. Springer, Heidelberg (2005)
7. Nguyen, H.T.: An Introduction to Random Sets. CRC Press, Boca Raton (2006)
8. Salamanca, J.J.: On the connectedness of random sets of $\mathbb{R}$. Int. J. Uncertain. Fuzziness Knowl.-Based Syst. (2020)
9. Salamanca, J.J., Herrera, J., Rubio, R.M.: On the connectedness of a random closed set in a Euclidean space (submitted)
10. Walley, P.: Peter: Statistical Reasoning with Imprecise Probabilities. Chapman and Hall, London (1991)

An Efficient Computation
of Dempster-Shafer Theory of Evidence
Based on Native GPU Implementation

Noelia Rico[1]([✉])[ID], Luigi Troiano[2][ID], and Irene Díaz[1][ID]

[1] Computer Science Department, University of Oviedo, Oviedo, Spain
{noeliarico,sirene}@uniovi.es
[2] Department of Innovation Systems, University of Salerno, Fisciano, Italy
ltroiano@unisa.it

Abstract. Interest in Dempster-Shafer's theory of evidence has often run up against problems associated with the inherently exponential complexity of the calculation of the Belief and Plausibility measures. This can be mitigated by looking at the technological possibilities offered by GPU computing. Some preliminary attempts have been oriented towards parallelization of the computation, but none of them natively use the support offered by lower-level GPUs. In this paper, we introduce a set of Python functions for operations related to the Dempster-Shafer theory and outline its implementation based natively on GPU computing, highlighting the speedup possibilities in relation to a CPU-based or CPU-derived implementation.

Keywords: Parallel computing · GPU · Computational complexity

1 Introduction

The Dempster-Shafer theory of evidence (DST) [3,6] provides an expressive framework for reasoning with uncertainty. It includes probability theory as a special case and is able to express imprecise probabilities. Given a universe set Ω, referred to as *frame of discernment* in the DST context, basic probabilities are allocated to its subsets, instead of being allocated to single elements (on the contrary to what happens in the case of probability theory). The derived measures of Plausibility and Belief for a subset, which determine a probability interval assigned to that subset, are calculated from the relationships between the subset and the basic assignment of probabilities on all subsets of Ω. This theory is broadly applied in many fields. For example, in [7,8] a reference model is proposed to analyze the relationships between users and media content that is based on Dempster-Shafer's theory. In [4], it is studied the next word prediction model based on Dempster-Shafer's combination rule. This approach manages high conflicting evidence.

For any application of this theory, the need to deal with the power set 2^Ω, i.e., the collection of all potential subsets of a universe set Ω, poses significant

© Springer Nature Switzerland AG 2021
T. Denœux et al. (Eds.): BELIEF 2021, LNAI 12915, pp. 291–299, 2021.
https://doi.org/10.1007/978-3-030-88601-1_29

problems in terms of scalability, being this a problem for the DST computation and consequently its use in real contexts. It is well known [1] that it is not possible to directly implement the computation of the Belief and Plausibility measures, as the time complexity is exponential on the number of elements in Ω. In addition, Dempster's rule of combination is #P-complete [5]. Some studies have focused on this issue proposing different solutions, mostly based on restrictions that limit their practical use. In [2] it is proposed a parallel computing approach for Dempster's rule of combination based on the concept of conquer and divide algorithms, inspired from the spawning technique of a single-celled bacterium.

In this paper we present Python functions for doing DST-related operations. First of all, an implementation scheme to compute the Belief and Plausibility measures defined by DST is given. Furthermore, a function to combine basic probability assignments from different sets as defined by DST is also provided. The functions of the package have been developed using parallel computing by means of a GPU, which allows to guarantee execution time even for large numbers of elements. Experimental results to support the benefits of the approach showing the tractability of the execution time are also given in this work. Moreover, these functions have been made open source so they can be used by anyone.

The remainder of this paper is organized as follows. In Sect. 2 the concepts of the Dempster-Shafer theory that are implemented are detailed. Section 3 highlights the problem regarding the computation of the Belief and Plausibility according to this theory and the parallel algorithm proposed to solve this problem is explained. In Sect. 4, the performance of the GPU functions is evaluated, showing that the reduction of the execution time for computing the Belief and Plausibility is such that, in some cases, goes from hours to less than one second. In the last section, the advances proposed in this paper are summarized and future research on this topic is outlined.

2   Preliminaries

In this section, the basic concepts related to Dempster-Shafer theory [3,6] are presented. In the following we denote the *frame of discernment* (i.e. the set of all the hypotheses) by Ω. Belief function represents the degree of belief to which the evidence supports and Plausibility function refers to the degree of belief to which a set is feasible.

Definition 1. *A function* $m : 2^{\Omega} \longrightarrow [0,1]$ *over* Ω *is called a* basic probability assignment *iff*

$$m(\emptyset) = 0 \quad and \quad \sum_{S \in 2^{\Omega}} m(S) = 1$$

Definition 2. *Any* $S \in 2^{\Omega}$ *is a focal element iff* $m(S) > 0$. *In addition, we name focal set the collection of focal elements* $F_m(\Omega) = \{S \subseteq \Omega \mid m(S) > 0\} \subseteq 2^{\Omega}$.

Definition 3. *The Belief measure of $A \subseteq \Omega$ induced by the basic probability assignment function m is defined as*

$$Bel(A) = \sum_{S \subseteq A} m(S) \tag{1}$$

Definition 4. *The Plausibility measure of $A \subseteq \Omega$ induced by the basic probability assignment function m is defined as*

$$Pl(A) = \sum_{S \cap A \neq \emptyset} m(S) \tag{2}$$

Definition 5. *Let m_1 and m_2 be two basic probability assignments, the joint basic probability assignment, i.e. the Dempster's combination rule (DCR), is computed as*

$$m_{1,2}(A) = \frac{1}{1-Z} \sum_{B \cap C = A} m_1(B) \cdot m_2(C) \tag{3}$$

where

$$Z = \sum_{B \cap C = \emptyset} m_1(B) \cdot m_2(C) \tag{4}$$

is a measure of conflict between the two basic probability assignment sets. In addition, $m_{1,2}(\emptyset) = 0$ by definition.

A reference algorithm to compute the Belief and Plausibility over all the possible subsets of Ω is given by Algorithm 1. This algorithm explores the lattice 2^Ω, which contains all the possible combinations of elements $A \subseteq \Omega$ for which is necessary to compute the Belief and Plausibility. This operation is done by checking the overlapping between the node and all the focal elements $S \in F_m(\Omega)$.

Algorithm 1: Compute Belief and Plausibility given Ω and $F_m(\Omega)$

Result: Two arrays Belief *Bel* and Plausibility *Pl* with these values for each element of the lattice $A \in 2^\Omega$.

Input: Set Ω. $F_m(\Omega)$ and their basic probability assignment.

1 Initialize an array *Bel* with one element $Bel(A) = 0$ for each node $A \in 2^\Omega$;
2 Initialize an array *Pl* with one element $Pl(A) = 0$ for each node $A \in 2^\Omega$;
3 **foreach** $A \in 2^\Omega$ **do**
4   **foreach** $S \in F_m(\Omega)$ **do**
5   **if** $S \cap A \neq \emptyset$ **then**
6 | $Pl(A) = Pl(A) + m(S)$;
7   **end**
8   **if** $S \subseteq A$ **then**
9 | $Bel(A) = Bel(A) + m(S)$;
10   **end**
11   **end**
12 **end**

Notice that, if this algorithm is developed by means of a sequential implementation and executed in a CPU, the execution time is governed by both the number of elements in Ω (which determines the size of the lattice of nodes to explore, making the execution time growing exponentially) and the number of focal elements.

The main aim of this work is to provide an alternative implementation that takes advantage of the parallel execution in a GPU in order to reduce the execution time required to compute the Belief and Plausibility of all the nodes in the lattice 2^{Ω}. Moreover, we also provide a GPU implementation for the combination of two different sets of focal elements in order to obtain a joint mass assignment as described in Definition 5.

3   The Computation Problem of Belief, Plausibility and Dempster's Combination Rule

There are broadly two types of programming languages, *compiled* and *interpreted*. On the one hand, compiled programming languages translate a program in a high-level language into a binary executable that contains instructions that the machine can execute. On the other hand, interpreted programming languages execute the code one line at a time as it is read from the source code. This makes compiled programming languages to be usually faster, as the result is a file that the operating system can run without further ado. For this reason, the execution time of any implementation depends on the programming language chosen.

3.1   A Python Implementation

The increasing popularity of the Python programming language in the last years makes it a suitable language for practitioners from all fields. Unfortunately, this is an interpreted language that does not stand out for being fast, which have an effect on the execution time of methods with heavy computation tasks. Numba is a Just-in-Time (JIT) compiler that allows the translation of the Python code to CPU efficient code by using the LLVM (Low Level Virtual Machine) compiler, which allows to speed up the execution time of the implementation. In order to do so, first, it analyses the Python code and turns it into an *LLVM IR* (intermediate representation), then it creates the corresponding bytecode for the selected architecture in which the function is going to be executed, which corresponds with the architecture where the host Python runtime is running on, making its execution faster.

3.2   Parallelization Based on GPU Computing

In recent years, GPUs have moved from graphic processing purpose to general purpose, as they have gain popularity for being powerful devices to execute parallel computing algorithms. The architecture of the GPUs differ from CPUs, which requires the algorithms to be designed using a different philosophy depending on how they are going to be executed.

In the last decade, CUDA has emerged as a tool for GPU code development, as it includes full extensions for writing code suitable to be executed in a GPU using classic programming languages such as C/C++ and Python.

In the previous section, we pointed out the drawbacks of a sequential implementation of Algorithm 1. The aim of this work is to improve the execution time of such kind of implementation by providing a parallel implementation that can be executed in a NVIDIA GPU.

In order to make the code efficient, for the GPU approach we represent each node of the lattice by a unique integer number. Therefore, given a set Ω containing n elements, the lattice formed by 2^n elements to explore requires numbers codified with n bits to represent all the possible combinations $A \subseteq \Omega$. Under this understanding, the element 0 represents for any number of bits the empty set and, consequently, the number $2^n - 1$ represents the full set Ω. Then, each node of the lattice is assigned exclusively to one of the threads in the GPU that are executed in parallel, matching the index of the set with the integer representing the node. The corresponding thread takes care of computing the Belief and Plausibility associated with the node by exploring its overlapping with all the focal elements $S \in F_m(\Omega)$. Therefore, following Algorithm 1, in the GPU implementation the loop in line 3 is parallelized in different threads, and each thread takes care of the inner loop over $F_m(\Omega)$ to compute the Belief and Plausibility associated with its corresponding node. By doing this, the execution time is relaxed as now it relies only on the number of focal elements, because the number of elements can be done in parallel whenever the GPU allows such number of threads.

As an example, let us define the set $\Omega = \{A, B, C, D\}$ with a focal set $F_m(\Omega)$ = $\{C, BC, AD\}$ with basic probability assignments $\{0.5, 0.4, 0.1\}$. Figure 1 shows the binary representation associated to each node, as well as its corresponding integer number, which is also the identifier of its associated thread. Each thread goes individually over the set of focal elements and computes the Belief and Plausibility of its associated node.

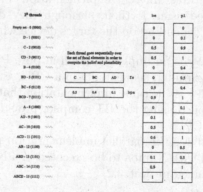

Fig. 1. Example of the parallel computation of the belief and plausibility given the set of items $F_m(\Omega) = A, B, C, D$ and the focal set $F_m(\Omega) = \{C, BC, AD\}$ with corresponding basic probability assignments $m(C) = 0.5, m(BC) = 0.4, m(AD) = 0.1$.

3.3   The Computation of DCR

A function to compute the joint basic probability assignment obtained from two focal set of elements $F_{m_1}(\Omega)$ and $F_{m_2}(\Omega)$ is also provided. This function has a native implementation based on the following steps:

1. Each thread evaluates, in parallel, one of the different pairs of focal elements $\{(B, C) \mid B \in F_{m_1}(\Omega) \wedge C \in F_{m_2}(\Omega)\}$, computing its intersection as well as the corresponding result obtained from $m_1(B) \cdot m_2(C)$.
2. Then, one single thread goes sequentially over all the obtained intersections, grouping the ones that represent the same focal element in order to create the resulting focal set $F_3(\Omega)$ and the corresponding basic probabilities assignment of the focal elements.
3. The threads are again used to divide in parallel the summed masses obtained for all $A \in F_3(\Omega)$ by $1 - Z$ (see Definition 5).

4   Empirical Proof of Time Reduction Using a GPU

The functions described in this work can be found at https://github.com/ noeliarico/belief. The execution times shown in this section have been measured using a Google Colab, which is freely available and provides a Tesla T4 for GPU execution and an Intel(R) Xeon(R) CPU @ 2.20GHzm, and 13GB of RAM memory.

In order to measure the execution time saved from using a GPU instead of a CPU to solve the computation of the Belief and Plausibility, we have benchmarked three different implementations of Algorithm 1. All of them have been developed using Python, although they differ in how they are compiled and written depending on the target device where they will be executed:

- AlgIPy is an implementation in pure Python 3.7 as an interpreted language and executed sequentially in a CPU.
- AlgCPy is the same implementation that AlgIPy but, in this case, it has been compiled using the JIT compiler provided by Numba, and it is also executed in a CPU.
- AlgGPU is the parallel implementation proposed in this work, which is compiled with Numba to be executed in a NVIDIA GPU with the considerations explained in Sect. 3.2.

Combinations of different sizes of Ω (which determines the number of nodes to explore in 2^{Ω}) and number of focal elements have been tested in order to observe the behaviour of the different implementations. Table 1 shows the execution times obtained for each combination and implementation.

Table 1. Execution time (in seconds) of the different implementations tested in order to compute the Belief and Plausibility for a set Ω given the focal elements $F_m(\Omega)$ and their corresponding basic probability assignments.

AlgIPy items/focals	2^2	2^4	2^9	2^{14}	2^{19}
5	0.0002	0.0006	–	–	–
10	0.0046	0.0182	0.5755	–	–
15	0.1683	0.5773	17.9401	563.0568	–
20	4.7980	18.3787	549.6553	18136.5335	–
25	–	–	–	–	–
26	–	–	–	–	–

AlgCPy items/focals	2^2	2^4	2^9	2^{14}	2^{19}
5	0.0000	0.0000	–	–	–
10	0.0000	0.0000	0.0015	–	–
15	0.0003	0.0010	0.0477	1.5555	–
20	0.0069	0.0295	1.5593	50.3568	1667.2546
25	–	–	–	–	–
26	–	–	–	–	–

AlgGPU items/focals	2^2	2^4	2^9	2^{14}	2^{19}
5	0.0033	0.0012	–	–	–
10	0.0025	0.0019	0.0015	–	±
15	0.0022	0.0143	0.0148	0.0135	–
20	0.0015	0.0070	0.0149	0.2293	7.2960
25	0.0044	0.0106	0.2449	7.0875	239.0286
26	0.0076	0.0185	0.4457	14.1326	478.5815

The `AlgIPy` shows how the time is incremented both in terms of the number of nodes to explore as well as the number of focal elements. The execution of this algorithm becomes intractable as the number of nodes in the lattice increases. Moving to the Numba[1] compiled `AlgCPy` implementation, it can be observed that it produces a reduction of the time in relation to `AlgIPy` implementation for all the cases studied by means of the compilation of the code.

CPU implementations produce a memory outbound for lattices with size greater than 2^{20} nodes and even for that size when trying to use a high number of focal elements. Table 1 shows that the maximum combination of number of nodes and focal elements is achieved in this experiments with a lattice of 2^{26} nodes and a number of focal elements equal to 2^{19} for the GPU. For the largest combination reached by the CPU, with a lattice with 2^{20} nodes and 2^{14} focal elements, the execution time of the original Python algorithm takes 5 h to complete the task. This time is greatly reduced by the compiled version which takes about 50 min, and extremely improved by the GPU version, which takes less than one second to solve the same task.

The GPU algorithm shows how the number of focal elements dominates now the execution time. For example, when the number of focal elements is 2^4, the execution time can be done in milliseconds, no matter the number of elements in the lattice. Up to 20 items and 2^{14} focal elements, the GPU takes similar execution times. The increase in bigger sizes is due to the need of reusing threads when the number of combinations increases.

Notice how for a lattice size greater than 2^{26} and a number of focal elements equal to 2^{19}, the execution time of `AlgGPU` is even lower than the time taken by the `AlgIPy` algorithm for a lattice of size 2^{15} and 2^{14} focal elements. This makes another point about the remarkable advantage of the GPU implementation in relation to the CPU versions.

Table 2. Execution time obtained with the `pyds` package (left) and speedup showing how much faster the results obtained with the GPU proposed algorithm in this work are (right).

nitems/nfocals	2^2	2^4	2^9	2^{14}	nitems/nfocals	2^2	2^4	2^9	2^{14}
5	0.0005	0.0008	–	–	5	6.6%	1.49 %	–	–
10	0.0013	0.0097	0.2108	–	10	1.93%	5.26%	142.86%	–
15	0.1572	0.4394	8.1160	268.4909	15	71.45%	33.33%	1000%	20000%
20	2.1417	16.8437	325.0773	–	20	1426.66%	2500%	1428.571%	–

---

[1] https://numba.pydata.org.

The execution time as well as the speed up of the GPU code proposed in this work in relation to the current available Python package pyds[2] that implements functions for Belief and Plausibility are shown in Table 2. This implementation suffers a memory outbound for 20 items when the number of elements is 2^{14}.

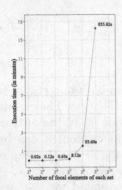

Fig. 2. Execution time of the method for combining two focal sets. x axis shows the size of each set of focal elements being combined.

In addition to the computation of the belief and plausibility, the execution times obtained for combining two different sets of focal elements given for a lattice of 2^{20} nodes (i.e. a set of 20 items) is given in Fig. 2 showing the time for different number of focal elements in the x axis. In order to test the function, large numbers of focal elements have been tested. In practical problems, it is likely to assume that the size of the focal set will not be as large, and it is shown that for focal sets smaller than 2^8 the problem can be solved in seconds.

5   Conclusion and Future Work

In this work we introduce a set of Python GPU functions to reduce the execution time of the operations related with the Dempster-Shafer Theory. Empirical results show the advantage of this kind of computation, making the methods feasible for large sets. Future work in this line will provide more functions to compute the joint probability assignments and also to explore further reductions of the execution time by the introduction of the equivalence classes concept into the parallel algorithms.

References

1. Barnett, J.A.: Computational methods for a mathematical theory of evidence. In: Proceedings of the 7th International Joint Conference on Artificial Intelligence, IJCAI'81, vol. 2, pp. 868–875. Morgan Kaufmann Publishers Inc., San Francisco (1981)
2. Benalla, M., Achchab, B., Hrimech, H.: On the computational complexity of Dempster's rule of combination, a parallel computing approach. J. Comput. Sci. **50**, 101283 (2021)
3. Dempster, A.P.: Upper and lower probabilities induced by a multivalued mapping. Ann. Math. Stat. **38**, 325–339 (1967)
4. Lal Prajapati, G., Saha, R.: Reeds: relevance and enhanced entropy based dempster Shafer approach for next word prediction using language model. J. Comput. Sci. **35**, 1–11 (2019)

---

[2] https://pypi.org/project/pyds/.

5. Orponen, P.: Dempster's rule of combination is #p-complete. Artif. Intell. **44**(1), 245–253 (1990)
6. Shafer, G.: A Mathematical Theory of Evidence. Princeton University Press, Princeton (1976)
7. Troiano, L., Díaz, I., Gaglione, C.: Matching media contents with user profiles by means of the Dempster-Shafer theory. In: 2017 IEEE International Conference on Fuzzy Systems, FUZZ-IEEE 2017, Naples, Italy, 9–12 July 2017, pp. 1–6. IEEE (2017)
8. Troiano, L., Rodríuez-Muñiz, L.J., Díaz, I.: Discovering user preferences using Dempster-Shafer theory. Fuzzy Sets Syst. **278**, 98–117 (2015)

QLEN: Quantum-Like Evidential Networks for Predicting the Decision in Prisoner's Dilemma

Jixiang Deng[1] and Yong Deng[1,2,3]

[1] Institute of Fundamental and Frontier Science, University of Electronic Science
and Technology of China, Chengdu 610054, China
dengjixiang@std.uestc.edu.cn, dengentropy@uestc.edu.cn
[2] School of Education, Shannxi Normal University, Xi'an 710062, China
[3] School of Knowledge Science, Japan Advanced Institute of Science and Technology,
Nomi, Ishikawa 923-1211, Japan

Abstract. For predicting the decision in prisoner's dilemma game, researchers have developed lots of models. However, the existing models only consider the networks based on quantum probability amplitude and the accuracy can still be improved. Thus, in this paper, quantum-like evidential network (QLEN) is proposed, which is the evidential extension of the original quantum-like Bayesian network. A QLEN consists of a directed acyclic graph associated with quantum mass function. In addition, a full joint quantum mass function can be derived from QLEN which can be applied in decision making and inference. Moreover, based on QLEN, this paper presents a decision model for predicting the players' decision in prisoner's dilemma. The results show that, compared with the existing models, the proposed model is more efficient and accurate to make predictions in prisoner's dilemma.

Keywords: Dempster-Shafer evidence theory · Prisoner's dilemma game · Bayesian networks · Quantum-like evidential networks · Quantum mass function

1 Introduction

Since Dempster-Shafer evidence theory (evidence theory) was firstly proposed by Dempster in 1967 [4] and later developed by Shafer in 1976 [12], uncertain information processing based on evidence theory becomes more and more popular, and many researchers have been promoting the development of evidence theory. For example, Smets proposed Pignistic Transformation [13], which can transform probability into mass function. Cuzzolin proposed a geometric approach for dealing with the uncertainty in evidence theory [2]. For measuring the uncertainty of mass function, Deng presented Deng entropy [6], which is a generalization of Shannon entropy [8] and is further developed into information volume [5,7]. Gao and Deng introduced quantum model of mass function [9], which extends classical mass function into its quantum counterpart.

© Springer Nature Switzerland AG 2021
T. Denœux et al. (Eds.): BELIEF 2021, LNAI 12915, pp. 300–308, 2021.
https://doi.org/10.1007/978-3-030-88601-1_30

The violation of Sure Thing Principle in prisoner's dilemma is an interesting phenomenon [11]. Various network models have been proposed for modeling the mechanism of this phenomenon. In 2016, Moreira and Wichert proposed quantum-like Bayesian networks (QLBN) [11], which replaces the classical probability in the classical Bayesian Networks model [10] by quantum probability amplitude. In 2020, Dai *et al.* presented a heuristic model to determine the interference effect in QLBN [3]. Apart from the network models, Benavoli *et al.* proposed a quantum gambling system in 2017 [1], which provides a potential model for explaining the violation of Sure Thing Principle in prisoner's dilemma.

However, the existing network models still have room for improving the accuracy of modeling the decision in prisoner's dilemma based on different techniques, such as modifying the way to determine interference and introducing some novel methods into the network. Moreover, how to predict the decision in prisoner's dilemma is still an open issue. To be specific, the existing network models only take quantum probability amplitude into consideration, which can be further generalized. By setting a connection between quantum theory and evidence theory, quantum mass function is an efficient tool for representing uncertainty [9]. It is reasonable to introduce quantum mass function into the network model to enhance the performance of predicting decision in prisoner's dilemma.

To address the issues mentioned above, a novel network model, named as quantum-like evidential network (QLEN), is proposed in this paper, which is the quantum evidential extension of the original quantum-like Bayesian network. A QLEN consists of a directed acyclic graph and its associated quantum mass function. Besides, a full joint quantum mass function can be derived from QLEN which can be applied in decision making and inference. In addition, based on QLEN, we present a decision model for predicting the players' decision in prisoner's dilemma. The results show that the proposed model can make predictions more efficiently and accurately than the existing models.

The rest of this paper is organized as follows. Section 2 briefly reviews some preliminaries. In Sect. 3, we propose quantum-like evidential networks (QLEN). In Sect. 4, a decision making model based on QLEN is presented for predicting the decision in prisoner's dilemma. Section 5 makes a brief conclusion.

2 Preliminaries

2.1 Dempster-Shafer Evidence Theory

Several conceptions about Dempster-Shafer evidence theory [4,12] are summarized as follows. Frame of discernment (FOD) is an exhaustive nonempty set of N hypotheses: $\Theta = \{\theta_1, \theta_2, \theta_3, \cdots, \theta_N\}$, where N elements are mutually exclusive. 2^Θ denotes the power set of Θ. It has 2^N elements which are all possible subsets of Θ. Basic probability assignment (BPA), or mass function, is a mapping function $\mathbf{m} : 2^\Theta \to [0,1]$, which is constrained by $\sum_{A \in 2^\Theta} \mathbf{m}(A) = 1$ and $\mathbf{m}(\emptyset) = 0$. For $A, B \subseteq \Theta$, the conditional mass function is defined as [14]:

$$\mathbf{m}(B\|A) = \begin{cases} \sum_{X \subseteq \bar{A}} \mathbf{m}(B \cup X) & \text{if } B \subseteq A \subseteq \Theta, \\ 0 & \text{otherwise.} \end{cases} \tag{1}$$

Given two variables $X, Y \subseteq \Theta$, the joint mass function is expressed as [14]:

$$\mathbf{m}_{X \cup Y}(x, y) = \prod_{y \subseteq Y} \mathbf{m}_X(x||y) = \prod_{x \subseteq X} \mathbf{m}_Y(y||x). \tag{2}$$

2.2 Quantum Mass Function

Quantum mass function is a quantum extension of classical mass function [9]. Quantum frame of discernment (QFOD) is the set of N mutually exhaustive events indicated by $|\Theta\rangle = \{|\theta_1\rangle, |\theta_2\rangle, |\theta_3\rangle, \cdots, |\theta_N\rangle\}$. The power set of QFOD is given by $2^{|\Theta\rangle} = \{|A_1\rangle, |A_2\rangle, |A_3\rangle, \cdots, |A_{2^N}\rangle\}$, which contains all the possible subsets of QFOD. On QFOD, quantum mass function (QMF) is defined as

$$\mathbb{M}(|A_i\rangle) = \psi_i e^{j\theta_i} \tag{3}$$

which is constrained by $\sum_{|A_i\rangle \in 2^{|\Theta\rangle}} |\mathbb{M}(|A_i\rangle)|^2 = \sum_{i=1}^{2^N} \psi_i^2 = 1$ and $\mathbb{M}(\emptyset) = 0$.

2.3 Quantum-Like Bayesian Networks

Quantum-like Bayesian Network is a directed acyclic graph [11]. Let the list of variables be $X = \{X_1, X_2, ..., X_N\}$. The full joint distribution of a quantum-like Bayesian Network (QLBN) is defined as follows [11]:

$$Pr(X_1, X_2, ..., X_N) = \left| \prod_{i=1}^{N} \psi(X_i|\pi_i) \right|^2 \tag{4}$$

where $\psi(X_i) = |\psi(X_i)| e^{j\theta_i}$ is the quantum probability amplitude and π_i represents all the parent nodes of X_i.

3 QLEN: Quantum-Like Evidential Networks

In this section, firstly, quantum-like evidential networks (QLEN) are proposed. Next, the joint mass function and the marginal mass function derived from QLEN are proposed.

Definition 1 (Quantum-like evidential networks). *Let the list of variables be $X = \{X_1, X_2, ..., X_M\}$. All variables are defined on the quantum frame of discernment $|\Theta\rangle = \{|\theta_1\rangle, |\theta_2\rangle, \cdots, |\theta_N\rangle\}$ and the value x_i for each variable X_i is in the power set $2^{|\Theta\rangle} = \{|\phi_1\rangle, |\phi_2\rangle, \cdots, |\phi_{2^N}\rangle\}$. A quantum-like evidential network (QLEN) is defined as:*

$$QLEN = \langle G, \mathbb{M} \rangle \tag{5}$$

where $G = \langle X, E \rangle$ is a directed acyclic graph, in which the nodes denote variables X_i and the edges denote dependencies from parent nodes to child nodes. $\mathbb{M}$

indicates a set of parameters for QLEN, and each element of $\mathbb{M}$ *is a conditional quantum mass function associated with each variable* X_i:

$$\mathbb{M}_{X_i}(x_i||\pi_i) = \sqrt{\boldsymbol{m}_{X_i}(x_i||\pi_i)}e^{j\theta_i} \tag{6}$$

where $\boldsymbol{m}_{X_i}(x_i||\pi_i)$ *is the classical conditional mass function and* π_i *are all the parent nodes of* X_i.

Accordingly, the full joint quantum mass function and its corresponding full joint mass function can be derived from QLEN:

$$\mathbb{M}_{\bigcup_{i=1}^M X_i}(x_1, x_2, ..., x_M) = \prod_{i=1}^M \mathbb{M}_{X_i}(x_i||\pi_i) \tag{7}$$

$$\boldsymbol{m}_{\bigcup_{i=1}^M X_i}(x_1, x_2, ..., x_M) = \left|\mathbb{M}_{\bigcup_{i=1}^M X_i}(x_1, x_2, ..., x_M)\right|^2 \tag{8}$$

For some query $X^q = \{X_1^q, X_2^q, ..., X_Q^q\} \subseteq X$, the marginal mass function can be calculated based on Born's rule:

$$\boldsymbol{m}_{\bigcup_{i=1}^Q X_i^q}(x_1, x_2, ..., x_Q||e) = \alpha \left|\sum_{y \in 2^{\langle\Theta\rangle}} \prod_{i=1}^Q \mathbb{M}_{X_i^q}(x_i||\pi_i, e, y)\right|^2 \tag{9}$$

$$\alpha = \frac{1}{\sum_{x_1...x_Q \in 2^{\langle\Theta\rangle}} \left|\sum_{y \in 2^{\langle\Theta\rangle}} \prod_{i=1}^Q \mathbb{M}_{X_i^q}(x_i||\pi_i, e, y)\right|^2} \tag{10}$$

where e is the value of observed variables, y is the value of unobserved variables, and α is the normalization factor.

Interference term will emerge by expanding Eq. 9 as follows:

$$\boldsymbol{m}_{\bigcup_{i=1}^Q X_i^q}(x_1, x_2, ..., x_Q||e) = \alpha \left(\sum_{y \in 2^{\langle\Theta\rangle}} \left|\prod_{i=1}^Q \mathbb{M}_{X_i^q}(x_i||\pi_i, e, y)\right|^2 + 2 \cdot INT\right) \tag{11}$$

$$INT = \sum_{a=1}^{2^N-1} \sum_{b=a+1}^{2^N} \left|\prod_{i=1}^Q \mathbb{M}_{X_i^q}(x_i||\pi_i, e, y = |\phi_a\rangle)\right| \cdot \left|\prod_{i=1}^Q \mathbb{M}_{X_i^q}(x_i||\pi_i, e, y = |\phi_b\rangle)\right| \cdot \cos(\theta_a - \theta_b) \tag{12}$$

where INT is the interference term and $\cos(\theta_a - \theta_b)$ is the interference degree.

4   QLEN-Based Decision Making Model and Its Application in Prisoner's Dilemma Game

In this section, a decision making model is proposed based on QLEN. Then, the proposed model is applied in prisoner's dilemma for predicting the decision.

Table 1. Results of prisoner's dilemma from different literature summarized by [11]

Literature	Known to defect[a]	Known to collaborate[b]	Observed $Pr^O(D)$[c]	Classical $Pr^C(D)$[d]
Shafir and Tversky, 1992	0.9700	0.8400	0.6300	0.9050
Li and Taplin, 2002[e]	0.8200	0.7700	0.7200	0.7950
Busemeyer *et al.*, 2006	0.9100	0.8400	0.6600	0.8750
Hristova and Grinberg, 2008	0.9700	0.9300	0.8800	0.9500
Game 1[f]	0.7333	0.6670	0.6000	0.7002
Game 2	0.8000	0.7667	0.6300	0.7834
Game 3	0.9000	0.8667	0.8667	0.8834
Game 4	0.8333	0.8000	0.7000	0.8167
Game 5	0.8333	0.7333	0.7000	0.7833
Game 6	0.7767	0.8333	0.8000	0.8050
Game 7	0.8867	0.7333	0.7667	0.8100

The results summarized by [11] show the probability that the second player (X_2) chooses to **defect** under the condition of knowing the choice of the first player (X_1) is: to defect[a], to collaborate[b], and unknown[c]. In the rest of this paper, [a] is denoted as $Pr(X_2 = D|X_1 = D)$ and [b] is denoted as $Pr(X_2 = D|X_1 = C)$. [c] Observed $Pr^O(D)$ is short for $Pr^O(X_2 = D)$. [d] Classical $Pr^C(D)$ is short for $Pr^C(X_2 = D)$ which is obtained based on total probability formula. [e] These results correspond to the average results of Game 1 to 7. [f] Game 1 to 7 are seven experiments reported in Li and Taplin, 2002.

4.1   Problem Statement

Prisoner's dilemma game describes the problem of collaboration. In this game, there are two players. Each player has two choices, namely, collaborating with or defecting to the other player. There are different payoffs for the players depending on the joint actions of the two players, which lead players to choose differently when knowing different prior information. Sure Thing Principle is essential in probability theory, which describes that if a person prefers to do X rather than do Y under the situation S, and if the person also would rather do X than do Y under the complementary situation $\bar{S}$, then the person should always prefer X to Y even when the situation is unspecified [11]. In 2016, Moreira *et al.* [11] summarized the results of prisoner's dilemma from different literature, which are shown in Table 1. The experimental results indicate the violations of Sure Thing Principle, which further lead to the violations of total probability formula, *i.e.*, there are vast differences between observed $Pr^O(D)$ and classical $Pr^C(D)$ [11]. Various models have been developed to explore the underlying mechanism and predict the observed $Pr^O(D)$ by different techniques [3,11]. However, how to model the decision in prisoner's dilemma is still an open issue.

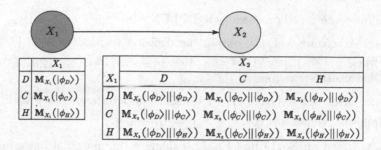

Fig. 1. The constructed QLEN model of prisoner's dilemma game and the associated tables of conditional quantum mass function for all the nodes (X_1 and X_2).

4.2 The Proposed QLEN-Based Decision Making Model

In this subsection, a decision making model based on QLEN is proposed for modeling the decision in prisoner's dilemma game.

Let the list of variables be $X = \{X_1, X_2\}$ where X_1 and X_2 respectively represent the first and the second player in prisoner's dilemma. Let the quantum frame of discernment be $|\Theta\rangle = \{|\theta_D\rangle, |\theta_C\rangle\}$ where $|\theta_D\rangle$ and $|\theta_C\rangle$ respectively denote the two choices of the players: **D**efect and **C**ollaboration. The value for each variable X_i is in the power set $2^{|\Theta\rangle} = \{\emptyset, |\phi_D\rangle, |\phi_C\rangle, |\phi_H\rangle\}\} = \{\emptyset, \{|\theta_D\rangle\}, \{|\theta_C\rangle\}, \{|\theta_D\rangle, |\theta_C\rangle\}\}$ where $|\phi_H\rangle$ denotes **H**esitancy which means that the player is hesitant about the choices. It should be pointed out that, since both mass function and quantum mass function of the empty set $\emptyset$ are constrained by zero, the proposed model just focus on $|\theta_D\rangle, |\theta_C\rangle$, and $|\theta_H\rangle$. The aim is to predict the observed probability $Pr^O(X_2 = D)$ based on QLEN. The proposed model proceeds as the following steps.

Step 1: Construct QLEN based on the list of variables. Since there are two players in prisoner's dilemma, QLEN contains two nodes, namely X_1 and X_2, which are illustrated in Fig. 1.

Step 2: Let $m, n \in \{D, C, H\}$. Input the probabilities in Table 1 and convert them into their associated mass functions:

$$\mathbf{m}_{X_1}(|\phi_m\rangle) = \begin{cases} w_m Pr(X_1 = m) & if\ m \in \{D, C\} \\ 1 - \sum_{i \in \{D,C\}} w_i Pr(X_1 = i) & if\ m = H \end{cases} \quad (13)$$

$$\mathbf{m}_{X_2}(|\phi_m\rangle \,|||\phi_n\rangle) = \begin{cases} x_{mn} Pr(X_2 = m|X_1 = n) & if\ m \in \{D, C\} \\ 1 - \sum_{i \in \{D,C\}} x_{in} Pr(X_2 = i|X_1 = n) & if\ m = H \end{cases} \quad (14)$$

where $Pr(X_2 = m|X_1 = H) \triangleq \sum_{j \in \{D,C\}} Pr(X_2 = m|X_1 = j) Pr(X_1 = j)$ for $m \in \{D, C\}$ are the classical probabilities based on total probability formula, and w_m, x_{mn} for $m \in \{D, C\}, n \in \{D, C, H\}$ are eight parameters to be determined.

Step 3: Convert the mass functions into the quantum mass functions (QMF): $\mathbb{M}_{X_1}(|\phi_n\rangle) = \sqrt{\mathbf{m}_{X_1}(|\phi_n\rangle)}e^{j\theta_1}, \mathbb{M}_{X_2}(|\phi_m\rangle \,|||\phi_n\rangle) = \sqrt{\mathbf{m}_{X_2}(|\phi_m\rangle \,|||\phi_n\rangle)}e^{j\theta_2}$, and assign the quantum mass functions to QLEN as shown in Fig. 1

Step 4: Inference based on constructed QLEN.

Step 4-1: According to Eq. 7, obtain the decision vectors for $K \in \{D, C, H\}$:
$V_K = \left[\mathbb{M}_{X_1 \cup X_2} (|\phi_D\rangle, |\phi_K\rangle), \mathbb{M}_{X_1 \cup X_2} (|\phi_C\rangle, |\phi_K\rangle), \mathbb{M}_{X_1 \cup X_2} (|\phi_H\rangle, |\phi_K\rangle) \right]^T$,
where T means the transpose of the vector.

Step 4-2: Given $a, b \in \{D, C, H\}$, determine the interference degree in Eq. 12 by the following equations: $\cos (\theta_a - \theta_b) \triangleq -\frac{V_a^T V_b}{|V_a| \cdot |V_b|}$.

Step 4-3: Based on Eq. 11 and Eq. 12, calculate the marginal mass functions for $m \in \{D, C, H\}$:

$$\mathbf{m}_{X_2} (|\phi_m\rangle) = \alpha \left(\sum_{n \in \{D,C,H\}} |\mathbb{M}_{X_2} (|\phi_m\rangle \,||\, |\phi_n\rangle) \cdot \mathbb{M}_{X_1} (|\phi_n\rangle)|^2 + 2 \cdot INT \right)$$
(15)

where α and INT are respectively the normalization factor and the interference term, which are as follows:

$$\alpha = \frac{1}{\sum_{m \in \{D,C,H\}} \left(\sum_{n \in \{D,C,H\}} |\mathbb{M}_{X_2} (|\phi_m\rangle \,||\, |\phi_n\rangle) \cdot \mathbb{M}_{X_1} (|\phi_n\rangle)|^2 + 2 \cdot INT \right)}$$
(16)

$$INT = \sum_{\substack{a,b \in \{D,C,H\} \\ a \neq b}} |\mathbb{M}_{X_1 \cup X_2} (|\phi_m\rangle, |\phi_a\rangle)| \cdot |\mathbb{M}_{X_1 \cup X_2} (|\phi_m\rangle, |\phi_b\rangle)| \cdot \left(-\frac{V_a^T V_b}{|V_a| \cdot |V_b|} \right).$$

Step 5: Based on Pignistic Transformation [13], convert the marginal mass functions into predicted probability for $i \in \{D, C\}$: $Pr^P (X_2 = i) = \mathbf{m}_{X_2} (|\phi_i\rangle) + \frac{1}{||\Theta\rangle|} \mathbf{m}_{X_2} (|\phi_H\rangle)$ where $||\Theta\rangle|$ is the cardinality of $|\Theta\rangle$ which is 2. Finally, output $Pr^P (X_2 = i)$ for predicting the decision in prisoner's dilemma.

4.3   Experiment and Discussion

In this subsection, the proposed QLEN-based model is applied in prisoner's dilemma for predicting the observed probability $Pr^O (X_2 = D)$.

Firstly, both $Pr (X_1 = D)$ and $Pr (X_1 = C)$ are assumed to be 0.5, since the first player (X_1) in QLEN has no parent nodes so that he/she cannot receive any prior information. Then, to determine the eight parameters w_m, x_{mn} for $m \in \{D, C\}, n \in \{D, C, H\}$, the proposed model is trained on the data in Table 1 by minimizing the loss function: $\mathcal{L} = \frac{1}{\#data} \sum_{k=1}^{\#data} \left[Pr_k^P (X_2 = D) - Pr_k^O (X_2 = D) \right]^2$. Next, calculate the fit error of the predicted probability $Pr^P (X_2 = D)$ and the observed probability $Pr^O (X_2 = D)$ based on $\frac{|Pr^P (X_2 = D) - Pr^O (X_2 = D)|}{Pr^O (X_2 = D)}$. Finally, the predicted probabilities generated by the proposed model and their associated fit errors are respectively summarized in the 7-th and 8-th column of Table 2.

Two models, namely Moreira et al.'s QLBN-based model (model 1) [11] and Dai et al.'s heuristic model (model 2) [3], are selected to compare with the

Table 2. Comparison of the results generated by different models

Literature	Observed $Pr^O(D)$	$Pr^P(D)$ model 1[b]	Fit error[a] model 1	$Pr^P(D)$ model 2[c]	Fit error model 2	$Pr^P(D)$ model 3[d]	Fit error model 3
S.&T., 1992	0.6300	0.6408	0.0171	0.8001	0.2700	0.6302	**0.0003**
L.&T., 2002	0.7200	0.7122	**0.0108**	0.6504	0.0967	0.6761	0.0609
B. et al., 2006	0.6600	0.7995	0.2113	0.7426	**0.1251**	0.7491	0.1350
H.&G., 2008	0.8800	0.8968	0.0191	0.8749	0.0058	0.8820	**0.0023**
G. 1	0.6000	0.6313	0.0522	0.5820	0.0299	0.6001	**0.0002**
G. 2	0.6300	0.7011	0.1129	0.6399	**0.0157**	0.6684	0.0610
G. 3	0.8667	0.8113	**0.0639**	0.7529	0.1313	0.7687	0.1130
G. 4	0.7000	0.7341	0.0487	0.6709	0.0415	0.7002	**0.0003**
G. 5	0.7000	0.7006	**0.0009**	0.6411	0.0841	0.6764	0.0337
G. 6	0.8000	0.7169	0.1039	0.6598	0.1752	0.8006	**0.0008**
G. 7	0.7667	0.7159	0.0663	0.6688	0.1277	0.7670	**0.0003**
Average error	–	–	0.0643	–	0.1003	–	**0.0371**

[a] Fit error is calculated by $\frac{|Pr^P(X_2=D)-Pr^O(X_2=D)|}{Pr^O(X_2=D)}$. Fit error in bold means the lowest.
[b] Model 1 is Moreira et al.'s QLBN-based model [11]. [c] Model 2 is Dai et al.'s heuristic model [3].
[d] Model 3 is the proposed QLEN-based model.

proposed QLEN-based model (model 3). Compared with model 1, the fit error of the proposed model is the lowest in 8 out of 11 experiments. In comparison of model 2, the proposed model's fit error is the lowest in 9 out of 11 experiments. This shows the stability of the proposed model is better than that of model 1 and model 2, namely: with respect to most of the experiments, the performance of the proposed model is the best compared with the other two models. Meanwhile, the average error of the proposed model is 3.71% which is much lower than that of model 1 (6.43%) and model 2 (10.03%). As a result, compared with the other two model, the proposed model is more accurate and effective to make predictions of the decision in prisoner's dilemma.

5   Conclusion

For modeling the phenomenon of violation of Sure Thing Principle in prisoner's dilemma game, lots of models have been proposed. However, the existing models only consider the networks based on quantum probability amplitude, which can be generalized into the networks based on quantum mass function. Besides, the accuracy of the existing models can still be improved. Therefore, this paper introduces quantum mass function into the original quantum-like Bayesian networks (QLBN) and proposes quantum-like evidential networks (QLEN). The main contributions of this paper are summarized as follows.

(i) Quantum-like evidential network (QLEN) is proposed, which is the evidential extension of QLBN. A QLEN consists of a directed acyclic graph and its associated quantum mass function.

(ii) A full joint quantum mass function can be derived from QLEN which can be applied in decision making and inference.

(iii) Based on QLEN, this paper presents a decision model for predicting the players' decision in prisoner's dilemma.

(iv) The results show that the proposed model is able to make predictions in prisoner's dilemma more efficiently and accurately than the existing models.

In future work, we will focus on applying QLEN into other fields. Also, modifications of QLEN for better performance are also worth exploring.

References

1. Benavoli, A., Facchini, A., Zaffalon, M.: A Gleason-type theorem for any dimension based on a gambling formulation of quantum mechanics. Found. Phys. **47**(7), 991–1002 (2017)
2. Cuzzolin, F.: The Geometry of Uncertainty. Springer, Cham (2021). https://doi.org/10.1007/978-3-030-63153-6
3. Dai, J., Deng, Y.: A new method to predict the interference effect in quantum-like Bayesian networks. Soft Comput. **24**, 1–8 (2020). https://doi.org/10.1007/s00500-020-04693-2
4. Dempster, A.P.: Upper and lower probabilities induced by a multivalued mapping. Ann. Math. Stat. **38**(2), 325–339 (1967)
5. Deng, J., Deng, Y.: Information volume of fuzzy membership function. Int. J. Comput. Commun. Control **16**(1), 4106 (2021)
6. Deng, Y.: Deng entropy. Chaos Solitons Fractals **91**, 549–553 (2016)
7. Deng, Y.: Information volume of mass function. Int. J. Comput. Commun. Control **15**(6), 3983 (2020)
8. Deng, Y.: Uncertainty measure in evidence theory. Sci. China Inf. Sci. **63**(11), 1–19 (2020). https://doi.org/10.1007/s11432-020-3006-9
9. Gao, X., Deng, Y.: Quantum model of mass function. Int. J. Intell. Syst. **35**(2), 267–282 (2020)
10. Jensen, F.V., et al.: An Introduction to Bayesian Networks, vol. 210. UCL Press, London (1996)
11. Moreira, C., Wichert, A.: Quantum-like Bayesian networks for modeling decision making. Front. Psychol. **7**, 11 (2016)
12. Shafer, G.: A Mathematical Theory of Evidence, vol. 1. Princeton University Press, Princeton (1976)
13. Smets, P.: Decision making in the TBM: the necessity of the pignistic transformation. Int. J. Approx. Reason. **38**(2), 133–147 (2005)
14. Xu, H., Smets, P.: Reasoning in evidential networks with conditional belief functions. Int. J. Approx. Reason. **14**(2–3), 155–185 (1996)

Author Index

Printed in the United States
by Baker & Taylor Publisher Services